Gel-Based Materials for Biomedical Engineering

Gel-Based Materials for Biomedical Engineering

Guest Editors

Daihua Fu
Jieyu Zhang

MDPI

Basel • Beijing • Wuhan • Barcelona • Belgrade • Novi Sad • Cluj • Manchester

Guest Editors
Daihua Fu
National Engineering Research Center for Biomaterials
Sichuan University
Chengdu
China

Jieyu Zhang
National Engineering Research Center for Biomaterials
Sichuan University
Chengdu
China

Editorial Office
MDPI AG
Grosspeteranlage 5
4052 Basel, Switzerland

This is a reprint of the Special Issue, published open access by the journal *Gels* (ISSN 2310-2861), freely accessible at: https://www.mdpi.com/journal/gels/special_issues/6Q4IP5LO4K.

For citation purposes, cite each article independently as indicated on the article page online and as indicated below:

Lastname, A.A.; Lastname, B.B. Article Title. *Journal Name* **Year**, *Volume Number*, Page Range.

ISBN 978-3-7258-3823-3 (Hbk)
ISBN 978-3-7258-3824-0 (PDF)
https://doi.org/10.3390/books978-3-7258-3824-0

Contents

Preface

This reprint, entitled "Gel-Based Materials for Biomedical Engineering", brings together ten selected papers that reflect the growing interest and innovation in the field of gel-based biomaterials. The aim of the Special Issue was to provide a platform for the presentation of recent research on the design, characterization, and biomedical applications of gels, particularly hydrogels, in areas such as tissue engineering, wound healing, and drug delivery.

The collected works highlight diverse approaches, from natural and synthetic hydrogels to nanocomposite and stimuli-responsive gel systems, showcasing their potential in addressing critical challenges in biomedical engineering. This reprint is intended for researchers, clinicians, and graduate students working in the fields of biomaterials, biomedical engineering, and regenerative medicine.

We would like to express our sincere gratitude to all the contributing authors for their excellent research and to the reviewers for their insightful comments. We also thank the editorial team at *Gels* for their continued support and professional assistance throughout the preparation of this Special Issue.

Daihua Fu and Jieyu Zhang

Guest Editors

gels

MDPI

Article

Bromelain- and Silver Nanoparticle-Loaded Polycaprolactone/Chitosan Nanofibrous Dressings for Skin Wound Healing

Yasaman Saghafi [1], Hadi Baharifar [1,2,3,*], Najmeh Najmoddin [1], Azadeh Asefnejad [1], Hassan Maleki [3,4], Sayed Mahmoud Sajjadi-Jazi [3,5], Alireza Bonkdar [6,7], Forough Shams [7] and Kamyar Khoshnevisan [3,6,7,*]

1 Department of Biomedical Engineering, Science and Research Branch, Islamic Azad University, Tehran 1477893855, Iran; yasaman.saghafi90@gmail.com (Y.S.); najmoddin@srbiau.ac.ir (N.N.)
2 Applied Biophotonics Research Center, Science and Research Branch, Islamic Azad University, Tehran 1477893855, Iran
3 Research and Development Team, Evolution Wound Dressing (EWD) Startup Co., Tehran 1983963113, Iran
4 Nano Drug Delivery Research Center, Health Technology Institute, Kermanshah University of Medical Sciences, Kermanshah 6715847141, Iran
5 Endocrinology and Metabolism Research Center, Endocrinology and Metabolism Clinical Sciences Institute, Tehran University of Medical Sciences, Tehran 1411713137, Iran
6 Medical Nanotechnology and Tissue Engineering Research Center, Shahid Beheshti University of Medical Sciences, Tehran 1983963113, Iran
7 Department of Tissue Engineering and Applied Cell Sciences, School of Advanced Technologies in Medicine, Shahid Beheshti University of Medical Sciences, Tehran 1983963113, Iran; forough.shamss@gmail.com
* Correspondence: baharifar.h@gmail.com (H.B.); khoshnevisan.k@sbmu.ac.ir (K.K.)

Abstract: A cutaneous wound is caused by various injuries in the skin, which can be wrapped with an efficient dressing. Electrospinning is a straightforward adjustable technique that quickly and continuously generates nanofibrous wound dressings containing antibacterial and anti-inflammatory agents to promote wound healing. The present study investigated the physicochemical and biological properties of bromelain (BRO)- and silver nanoparticle (Ag NPs)-loaded gel-based electrospun polycaprolactone/chitosan (PCL/CS) nanofibrous dressings for wound-healing applications. Electron microscopy results showed that the obtained nanofibers (NFs) had a uniform and homogeneous morphology without beads with an average diameter of 176 ± 63 nm. The FTIR (Fourier transform infrared) analysis exhibited the loading of the components. Moreover, adding BRO and Ag NPs increased the tensile strength of the NFs up to 4.59 MPa. BRO and Ag NPs did not significantly affect the hydrophilicity and toxicity of the obtained wound dressing; however, the antibacterial activity against *E. coli* and *S. aureus* bacteria was significantly improved. The in vivo study showed that the wound dressing containing BRO and Ag NPs improved the wound-healing process within one week compared to other groups. Therefore, gel-based PCL/CS nanofibrous dressings containing BRO and Ag NPs could be a promising solution for healing skin wounds.

Keywords: nanofibers; wound dressing; electrospinning; polycaprolactone; chitosan; silver nanoparticles; bromelain

Citation: Saghafi, Y.; Baharifar, H.; Najmoddin, N.; Asefnejad, A.; Maleki, H.; Sajjadi-Jazi, S.M.; Bonkdar, A.; Shams, F.; Khoshnevisan, K. Bromelain- and Silver Nanoparticle-Loaded Polycaprolactone/Chitosan Nanofibrous Dressings for Skin Wound Healing. *Gels* **2023**, *9*, 672. https://doi.org/10.3390/gels9080672

Academic Editors: Xin Zhao and Esmaiel Jabbari

Received: 25 June 2023
Revised: 7 August 2023
Accepted: 17 August 2023
Published: 19 August 2023

1. Introduction

The skin acts as a natural protector against external pathogenic microbial, chemical, mechanical, and thermal stresses and can therefore be exposed to various injuries such as wounds, trauma, and burns [1,2]. The protection of body fluids, electrolytes, and nutritional components of the body depends on the safekeeping of the skin against severe physicochemical damage and microbial invasion. Tissue engineering, as a method to replace or improve portions of or whole biological tissues such as the skin, bladder, bone, blood vessels, and cartilage, holds great potential for the treatment of human diseases [3,4]. In recent years, a growing deal of attention has been given to promoting wound healing by suitable dressings. An ideal wound dressing should be applied to protect the wound against external contaminants and pathogenesis, as well as facilitate the healing process [5]. The most powerful features of a perfect dressing include adequate tensile strength, high biocompatibility, intense antibacterial activity, and sufficient permeability for the enhancement of humidity and oxygen [6–8]. Infection is another major problem that hinders wound healing and can cause various life-threatening complications, such as multiple organ system defects, bacteremia, and sepsis [9]. The primary approach to overcoming this problem is using wound dressings that contain antibacterial agents on the surface, which can gradually release the agents into the wounded area [10,11]. Effective antibacterial wound dressings have recently attracted a great deal of attention from researchers because they can inhibit the growth of Gram-positive and Gram-negative bacteria in the wound area [12].

Fibrous-containing bioactive agents are one of the eligible candidates for dressings employed for the promotion of wound healing. The range of the fibers' diameter is from several micrometers down to a few nanometers. Their high surface area and high porosity subsequently resulted in effective controlled delivery of drugs and different biological molecules to the surrounding environment [10,11].

Researchers bestow a high value on natural polymers like chitin, chitosan (CS), and collagen due to their excellent antimicrobial and biological properties [13]. Innovative efforts have been made to produce nanofibrous dressings of multi-polymer mixtures—incredibly natural and synthetic polymer mixtures—for environmental use. In this case, most studies have investigated CS and polycaprolactone (PCL) for the fabrication of films, membranes [14], and electrospinning fibers [15,16]; however, a few studies have considered mixtures of these [17]. A one-layer wound dressing cannot meet all clinical needs at the same time because it does not have the characteristics of an ideal dressing, including surface wettability, antibacterial and anti-inflammatory activity, biodegradability, and biocompatibility [18]. Thus, bilayer dressings comprising two layers with various properties have received significant attention [19]. A high-density upper layer can protect the wound from infection and mechanical stress as well as prevent wound dehydration, and it can also provide a humid environment in the wound region [20]. Subsequently, the sub-layers come into contact directly with the wound area and should mimic the ECM structure to facilitate cell adherence and accelerate cell proliferation [21].

Among the biomaterials applied in manufacturing, CS, as an antibacterial wound dressing, significantly improves the antibacterial activity against Gram-positive and Gram-negative bacteria and limits infection in the wound area [21,22]. Also, it has been reported that positively charged CS interacts with negatively charged microbial cell membranes [23]. The negative charge of lipopolysaccharides present in Gram-negative bacteria membranes engages in an ionic type of binding with the positively charged amino groups of CS. Consequently, this interaction leads to impeded nutrient flow and eventual bacterial death [22]. Additionally, CS-based wound dressings provide positive effects on all wound-healing stages, such as the synthesis of hyaluronic acid and type II collagen [24]. CS is a biocompatible, biodegradable, and non-toxic polysaccharide with excellent properties, including adhesive capability and antimicrobial/antifungal activity [25]. Despite these benefits, it has drawbacks such as low dissolvability in water, poor mechanical efficiency, and poor dispensability consistency [26].

PCL is a common synthetic polymer in biomedical applications owing to its biocompatibility, safety, and resistance. Although electrospun PCL webs are structurally similar to the ECM in living tissues, their hydrophobicity causes flow cell loading and inhibits cell adhesion, proliferation, and differentiation [26,27]. PCL's hydrophobicity should be reduced to improve its biological properties. Further, PCL blending with other hydrophilic polymers can provide a fruitful approach due to enhancing the overall material properties [28,29]. Correspondingly, apart from highly required biological properties, the dressing should also have the appropriate mechanical properties depending on the tissue, which has been proposed with the highly mechanical properties of PCL [30].

CS provides moderate antimicrobial activity due to its cationic excess. The antibacterial activity of silver species has been well known since ancient times [31]. Studies have shown that a double-layer wound dressing consisting of silver nanoparticles (Ag NPs) and CS can significantly accelerate the healing rate of skin wounds [32]. The antibacterial and antifungal behavior of biosynthesized Ag NPs against various microorganisms has been widely noted [33]. Unlike antibiotics, Ag NPs are released after exposure to microbes and affect microorganisms [34]. In addition, other essential features of these NPs include their being non-toxic, biocompatible, heat-resistant, and very highly durable and their failure to create and increase resistance and compatibility in microorganisms. However, some researchers have suggested that blending CS and PCL can improve the disadvantages of CS, for instance, its spin ability and mechanical strength, during the electrospinning process [32,33]. The blending of PCL/CS introduced the advancement of dressing properties concerning an acceptable degradation rate, water retention, and vascularization during the wound-healing process. Thus, the idea of adding any additional layer of a suitable biopolymer onto CS/PCL matrices has been recommended to enhance PCL/CS's admissibility as a skin dressing [34,35].

Bromelain (BRO) is a major sulfhydryl proteolytic enzyme present in pineapple kernels and stems. This enzyme consists of a mixture of different thiol endopeptidases and non-protease components [36]. Studies have shown that BRO has multifarious therapeutic attributes, e.g., anti-oxidant, anti-inflammatory, immunomodulatory, anti-thrombotic, cardioprotective, fibrinolytic, and anti-cancer properties [37]. Hence, it has contributed to the treatment and relief of various diseases, such as angina pectoris, bronchitis, sinusitis, surgical trauma, osteoarthritis, diarrhea, and dermatological issues [38]. Furthermore, it has shown potential wound-healing activity by the efficient debridement of necrotic tissue and the modulation of inflammatory reactions [39].

Various biomimetic systems have been produced in recent years, comprising microporous gels and substrates with varying chemistry and topography. The ideal substrate for wound-healing applications should be biocompatible, biodegradable, and capable of supporting cell development in a manner comparable to the in vivo microenvironment. Although the effectiveness of microporous scaffolds was proved in some applications, they are not an actual replica of an ECM structure, because they influence cell binding. In addition, the bulk of ECM proteins are fibrous; however, nanofibrous dressings are more bio-mimicking [40]. Nanofibers (NFs) are principally interesting thanks to their simplicity of fabrication, high surface area to volume ratio, diversity of composition, customizable shape and physicochemical characteristics, bioactive compound loading potential, programmable release, and degradation kinetics [41,42].

Many natural and artificial polymers have been electrospun to create a three-dimensional (3D) ECM that resembles NFs [43]. Some recent research has suggested using merged PCL and CS for NF production due to the mechanical strength, processability, and biocompatibility of PCL and the ECM mimicking qualities of CS [44,45]. It is believed that CS-based dressings have been widely applied as wound dressings for the promotion of wound healing [46,47].

In the present study, biocompatible gel-based PCL/CS nanofibrous scaffolds were prepared using novel 3-nozzle electrospinning technology and then coated with Ag NPs and BRO by sprinkling these on the surface of the electrospun NFs. The present study aimed to investigate the synergic effects of Ag NPs and BRO on the antibacterial properties of composite NFs. The resultant physicochemical and in vitro properties in the presence and absence of an Ag NP and BRO coating layer on the wound dressing were scrutinized. An animal study containing a control, PCL/CS, and PCL/CS along with BRO and Ag NPs was conducted to exhibit the functionality of the developed products as a skin wound dressing.

2. Results and Discussion

2.1. Characterizations of Ag NPs

The diameter and shapes of Ag NPs are shown in Figure 1. In the SEM image, the average size of the NPs was attained at about 55 nm (Figure 1a). The particles' hydrodynamic size and distribution were examined by Particle Size Analyzer (VASCO, CORDOUAN, France). The particles are monodispersed, and their hydrodynamic size was about 135 nm, as represented in Figure 1b. DLS analysis requires a larger number of NPs compared to SEM, and it therefore provides significantly better statistics. However, in most cases, the hydrodynamic particle size obtained by DLS was larger than the values obtained by SEM [31].

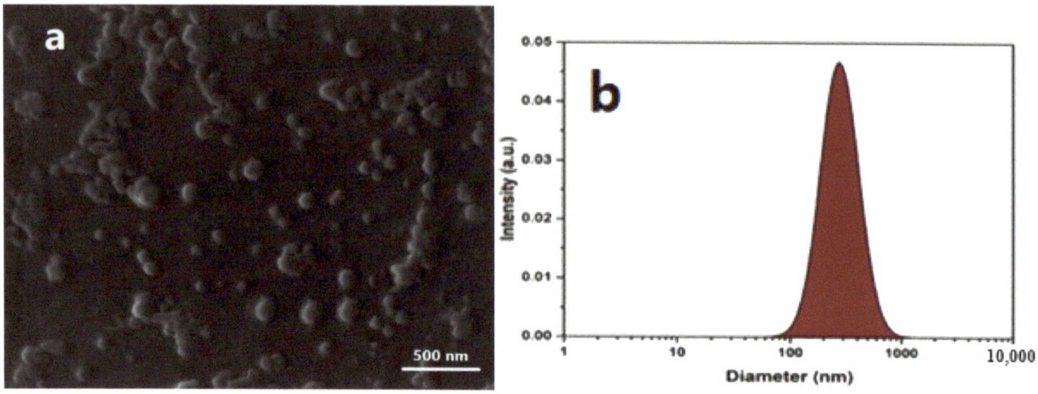

Figure 1. Scanning electron micrograph of Ag NPs at 20 kV (**a**) and hydrodynamic diameter and dispersity of the particles (**b**). In DLS analysis, the sample was used without dilution and was performed in triplicate.

2.2. Nanofibrous Scaffold Characterizations

NFs' shape, diameter, and morphology were observed and measured using SEM (Figure 2). Different polymer (i.e., PCL/Cs) ratios were investigated to obtain a consistent bead-less nanofibrous structure (Table 1 and Figure 2).

Table 1. Details of different electrospun scaffolds to select the proper blend ratio.

PCL/CS Ratio	Average Size ± (SD) nm	Bead Formation	Spinnability of the Solution
50:50	212 ± 25	High	Unsuccessful
60:40	123 ± 24	Rare	Successful
70:30	104 ± 21	Rare	Unsuccessful

The SEM micrographs of the PCL/CS (50-50) (average diameter 83 ± 27 nm) NFs exhibited spindle-like beads (Figure 2a). By increasing the PCL content (i.e., 60:40 PCL/CS), a uniform mesh and fine NFs with 126 ± 8 nm average diameter were acquired (Figure 2b). Further increasing the PCL ratio (70:30 PCL/CS) resulted in fibers with finer diameters (average diameter of 145 ± 73 nm); however, the chain entanglement was dropped, and the solution jet was broken during elongation and evaporation of the solvent (Figure 2c). Therefore, the 60:40 ratio was applied as an optimum mixture for further modification. Figure 2d shows the SEM image of PCL/CS containing Ag NPs. As detailed in the image above, the uniformity of the scaffolds increased, and a bead-less fiber was obtained (113 nm) [45]. The presence of conductive agents including Ag NPs produced enhanced electric charges carried out by the electrospinning jet [46], so the fiber diameter dropped significantly [45]. The SEM image showed white grains that may indicate the presence of

NPs in the substrate. By adding Bro, the fiber diameter was changed (i.e., 147 ± 16 nm), as shown in Figure 2e. The average diameter of the NFs along with both Ag NPs and Bro was obtained as 185 ± 24 nm, which is in the range of ECM collagen fibers (50 to 500 nm). Thus, it can be suggested that the produced nanofibers may mimic some structural features of the ECM and provide an environment for cell adhesion, proliferation, differentiation, and improved skin tissue regeneration at the wound site [48]. Figure 2f shows the SEM image of PCL/CS containing 60:40 Ag NPs + Bro (average diameter 176 ± 63 nm).

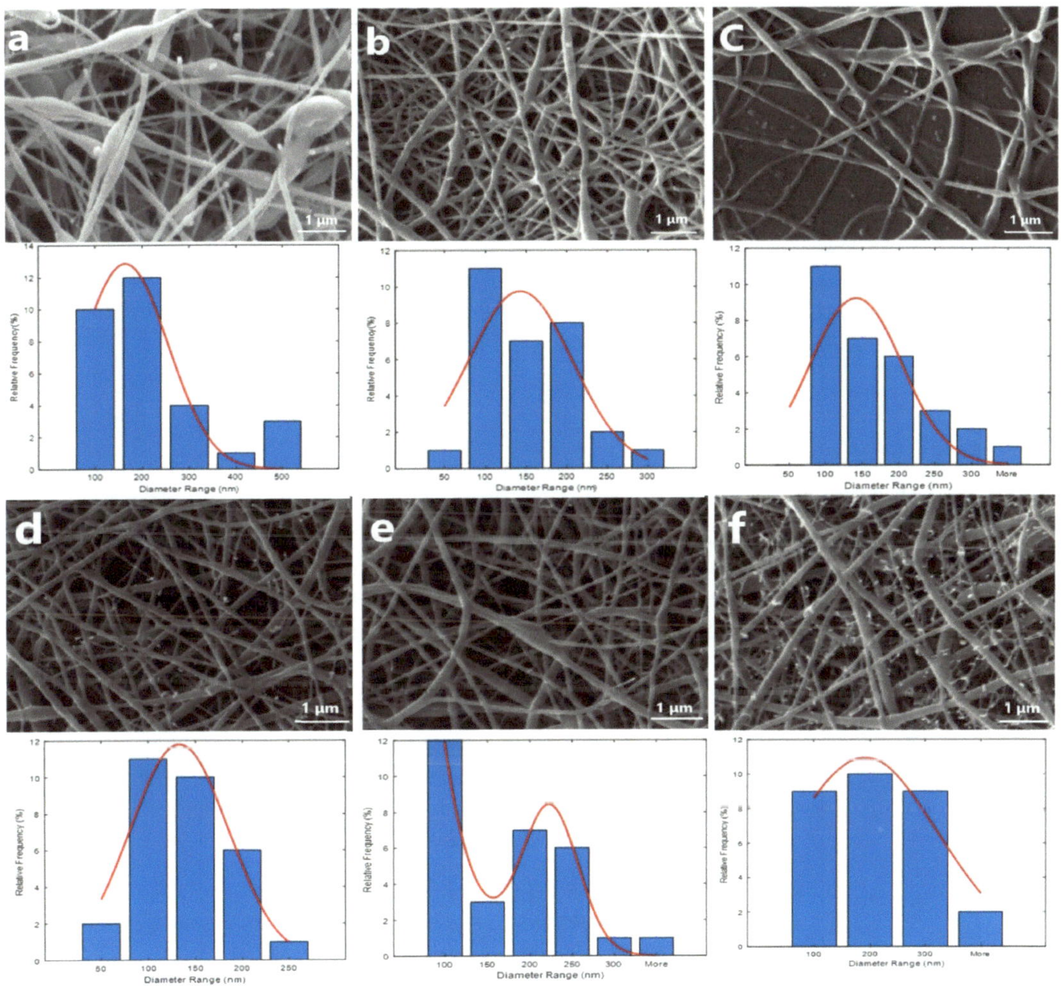

Figure 2. SEM micrographs of NFs show the morphology and random distribution of fibers: nanofibrous webs of PCL/CS with ratios of (**a**) 50:50, (**b**) 60:40, (**c**) 70:30, and (**d**) 60:40 containing Ag NPs, (**e**) 60:40 containing Bro, and (**f**) 60:40 containing Ag NPs + Bro. All samples were coated with a thin layer of gold to produce a conductive surface and were examined at 15 kV.

2.3. Fourier Transform Infrared Analysis

The IR spectra of all samples (CS/PCL, CS/PCL-Ag NPs, CS/PCL-BRO, and CS/PCL-Ag NPs-BRO) in Figure 3 confirmed the successful combination of CS and PCL in the electrospun NFs. Specific characteristic peaks at certain wavenumbers indicated the presence of PCL in the NFs (1723 cm^{-1}, 1293 cm^{-1}, and 1365 cm^{-1}). The spectra revealed the presence of CS, with characteristic peaks at 2864 cm^{-1}, 2943 cm^{-1}, and 1046 cm^{-1}, as well as a peak at 1723 cm^{-1} showing carboxylate adsorption [49–51].

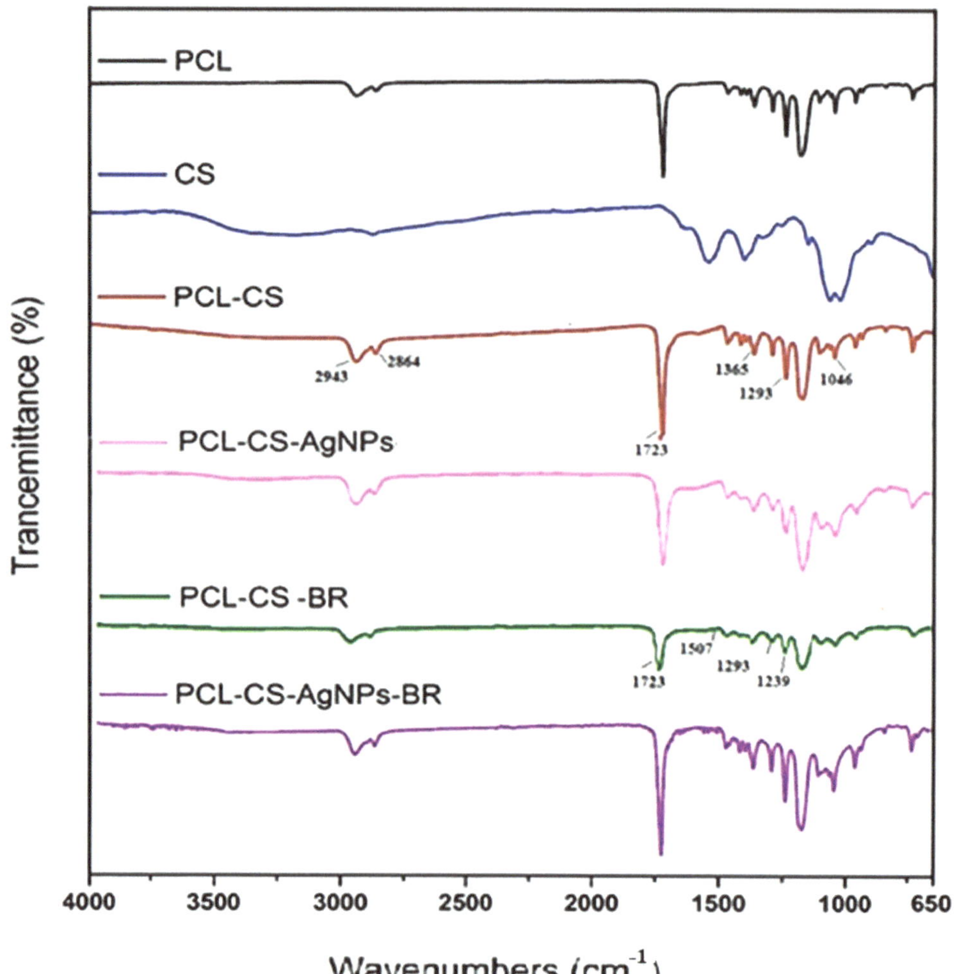

Figure 3. FTIR spectra of the gel-based PCL/CS, PCL/CS-BRO, PCL/CS-Ag NPs, and PCL/CS-BRO-Ag NPs scaffolds. For the FTIR, 1 mg of the samples was mixed with 300 mg of KBr under a vacuum. Then, IR spectra were recorded in the range of 400 and 4000 cm^{-1}.

In the FTIR spectrum of the PCL/CS/BRO scaffold, specific IR bands at 1239 cm^{-1} and 1293 cm^{-1} indicated the presence of guanidinium monoalkyl, and the peak at 1723 cm^{-1} confirmed the group C = O and the presence of amino acids like asparagine and glutamine [52]. For both PCL/CS-Ag NPs and PCL/CS-BRO-Ag NPs, the characteristic peaks observed at 3000–3650 cm^{-1} were attributed to symmetric vibrations of OH and NH. These peaks indicated interactions between Ag NPs and the OH and NH groups of CS, with the interaction decreasing as the amount of Ag NPs increased [53,54].

2.4. Mechanical Measurement

The tensile strength results of the obtained scaffolds (Figure 4) revealed that adding Ag NPs and BRO increased the tensile stress–strain curves due to chemical bonding with the base polymers [45]. The proposed NFs had even better tensile strength, attributed to the higher crosslinking of BRO with CS (2% *w*/*v*) compared to NFs with Ag NPs or both BRO and Ag NPs [55].

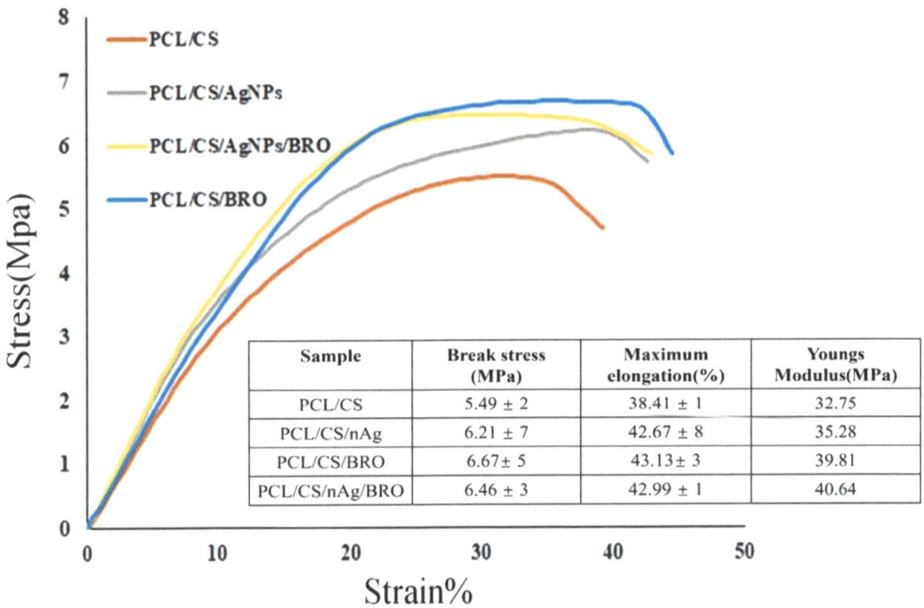

Sample	Break stress (MPa)	Maximum elongation(%)	Youngs Modulus(MPa)
PCL/CS	5.49 ± 2	38.41 ± 1	32.75
PCL/CS/nAg	6.21 ± 7	42.67 ± 8	35.28
PCL/CS/BRO	6.67± 5	43.13± 3	39.81
PCL/CS/nAg/BRO	6.46 ± 3	42.99 ± 1	40.64

Figure 4. The tensile stress–strain curves of gel-based electrospun nanofibrous scaffolds along with the related parameters. The samples were cut into 2 × 5 cm^2 pieces and were applied to the jaws at 1 mm/min and 500 N load cell. The test was performed three times for each sample, and the values were reported as mean ± standard deviation (SD).

2.5. Contact Angle of NFs

The hydrophilicity of nanofibrous dressings is a central factor for biocompatibility and cellular interactions such as cell adhesion, differentiation, and proliferation [48,56]. The contact angle measurement results obtained from gel based PCL/CS (a), PCL/CS-BRO (b), PCL/CS-Ag NPs (c), and PCL/CS-Ag NPs-BRO (d) NFs were recorded after 3 s (Figure 5). The PCL/CS NFs displayed inherent hydrophilicity with a contact angle of 60° due to the presence of CS as a rather hydrophilic polymer [53] (Figure 5a). As shown in Figure 5b, the PCL/CS-BRO NFs exposed a contact angle of 69°, which proves hydrophobic interaction due to non-polar chain amino acids in the BRO structure [54]. On the contrary, according to Figure 5c, by adding Ag NPs, the hydrophilicity of NFs (56°) improved due to an increase in crosslinking with the polymer structure. Similarly, the findings were consistent with the background hypothesis presented by Li and colleagues [57], who stated that Ag NPs might enhance membrane surface hydrophilicity. Following Figure 5d, where both Ag NPs and BRO are presented, the contact angle increased, and the hydrophilicity decreased. Therefore, adding both BRO and Ag NPs did not have much of an effect on the hydrophilicity and hydrophobicity of the initial obtained NFs. All samples have a contact angle between 40 and 70°, which is suitable for cell attachment and proliferation.

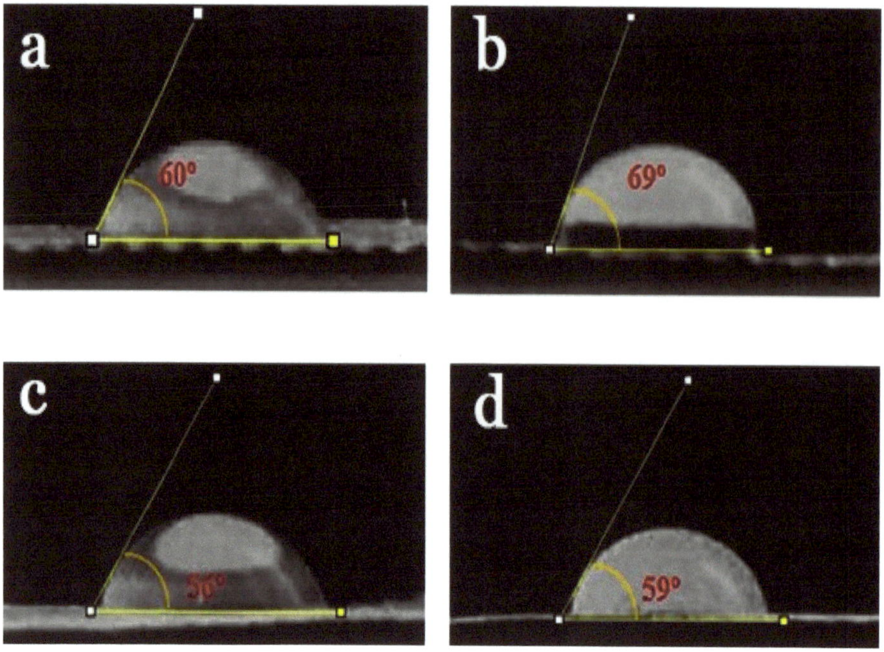

Figure 5. Contact angle images of PCL/CS (**a**), PCL/CS-BRO (**b**), PCL/CS-Ag NPs (**c**), PCL/CS-BRO-Ag NPs (**d**) NFs. The test dressings were cut to 15 × 20 mm^2, and a drop of distilled water (2 μL) was spilled on each sample surface. The droplet figure was photographed after 3 s.

2.6. BRO Activity

The results of an azocasein test indicated that upon adding bromelain to the nanofiber structure, its activity reached 87% of the initial value. This effect may be attributed to the electrospinning process as well as a possible disruption in the enzyme's release from the fiber structure [36,37].

2.7. Antibacterial Results

The antibacterial effects of electrospinning the NFs, BRO solution, and Ag NPs were determined against *Escherichia coli* (*E. coli*) (ATCC 25922) and *Staphylococcus aureus* (*S. aureus*) (ATCC 25923). The results of the antimicrobial test are displayed in Figure 6. According to the plate images, sample 2b, 1d, which is a gel-based PCL/CS NF, was not observed with a significant inhibition zone and did not provide inhibitory activity on the growth of both bacterial strains. Although CS's antibacterial activity has been proven in many studies [56], its antibacterial activity is affected by the molecular weight, degree of deacetylation, and concentration [58]. It was observed that (1b, 2d) Ag NPs exhibited the largest inhibition zone at about 24 ± 2 and 19 ± 1 mm against *E. coli* and *S. aureus*, respectively. Based on the articles and conducted experiments, the presence of Ag NPs provided antibacterial properties in the obtained NFs, because bacteria are affected by silver ions via their release into the bacterial cell wall, where they then destroy the bacteria [53,59]. Moreover, the ions can remove the bacterial respiratory enzyme that helps bacteria survive. It can be noted that the reduced size of Ag NPs renders a significant effect on the antibacterial property [53,60,61].

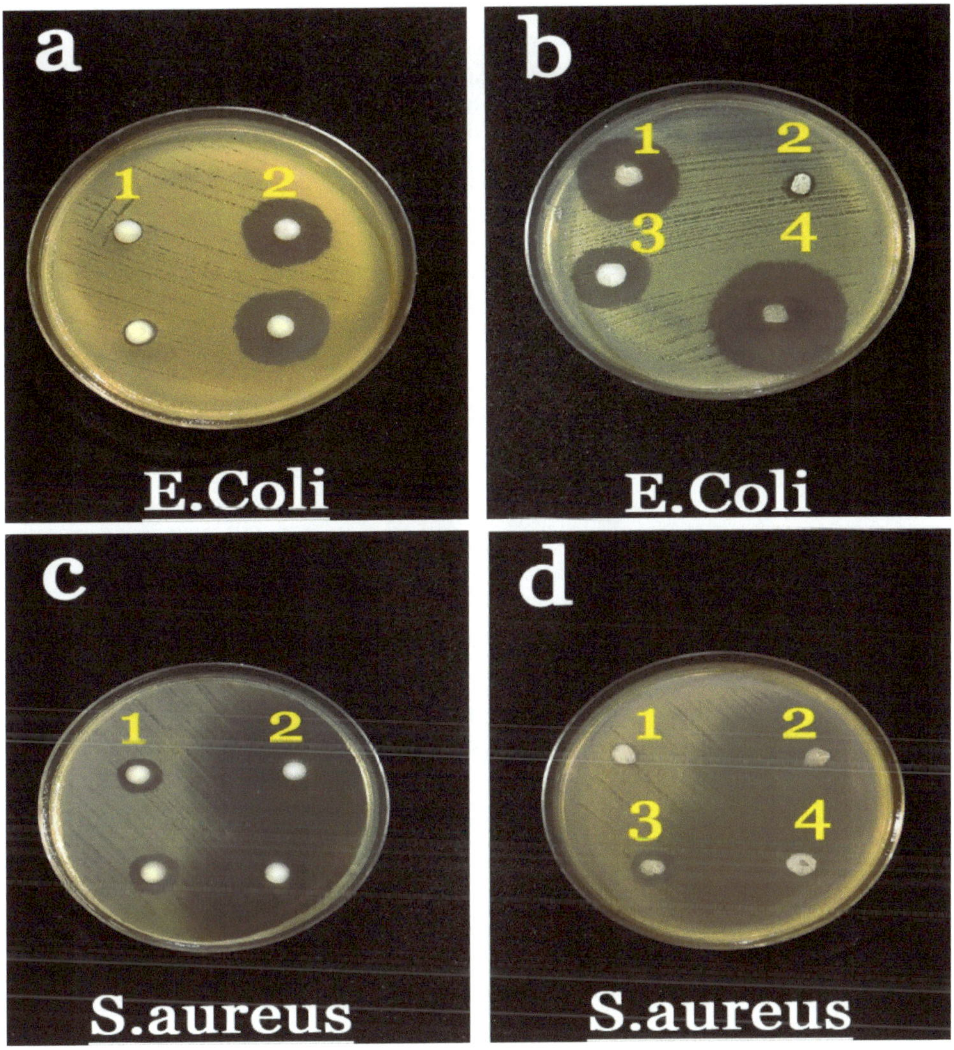

Figure 6. Antibacterial susceptibility by the zone inhibition against *E. coli* (**a**,**b**) and *S. aureus* (**c**,**d**): free BRO (1a, 1c), free Ag NPs (2a, 2c), PCL/CS (2b, 1d), PCL/CS-Ag NPs (1b, 2d), PCL/CS-BRO (3b, 3d), and PCL/CS-Ag NPs-BRO (4b, 4d).

In the sample 3b, 3d, which contained BRO, antibacterial activity with an inhibition zone was attained around 10 ± 2 mm on *E. coli*; however, the sample did not provide high inhibitory activity against *S. aureus* (6 ± 1 mm). The suppression of growth and growth inhibition of both strains was observed in the samples (4b, 4d). The obtained results revealed that the largest inhibition zone reached about 25 ± 2 and 12 ± 1 mm against *E. coli* and *S. aureus*, respectively.

Overall, the antibacterial activity of NFs against *S. aureus* was higher than that against *E. coli*. This observation is due to the fact that the cell wall structure of these bacteria is dissimilar. The outer membrane of Gram-negative bacteria contains lipopolysaccharides and phospholipids in its outer and inner cortex, respectively; however, Gram-positive bacteria do not contain a lipopolysaccharide layer.

2.8. MTT Assay Results

The in vitro viability activities of samples were assessed by MTT assay for 24 h. According to Figure 7, The sample containing gel-based PCL/CS NFs showed high cytotoxicity in mouse fibroblast cells within 24 h (Table 2). BRO NPs displayed higher cytotoxicity than that of Ag NPs. This is owing to the fact that Ag NPs can cause adverse biological effects such as cytotoxicity and genotoxicity effects, which are dose-dependent [32]. In several studies, the use of low concentrations of Ag NPs has been investigated for high survival and cell proliferation [59,62]. PCL/CS-Ag NPs-BRO NFs improved cell viability after 24 h because of the positive effect of Ag NPs and BRO on the gel-based PCL/CS NFs. The degree of cytotoxicity is negligible and close to that of the control sample. Thus, the proposed PCL/CS-Ag NPs-BRO mat could be applied as a suitable structure for cell adhesion, growth, and migration.

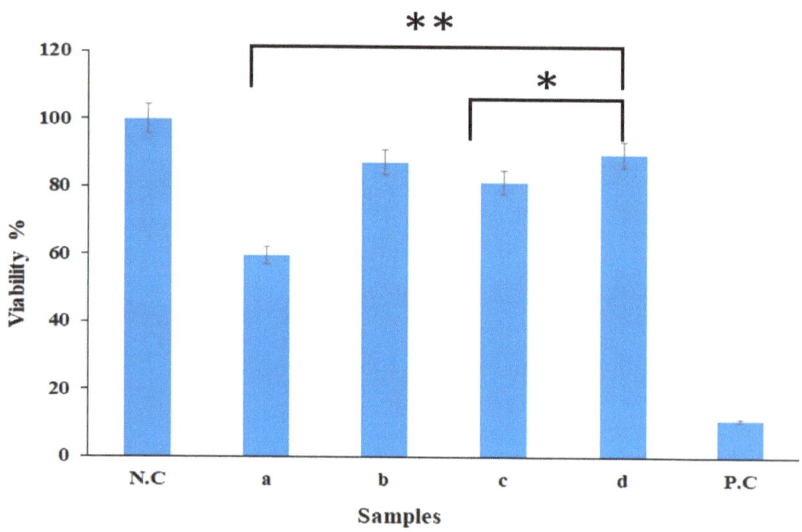

Figure 7. Cell viability by the MTT assay: (N.C) negative control, (a) PCL/CS, (b) PCL/CS-Ag NPs, (c) PCL/CS-BRO, and (d) PCL/CS-BRO-Ag NPs, (P.C) positive control after 24 h. * $p < 0.05$ and ** $p < 0.01$ using Mann–Whitney U test.

Table 2. Toxicity% by the MTT assay: (N.C) negative control, (a) PCL/CS, (b) PCL/CS-Ag NPs, (c) PCL/CS-BRO, and (d) PCL/CS-BRO-Ag NPs, (P.C) positive control after 24 h.

Samples	N.C	a	b	c	d	P.C
Toxicity %	0.10 ± 0.02	40.49 ± 3.21	12.89 ± 1.57	18.79 ±1.23	10.44 ± 1.17	88.98 ± 5.27

2.9. Animal Study

The results of the healing process from a second-degree burn model indicated that there was no significant difference between the groups treated with wound dressings and the control group on the first and third days (Figure 8). However, on the fifth day, both samples treated with wound dressings, namely, PCL/CS and PCL/CS-BRO-Ag NPs, exhibited a significant difference ($p < 0.05$) compared to the control, indicating an acceleration of the healing process. By the seventh day, the rate of wound healing in the control sample and the sample containing the PCL/CS wound dressing was similar. However, the group treated with NFs containing bromelain and Ag NPs showed a significant difference compared to the other groups. The presence of Ag NPs reduces the probability of infection at the wound site.

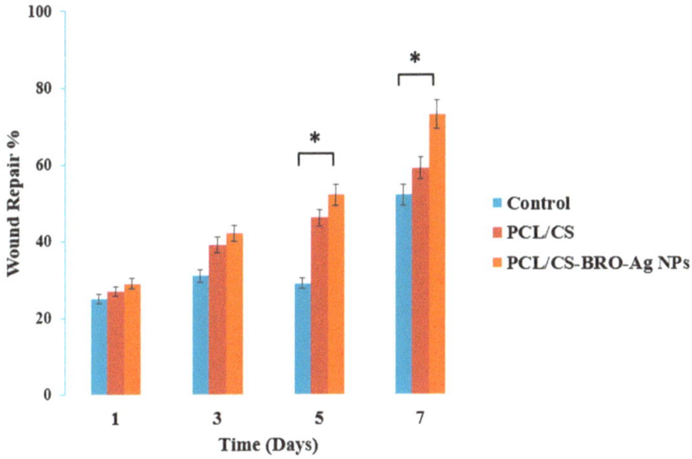

Figure 8. Wound repair % of 2nd degree burn rat model on different days. * $p < 0.05$ using ANOVA test.

Furthermore, Ag NPs effectively transformed fibroblasts into myofibroblasts, facilitating quick wound closure and promoting the proliferation and migration of keratinocytes [63,64]. Bromelain also reduces inflammation and swelling by suppressing kinin production, thereby accelerating wound healing [65]. Figure 9 shows that after 7 days, the skin layers in the group treated with PCL/CS-BRO-Ag NPs are more obviously formed, and the dermis layer is visible.

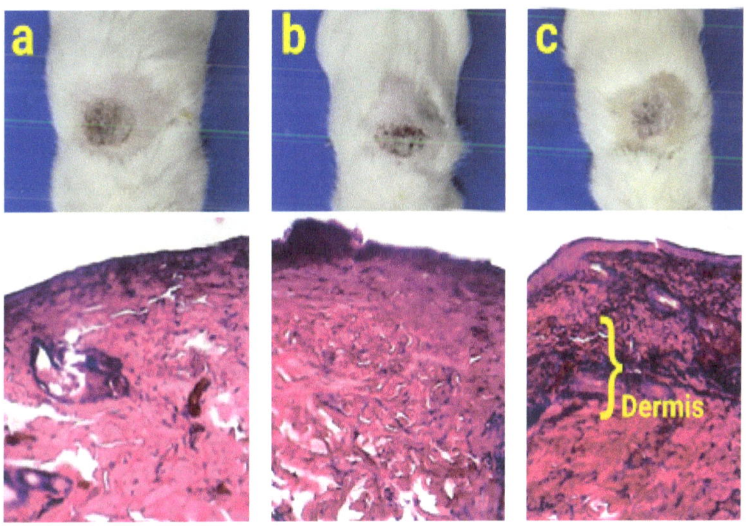

Figure 9. Rat skin wound repair and related H&E staining of wound-healing histology in different groups after 7 days: (**a**) control, (**b**) PCL/CS, (**c**) PCL/CS-BRO-Ag NPs.

3. Conclusions

We evaluated the preparation, characteristics, and antibacterial and cell viability properties of a gel-based electrospun PCL/CS dressing co-loaded with Bro and Ag NPs to promote wound healing. The optimized electrospun PCL/CS (60:40) nanofibrous mat was co-loaded with BRO and Ag NPs through the electrospray method, which led to the formation of homogenous NFs with a nano-scale diameter and uniform morphology. The relevant analyses specified the successful loading and structural stability of the components. Also, the results indicated the high mechanical strength and suitable hydrophilicity of the prepared mats. The gel-based PCL/CS-Ag NPs-BRO nanofibrous dressing provided remarkable antibacterial activities against both Gram-positive and -negative bacteria. Moreover, the dermal fibroblast viability measurement confirmed the mat's high biocompatibility and proper cell proliferation effects. The animal study displayed that the obtained mat enhanced the speed of wound healing over a week. Therefore, gel-based PCL/CS nanofibrous wound dressings co-loaded with BRO and Ag NPs can be exploited as an efficient scaffold for skin regeneration and wound-healing applications.

4. Materials and Methods

4.1. Materials

Chitosan (medium molecular weight, DD of 85%) and polycaprolactone (average Mn 80 kDa) were purchased from Sigma Aldrich, St. Louis, MO, USA. Formic acid, acetic acid, trisodium citrate, and silver nitrate were purchased from Merck Millipore, Darmstadt, Germany. Bromelain was purchased from Biozym, Hamburg, Germany, and other materials were provided from authentic centers unless otherwise specified.

4.2. Synthesis of Ag NPs

A simple reduction method was applied to fabricate Ag NPs [42]. Briefly, trisodium citrate (1% w/v) and silver nitrate (0.02% w/v) solutions were prepared in distilled water. Then, the silver nitrate solution was heated to boiling temperature using a magnetic heater stirrer. After that, the trisodium citrate solution was added dropwise to the boiling silver nitrate until the solution turned colored (pale yellowish-brown). Then, the heating was stopped, but dispersing was started by raising the stiller round. Then, the solution was placed on another stiller with a temperature of 25 °C for 15 min at 1000 rpm. The NPs obtained were stored in amber bottles at 4 °C. The formed colloidal Ag NPs were characterized and measured until day 14 after synthesis by DLS and SEM. These NPs were the reference Ag NPs for the study.

4.3. Fabrication of Free PCL/CS NFs

The fabrication of gel-based PCL/CS NFs started by mixing PCL (14% w/v) and CS (2% w/v) in a 3:1 ratio of formic acid: acetic acid solvent. After 2 h, the mix of CS and PCL with different blending ratios (Table 1) was kept on the magnetic stirrer at 250 rpm for 30 min. Then, the solution was electrospun under ambient conditions (a temperature of 22 to 25 °C and relative humidity of 45% to 52%) at a flow rate of 100 mL/h and with a 10 cm distance between the needle and collector, under 16 kV electrical potential.

4.4. Fabrication of PCL/CS-Ag NPs-BRO Mat

NFs were fabricated by electrospinning technique at room temperature. In the first step, the optimal polymer solution PCL/CS (60:40) (by mixing 14 % PCL w/v and 2% CS w/v in a 3:1 formic acid: acetic acid solvent system) was filled in a 1 mL syringe and equipped with a 29G needle. In addition, 1% w/v of Ag NPs and 2% w/v of BRO were prepared separately and then placed in individual syringes for the electrospinning process. In the next step, according to Figure 8, for fabrication of PCL/CS NFs including Ag NPs and BRO, the solutions of Ag NPs and BRO were separately electrosprayed on the surface of the base NFs. The solutions were drawn from the syringes with a flow rate of 0.1 mL/h using a syringe pump and were spun at a 10 cm distance between the needle and collector plate, under an electrical field of 16 kV potential. These processing parameters were standardized after several variations to obtain NFs of preferred quality. The resultant wound dressings (PCL/CS, PCL/CS-Ag NPs, PCL/CS-BRO, and PCL/CS-Ag NPs-BRO) were collected on 5×5 cm^2 drum that was pre-adhered to aluminum foil for further analysis according to Figure 10.

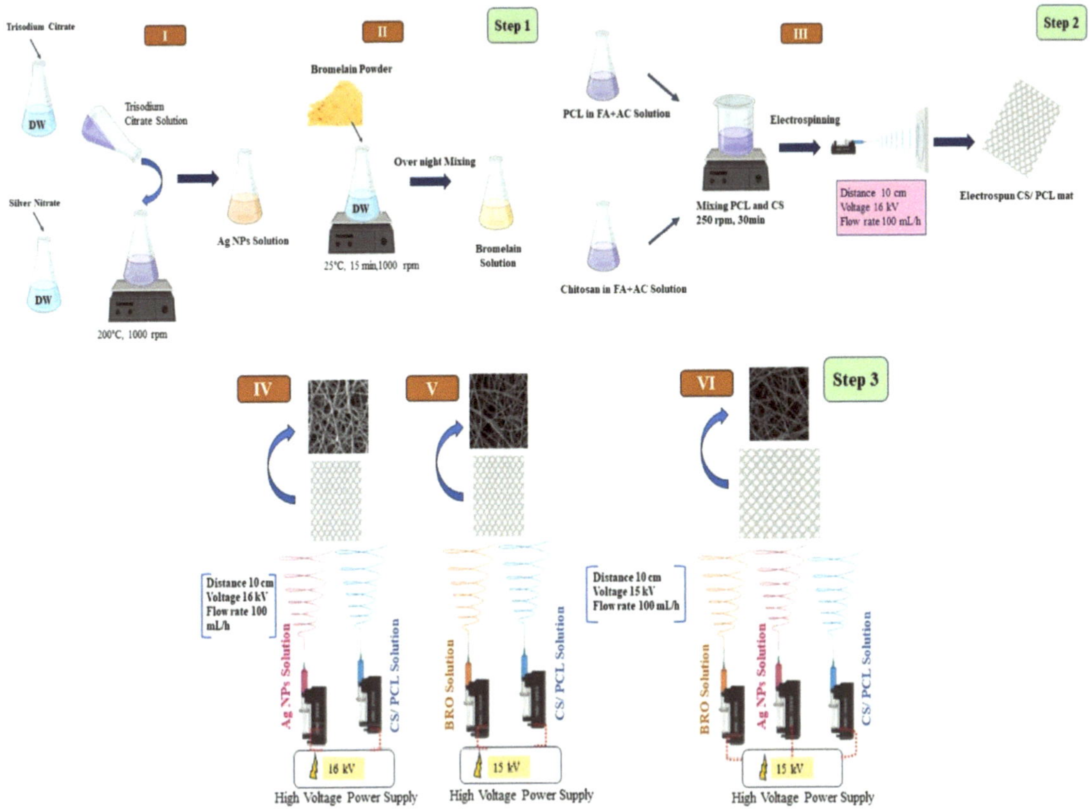

Figure 10. Fabrication processes of electrospinning. Step 1: Fabrication of Ag NPs (**I**) and BRO (**II**), step 2: fabrication of PCL/CS NFs (**III**), step 3: fabrication of PCL/CS-Ag NPs (**IV**), PCL/CS-BRO (**V**), and PCL/CS-Ag NPs-BRO (**VI**) scaffolds.

4.5. Washing and Sterilization of Nanofibrous Dressing

Before performing the extraction procedure, all attained gel-based NFs, including PCL/CS, PCL/CS Ag NPs, PCL/CS-BRO, and PCL/CS-BRO-Ag NPs mats, were washed at least three times with distilled water to ensure that the free materials were removed. Then, they were sterilized under a UV lamp (265 nm) for 4 h in a laminar flow cabinet (Pasteur Institute of Iran Class II Biological Safety Cabinet, Tehran, Iran) to keep the structural properties of the NFs intact.

4.6. Characterization of Nanofibrous Dressing

4.6.1. Morphology Observation

The morphology of all nanofiber scaffolds was observed by a scanning electron microscope (SEM) (AIS2100, SERON TECHNOLOGIES, Anseong, Korea). In this regard, samples were coated with a thin layer of gold to produce a conductive surface and reduce charging.

4.6.2. Fourier Transform Infrared Spectroscopy

The chemical characteristics of raw materials and the structures of samples were evaluated using Fourier transform infrared (FTIR) spectra (Nicolet Magna-IR 560) by the KBr method. As such, 1 mg of the materials and samples were mixed with 300 mg of KBr under a vacuum. Then, IR spectra were recorded in the range of 400 to 4000 cm^{-1}.

4.6.3. Tensile Testing

Tensile tests of the nanofiber membranes were carried out with a Universal Testing Machine (Instron TM—SM, England). The samples were cut into a rectangular shape with equal sizes of 2×5 cm^2 and then placed in the middle of a paper frame with dimensions of 5×5 cm^2 to prevent sample damage during the study. The jaws were spaced at a 1 mm/min speed by applying a 500 N load cell (according to the ASTM D 882 standard) until rupture occurred, and the stress–strain curve was plotted. The average values of tensile properties were achieved from the results of the four tests and expressed as the mean $\pm$ standard deviation (SD).

4.6.4. Contact Angle Determination Test

All gel-based NFs were cut to 15×20 mm^2 and were placed on a retentive base. A drop of distilled water (2 µL) was spilled onto each sample surface. The droplet image was photographed after 3 s, and the contact angle was gauged by ImageJ software.

4.6.5. BRO Activity

Enzyme activity was measured using the azocasein method (REF). Briefly, a 1.0% *w*/*v* solution of azocasein was prepared in 0.2 M Tris-HCL and 1 mM calcium chloride. The pH of the solution was then adjusted to 7.2. The enzyme activity was measured both before and after combination into the fiber structure. The test was conducted in triplicate at 4 °C.

4.6.6. Inhibition Zone Assay (Antibacterial Activity Measurements)

The antibacterial properties of the samples were evaluated according to the ISO 20645: 2004 standard Disc diffusion method. The samples were tested against *E. coli* (ATCC 25922) and *S. aureus* (ATCC 25923) as a model for Gram-negative and Gram-positive bacteria, respectively [49]. One hundred milliliters of *E. coli* and *S. aureus* were cultured on Luria–Bertani (LB) agar plates used for the antibacterial activity tests. Gel-based PCL/CS NFs were used as the control samples. The PCL/CS containing Ag NPs, BRO, and Ag NPs/BRO were applied as the test samples. The obtained dressings were cut into small circular pieces (5 mm) and placed over the solidified agar gel. Then, the plates were incubated for 24 h at 37 °C in a bacteriological incubator. Subsequently, the zone of bacterial inhibition was monitored.

4.6.7. Cytotoxicity Test (MTT Assay)

An MTT assay was carried out to determine the potential cytotoxicity and viability of fibroblast cells involved in cutaneous wound healing [66]. The L929 cell (NCBI C161) (Pasteur Institute, Iran) was exploited for the proliferation study. After defrosting the cells, they were moved to a flask containing DMEM medium with 10% FBS, and then the flask was incubated at 37 °C and 5% carbon dioxide [44]. The extraction process of the sterilized samples was carried out according to the ISO 10993-5:2009 standard.

All samples were cut into circular pieces and sterilized by exposing them to UV light at 254 nm for 4 h. Subsequently, the samples were positioned at the bottom of the wells in a 96-well microplate using sterilized silicon rubber washers. A total of 10,000 cells in 100 µL of DMEM containing 10% FBS were added to each well. The entire plate was then incubated at 37 °C for 24 h. Then, the culture medium was removed, and 100 µL of MTT solution (0.5 mg/mL) was added to each well, followed by incubation in the dark for 4 h. Finally, the MTT solution was removed, and 50 µL of isopropanol was added to dissolve the purple formazan crystals. The plate was placed on a shaker for 15 min. All assays were carried out in triplicate. The same method was applied to the control group, but the cells were not treated with the samples. Then, the absorption of each well was read out by a microplate reader (STAT FAX 2100, USA) at 545 nm. The cell viability of the samples was calculated using Equation (1):

$$\% \text{ Cell viability} = (\text{Mean OD of sample}/\text{Mean OD of control}) \times 100 \quad (1)$$

4.6.8. Animal Study

Male Wistar rats weighing approximately 250 g was used for the animal study. The animal experiments were carried out following the guidelines established by the Ethic Committee for Animal Experiments at Science and Research Branch, Islamic Azad University (Ethical code number: IR.IAU.SRB.REC.1401.235). The studies were also in accordance with the EU Directive 2010/63/EU for animal experiments and any subsequent amendments, or similar ethical standards. The animals were housed in separate cages with a 12 h light and 12 h dark cycle and had unrestricted access to food and water. To establish a burn model, the rats were randomly divided into three groups, each consisting of 5 rats. One group served as the control, another was treated with PCL/CS NFs, and the third received the PCL/CS-BRO-Ag NP wound dressing. Subsequently, the rats were anaesthetized by intraperitoneal injection of ketamine (80 mg/kg) and xylazine (10 mg/kg), and the hair on their backs was shaved. A circular metal piece with a 1 cm diameter was immersed in boiling water to induce second-degree burns. After 5 min, it was placed on the rats' back skin for 10 s [67].

The day of the burning procedure was designated as day zero. On the 3rd, 5th, and 7th days following the burns, images of the wounds were captured, and the wound area was calculated using image analysis software (ImageJ). The healing rate of the wounds was determined using Equation (2). On the seventh day, the skin at the wound site was excised, fixed in 10% formalin, and used to prepare a pathological slide. Hematoxylin-eosin staining was employed to examine the formation process of different skin layers.

$$Wound\ repair\ \% = (W_0 - W_1)/W_0 \times 100 \quad (2)$$

Here, W_0 is the initial area of the wound, and W_1 is the area of the wound on the day of the study.

4.7. Statistical Analysis

The experiments were conducted with a minimum of three repetitions, and the average values, along with their corresponding standard deviations, were described (mean $\pm$ SD). One-way ANOVA and Mann–Whitney U tests were employed to compare groups, considering a significance level of $p < 0.05$. Statistical analysis and graph generation were performed using Excel and SPSS (V. 16.0) software.

Author Contributions: Conceptualization, Y.S. and H.B.; Methodology, Y.S., H.B., N.N. and K.K.; Software, F.S. and K.K.; Validation, H.M. and K.K.; Formal analysis, Y.S., H.B., N.N. and H.M.; Investigation, Y.S., H.B., H.M., S.M.S.-J. and F.S.; Resources, H.B., S.M.S.-J. and A.B.; Data curation, K.K.; Writing—original draft, Y.S., A.A., H.M., S.M.S.-J., A.B., F.S. and K.K. Writing—review & editing, H.B., N.N. and K.K.; Visualization, A.A., F.S. and K.K.; Supervision, H.B.; Project administration, H.B., A.A., A.B. and F.S.; Funding acquisition, A.A., S.M.S.-J. and A.B. All authors have read and agreed to the published version of the manuscript.

Funding: This research received no external funding.

Institutional Review Board Statement: The animal experiments were carried out following the guidelines established by the Ethic Committee for Animal Experiments at Science and Research Branch, Islamic Azad University (Ethical code number: IR.IAU.SRB.REC.1401.235).

Informed Consent Statement: Not available.

Data Availability Statement: Data is available upon request.

Conflicts of Interest: The authors declare no conflict of interest.

References

1. Ahmed, J.; Mulla, M.; Arfat, Y.A.; Thai, T.L.A. Mechanical, Thermal, Structural and Barrier Properties of Crab Shell Chitosan/Graphene Oxide Composite Films. *Food Hydrocoll.* **2017**, *71*, 141–148. [CrossRef]

2. Chen, Q.; Xu, A.; Li, Z.; Wang, J.; Zhang, S. Influence of Anionic Structure on the Dissolution of Chitosan in 1-Butyl-3-Methylimidazolium-Based Ionic Liquids. *Green Chem.* **2011**, *13*, 3446–3452. [CrossRef]
3. Bajgai, M.P.; Kim, K.W.; Chandra Parajuli, D.; Yoo, Y.C.; Kim, W.D.; Khil, M.S.; Kim, H.Y. In Vitro Hydrolytic Degradation of Poly(ε-Caprolactone) Grafted Dextran Fibers and Films. *Polym. Degrad. Stab.* **2008**, *93*, 2172–2179. [CrossRef]
4. Loesberg, W.A.; Walboomers, X.F.; Van Loon, J.J.W.A.; Jansen, J.A. The Effect of Combined Hypergravity and Microgrooved Surface Topography on the Behaviour of Fibroblasts. *Cell Motil. Cytoskelet.* **2006**, *63*, 384–394. [CrossRef] [PubMed]
5. Lindahl, U.; Höök, M. Glycosaminoglycans and Their Binding to Biological Macromolecules. *Annu. Rev. Biochem.* **1978**, *47*, 385–417. [CrossRef]
6. Liao, S.; Li, B.; Ma, Z.; Wei, H.; Chan, C.; Ramakrishna, S. Biomimetic Electrospun Nanofibers for Tissue Regeneration. *Biomed. Mater.* **2006**, *1*, R45–R53. [CrossRef]
7. Martínez-Castañón, G.A.; Niño-Martínez, N.; Martínez-Gutierrez, F.; Martínez-Mendoza, J.R.; Ruiz, F. Synthesis and Antibacterial Activity of Silver Nanoparticles with Different Sizes. *J. Nanoparticle Res.* **2008**, *10*, 1343–1348. [CrossRef]
8. Hajji, S.; Khedir, S.B.; Hamza-Mnif, I.; Hamdi, M.; Jedidi, I.; Kallel, R.; Boufi, S.; Nasri, M. Biomedical Potential of Chitosan-Silver Nanoparticles with Special Reference to Antioxidant, Antibacterial, Hemolytic and in Vivo Cutaneous Wound Healing Effects. *Biochim. Biophys. Acta Gen. Subj.* **2019**, *1863*, 241–254. [CrossRef]
9. Gnanadesigan, M.; Anand, M.; Ravikumar, S.; Maruthupandy, M.; Syed Ali, M.; Vijayakumar, V.; Kumaraguru, A.K. Antibacterial Potential of Biosynthesised Silver Nanoparticles Using Avicennia Marina Mangrove Plant. *Appl. Nanosci.* **2012**, *2*, 143–147. [CrossRef]
10. Amoabediny, G.; Salehi-Nik, N.; Heli, B. The Role of Biodegradable Engineered Scaffold in Tissue Engineering. In *Biomaterials Science and Engineering*; InTech: Rijeka, Croatia, 2011.
11. Gunn, J.; Zhang, M. Polyblend Nanofibers for Biomedical Applications: Perspectives and Challenges. *Trends Biotechnol.* **2010**, *28*, 189–197. [CrossRef] [PubMed]
12. Arsad, N.K.; Samad, A.A.; Jamaluddin, H. Isolation of Proteolytic Enzyme from Pineapple Crown. *Pertanika J. Trop. Agri. Sci.* **2023**, *46*, 607–626. [CrossRef]
13. Chobotova, K.; Vernallis, A.B.; Majid, F.A.A. Bromelain's Activity and Potential as an Anti-Cancer Agent: Current Evidence and Perspectives. *Cancer Lett.* **2010**, *290*, 148–156. [CrossRef]
14. Chakraborty, A.J.; Mitra, S.; Tallei, T.E.; Tareq, A.M.; Nainu, F.; Cicia, D.; Dhama, K.; Emran, T.B.; Simal-Gandara, J.; Capasso, R. Bromelain a Potential Bioactive Compound: A Comprehensive Overview from a Pharmacological Perspective. *Life* **2021**, *11*, 317. [CrossRef]
15. Hikisz, P.; Bernasinska-Slomczewska, J. Beneficial Properties of Bromelain. *Nutrients* **2021**, *13*, 4313. [CrossRef] [PubMed]
16. Doostan, M.; Doostan, M.; Mohammadi, P.; Khoshnevisan, K.; Maleki, H. Wound Healing Promotion by Flaxseed Extract-Loaded Polyvinyl Alcohol/Chitosan Nanofibrous Scaffolds. *Int. J. Biol. Macromol.* **2023**, *228*, 506–516. [CrossRef]
17. Khalilimofrad, Z.; Baharifar, H.; Asefnejad, A.; Khoshnevisan, K. Collagen Type I Cross-Linked to Gelatin/Chitosan Electrospun Mats: Application for Skin Tissue Engineering. *Mater. Today Commun.* **2023**, *35*, 105889. [CrossRef]
18. Davarchi, S.M.; Kaffashi, B.; Torabinejad, B.; Zamanian, A. In-Vitro Investigation and Hydrolytic Degradation of Antibacterial Nanocomposites Based on PLLA/Triclosan/Nano-Hydroxyapatite. *Polymer* **2016**, *83*, 101–110. [CrossRef]
19. Maleki, H.; Khoshnevisan, K.; Sajjadi-Jazi, S.M.; Baharifar, H.; Doostan, M.; Khoshnevisan, N.; Sharifi, F. Nanofiber-Based Systems Intended for Diabetes. *J. Nanobiotechnol.* **2021**, *19*, 317. [CrossRef] [PubMed]
20. Kahdim, Q.S.; Abdelmoula, N.; Al-Karagoly, H.; Albukhaty, S.; Al-Saaidi, J. Fabrication of a Polycaprolactone/Chitosan Nanofibrous Scaffold Loaded with Nigella Sativa Extract for Biomedical Applications. *BioTech* **2023**, *12*, 19. [CrossRef]
21. Augustine, R.; Kalarikkal, N.; Thomas, S. Electrospun PCL Membranes Incorporated with Biosynthesized Silver Nanoparticles as Antibacterial Wound Dressings. *Appl. Nanosci.* **2016**, *6*, 337–344. [CrossRef]
22. Madhavan, R.V.; Rosemary, M.J.; Nandkumar, M.A.; Krishnan, K.V.; Krishnan, L.K. Silver Nanoparticle Impregnated Poly (ε-Caprolactone) Scaffolds: Optimization of Antimicrobial and Noncytotoxic Concentrations. *Tissue Eng. Part A* **2011**, *17*, 439–449. [CrossRef] [PubMed]
23. Taylor, P.; Sundaramurthi, D.; Krishnan, U.M. Electrospun Nanofibers as Scaffolds for Skin Tissue Engineering. *Polym. Rev.* **2014**, *54*, 348–376. [CrossRef]
24. Kalwar, K.; Sun, W.X.; Li, D.L.; Zhang, X.J.; Shan, D. Coaxial Electrospinning of Polycaprolactone@chitosan: Characterization and Silver Nanoparticles Incorporation for Antibacterial Activity. *React. Funct. Polym.* **2016**, *107*, 87–92. [CrossRef]
25. Yang, X.; Chen, X.; Wang, H. Acceleration of Osteogenic Differentiation of Preosteoblastic Cells by Chitosan Containing Nanofibrous Scaffolds. *Biomacromolecules* **2009**, *10*, 2772–2778. [CrossRef]
26. Chanda, A.; Adhikari, J.; Ghosh, A.; Chowdhury, S.R.; Thomas, S.; Datta, P.; Saha, P. Electrospun Chitosan/Polycaprolactone-Hyaluronic Acid Bilayered Scaffold for Potential Wound Healing Applications. *Int. J. Biol. Macromol.* **2018**, *116*, 774–785. [CrossRef]
27. Luna-Hernández, E.; Cruz-Soto, M.E.; Padilla-Vaca, F.; Mauricio-Sánchez, R.A.; Ramirez-Wong, D.; Muñoz, R.; Granados-López, L.; Ovalle-Flores, L.R.; Menchaca-Arredondo, J.L.; Hernández-Rangel, A.; et al. Combined Antibacterial/Tissue Regeneration Response in Thermal Burns Promoted by Functional Chitosan/Silver Nanocomposites. *Int. J. Biol. Macromol.* **2017**, *105*, 1241–1249. [CrossRef]

28. Angadi, S.C.; Manjeshwar, L.S.; Aminabhavi, T.M. Stearic Acid-Coated Chitosan-Based Interpenetrating Polymer Network Microspheres: Controlled Release Characteristics. *Ind. Eng. Chem. Res.* **2011**, *50*, 4504–4514. [CrossRef]
29. Maurer, H.R. Bromelain: Biochemistry, Pharmacology and Medical Use. *Cell. Mol. Life Sci.* **2001**, *58*, 1234–1245. [CrossRef]
30. Kim, S.; Choi, J.E.; Choi, J.; Chung, K.H.; Park, K.; Yi, J.; Ryu, D.Y. Oxidative Stress-Dependent Toxicity of Silver Nanoparticles in Human Hepatoma Cells. *Toxicol. Vitr.* **2009**, *23*, 1076–1084. [CrossRef]
31. Ramli, A.N.M.; Manas, N.H.A.; Hamid, A.A.A.; Hamid, H.A.; Illias, R.M. Comparative Structural Analysis of Fruit and Stem Bromelain from Ananas Comosus. *Food Chem.* **2018**, *266*, 183–191. [CrossRef]
32. Bayat, S.; Amiri, N.; Pishavar, E.; Kalalinia, F.; Movaffagh, J.; Hahsemi, M. Bromelain-Loaded Chitosan Nanofibers Prepared by Electrospinning Method for Burn Wound Healing in Animal Models. *Life Sci.* **2019**, *229*, 57–66. [CrossRef] [PubMed]
33. Takahashi, T.; Imai, M.; Suzuki, I.; Sawai, J. Growth Inhibitory Effect on Bacteria of Chitosan Membranes Regulated with Deacetylation Degree. *Biochem. Eng. J.* **2008**, *40*, 485–491. [CrossRef]
34. Li, J.H.; Shao, X.S.; Zhou, Q.; Li, M.Z.; Zhang, Q.Q. The Double Effects of Silver Nanoparticles on the PVDF Membrane: Surface Hydrophilicity and Antifouling Performance. *Appl. Surf. Sci.* **2013**, *265*, 663–670. [CrossRef]
35. No, H.K.; Young Park, N.; Ho Lee, S.; Meyers, S.P. Antibacterial Activity of Chitosans and Chitosan Oligomers with Different Molecular Weights. *Int. J. Food Microbiol.* **2002**, *74*, 65–72. [CrossRef]
36. Thompson, E.A.; Graham, E.; Macneill, C.M.; Young, M.; Donati, G.; Wailes, E.M.; Jones, B.T.; Levi-Polyachenko, N.H. Differential Response of MCF7, MDA-MB-231, and MCF 10A Cells to Hyperthermia, Silver Nanoparticles and Silver Nanoparticle-Induced Photothermal Therapy. *Int. J. Hyperth.* **2014**, *30*, 312–323. [CrossRef]
37. Li, X.; Xie, J.; Liao, L.; Jiang, X.; Fu, H. UV-Curable Polyurethane Acrylate–Ag/TiO_2 Nanocomposites with Superior UV Light Antibacterial Activity. *Int. J. Polym. Mater. Polym. Biomater.* **2017**, *66*, 835–843. [CrossRef]
38. Song, K.C.; Lee, S.M.; Park, T.S.; Lee, B.S. Preparation of Colloidal Silver Nanoparticles by Chemical Reduction Method. *Korean J. Chem. Eng.* **2009**, *26*, 153–155. [CrossRef]
39. Liu, X.; Lee, P.Y.; Ho, C.M.; Lui, V.C.H.; Chen, Y.; Che, C.M.; Tam, P.K.H.; Wong, K.K.Y. Silver Nanoparticles Mediate Differential Responses in Keratinocytes and Fibroblasts during Skin Wound Healing. *ChemMedChem* **2010**, *5*, 468–475. [CrossRef]
40. Chinnasamy, G.; Chandrasekharan, S.; Koh, T.W.; Bhatnagar, S. Synthesis, characterization, antibacterial and wound healing efficacy of silver nanoparticles from Azadirachta indica. *Front. Microbiol.* **2021**, *12*, 611560. [CrossRef]
41. Vijayakumar, V.; Samal, S.K.; Mohanty, S.; Nayak, S.K. Recent advancements in biopolymer and metal nanoparticle-based materials in diabetic wound healing management. *Int. J. Biol. Macromol.* **2019**, *122*, 137–148. [CrossRef]
42. Soheilifar, S., Bidgoli, M.; Hooshyarfard, A.; Shahbazi, A.; Vahdatinia, F.; Khoshkhooie, F. Effect of oral bromelain on wound healing, pain, and bleeding at donor site following free gingival grafting: A clinical trial. *J. Dent.* **2018**, *15*, 309.
43. Cai, E.Z.; Ang, C.H.; Raju, A.; Tan, K.B.; Hing, E.C.; Loo, Y.; Wong, Y.C.; Lee, H.; Lim, J.; Moochhala, S.M.; et al. Creation of consistent burn wounds: A rat model. *Arch. Plast. Surg.* **2014**, *41*, 317–324. [CrossRef]
44. Yao, C.H.; Lee, C.Y.; Huang, C.H.; Chen, Y.S.; Chen, K.Y. Novel Bilayer Wound Dressing Based on Electrospun Gelatin/Keratin Nanofibrous Mats for Skin Wound Repair. *Mater. Sci. Eng. C* **2017**, *79*, 533–540. [CrossRef] [PubMed]
45. Zhang, X.; Jia, C., Qiao, X.; Liu, T.; Sun, K. Porous Poly(Glycerol Sebacate) (PGS) Elastomer Scaffolds for Skin Tissue Engineering. *Polym. Test.* **2016**, *54*, 118–125. [CrossRef]
46. Nadim, A.; Khorasani, S.N.; Kharaziha, M.; Davoodi, S.M. Design and Characterization of Dexamethasone-Loaded Poly (Glycerol Sebacate)-Poly Caprolactone/Gelatin Scaffold by Coaxial Electro Spinning for Soft Tissue Engineering. *Mater. Sci. Eng. C* **2017**, *78*, 47–58. [CrossRef] [PubMed]
47. Shah, S.A.; Sohail, M.; Khan, S.; Minhas, M.U.; de Matas, M.; Sikstone, V.; Hussain, Z.; Abbasi, M.; Kousar, M. Biopolymer-Based Biomaterials for Accelerated Diabetic Wound Healing: A Critical Review. *Int. J. Biol. Macromol.* **2019**, *139*, 975–993. [CrossRef] [PubMed]
48. Michailidou, G.; Christodoulou, E.; Nanaki, S.; Barmpalexis, P.; Karavas, E.; Vergkizi-Nikolakaki, S.; Bikiaris, D.N. Super-Hydrophilic and High Strength Polymeric Foam Dressings of Modified Chitosan Blends for Topical Wound Delivery of Chloramphenicol. *Carbohydr. Polym.* **2019**, *208*, 1–13. [CrossRef] [PubMed]
49. Vig, K.; Chaudhari, A.; Tripathi, S.; Dixit, S.; Sahu, R.; Pillai, S.; Dennis, V.A.; Singh, S.R. Advances in Skin Regeneration Using Tissue Engineering. *Int. J. Mol. Sci.* **2017**, *18*, 789. [CrossRef]
50. Najafabadi, S.A.A.; Mohammadi, A.; Kharazi, A.Z. Polyurethane Nanocomposite Impregnated with Chitosan-Modified Graphene Oxide as a Potential Antibacterial Wound Dressing. *Mater. Sci. Eng. C* **2020**, *115*, 110899. [CrossRef]
51. Kausar, R.; Khan, A.; Jamil, B.; Shahzad, Y.; ul-Haq, I. Development and Pharmacological Evaluation of Vancomycin Loaded Chitosan Films. *Carbohydr. Polym.* **2021**, *256*, 117565. [CrossRef]
52. Dart, A.; Bhave, M.; Kingshott, P. Antimicrobial Peptide-Based Electrospun Fibers for Wound Healing Applications. *Macromol. Biosci.* **2019**, *19*, e1800488. [CrossRef]
53. Lumbreras-Aguayo, A.; Meléndez-Ortiz, H.I.; Puente-Urbina, B.; Alvarado-Canché, C.; Ledezma, A.; Romero-García, J.; Betancourt-Galindo, R. Poly(Methacrylic Acid)-Modified Medical Cotton Gauzes with Antimicrobial and Drug Delivery Properties for Their Use as Wound Dressings. *Carbohydr. Polym.* **2019**, *205*, 203–210. [CrossRef]
54. Fan, Z.; Liu, B.; Wang, J.; Zhang, S.; Lin, Q.; Gong, P.; Ma, L.; Yang, S. A Novel Wound Dressing Based on Ag/Graphene Polymer Hydrogel: Effectively Kill Bacteria and Accelerate Wound Healing. *Adv. Funct. Mater.* **2014**, *24*, 3933–3943. [CrossRef]
55. Li, L.; Hsieh, Y. Lo Chitosan Bicomponent Nanofibers and Nanoporous Fibers. *Carbohydr. Res.* **2006**, *341*, 374–381. [CrossRef]

56. Sahoo, S.; Sasmal, A.; Nanda, R.; Phani, A.R.; Nayak, P.L. Synthesis of Chitosan-Polycaprolactone Blend for Control Delivery of Ofloxacin Drug. *Carbohydr. Polym.* **2010**, *79*, 106–113. [CrossRef]
57. Malheiro, V.N.; Caridade, S.G.; Alves, N.M.; Mano, J.F. New Poly(ε-Caprolactone)/Chitosan Blend Fibers for Tissue Engineering Applications. *Acta Biomater.* **2010**, *6*, 418–428. [CrossRef]
58. Neves, S.C.; Moreira Teixeira, L.S.; Moroni, L.; Reis, R.L.; Van Blitterswijk, C.A.; Alves, N.M.; Karperien, M.; Mano, J.F. Chitosan/Poly(ε-Caprolactone) Blend Scaffolds for Cartilage Repair. *Biomaterials* **2011**, *32*, 1068–1079. [CrossRef]
59. Gholipour-Kanani, A.; Bahrami, S.H.; Joghataie, M.T.; Samadikuchaksaraei, A. Nanofibrous Scaffolds Based on Poly (Caprolactone)/Chitosan/Poly (Vinyl Alcohol) Blend for Skin Tissue Engineering. *Iran. J. Polym. Sci. Technol.* **2013**, *26*, 159–170.
60. Eskandarinia, A.; Kefayat, A.; Agheb, M.; Rafienia, M.; Amini Baghbadorani, M.; Navid, S.; Ebrahimpour, K.; Khodabakhshi, D.; Ghahremani, F. A Novel Bilayer Wound Dressing Composed of a Dense Polyurethane/Propolis Membrane and a Biodegradable Polycaprolactone/Gelatin Nanofibrous Scaffold. *Sci. Rep.* **2020**, *10*, 3063. [CrossRef] [PubMed]
61. Franco, R.A.; Min, Y.K.; Yang, H.M.; Lee, B.T. Fabrication and Biocompatibility of Novel Bilayer Scaffold for Skin Tissue Engineering Applications. *J. Biomater. Appl.* **2013**, *27*, 605–615. [CrossRef]
62. Hinrichs, W.L.; Lommen, E.J.; Wildevuur, C.R.; Feijen, J. Fabrication and Characterization of an Asymmetric Polyurethane Membrane for Use as a Wound Dressing. *J. Appl. Biomater.* **1992**, *3*, 287–303. [CrossRef]
63. Zhang, L.; Webster, T.J. Nanotechnology and Nanomaterials: Promises for Improved Tissue Regeneration. *Nano Today* **2009**, *4*, 66–80. [CrossRef]
64. Tsao, C.T.; Chang, C.H.; Lin, Y.Y.; Wu, M.F.; Wang, J.L.; Han, J.L.; Hsieh, K.H. Antibacterial Activity and Biocompatibility of a Chitosan-γ- Poly(Glutamic Acid) Polyelectrolyte Complex Hydrogel. *Carbohydr. Res.* **2010**, *345*, 1774–1780. [CrossRef] [PubMed]
65. Tamara, F.R.; Lin, C.; Mi, F.L.; Ho, Y.C. Antibacterial Effects of Chitosan/Cationic Peptide Nanoparticles. *Nanomaterials* **2018**, *8*, 88. [CrossRef] [PubMed]
66. Wang, Y.; Wang, C.; Xie, Y.; Yang, Y.; Zheng, Y.; Meng, H.; He, W.; Qiao, K. Highly Transparent, Highly Flexible Composite Membrane with Multiple Antimicrobial Effects Used for Promoting Wound Healing. *Carbohydr. Polym.* **2019**, *222*, 114985. [CrossRef] [PubMed]
67. López-Iglesias, C.; Barros, J.; Ardao, I.; Monteiro, F.J.; Alvarez-Lorenzo, C.; Gómez-Amoza, J.L.; García-González, C.A. Vancomycin-Loaded Chitosan Aerogel Particles for Chronic Wound Applications. *Carbohydr. Polym.* **2019**, *204*, 223–231. [CrossRef]

gels

Article

Biomimetic Gradient Hydrogels with High Toughness and Antibacterial Properties

Mingzhu Zeng [1,2,3], Zhimao Huang [2,3], Xiao Cen [2,3], Yinyu Zhao [2,3], Fei Xu [1], Jiru Miao [2,3], Quan Zhang [1,*] and Rong Wang [2,3,*]

1 Institute of Smart Biomedical Materials, School of Materials Science and Engineering, Zhejiang Sci-Tech University, Hangzhou 310018, China
2 Zhejiang International Scientific and Technological Cooperative Base of Biomedical Materials and Technology, Institute of Biomedical Engineering, Ningbo Institute of Materials Technology and Engineering, Chinese Academy of Sciences, Ningbo 315201, China
3 Ningbo Cixi Institute of Biomedical Engineering, Ningbo 315300, China
* Correspondence: quanzhang@zstu.edu.cn (Q.Z.); rong.wang@nimte.ac.cn (R.W.); Tel.: +86-0571-8684-0651 (Q.Z.)

Abstract: Traditional hydrogels, as wound dressings, usually exhibit poor mechanical strength and slow drug release performance in clinical biomedical applications. Although various strategies have been investigated to address the above issues, it remains a challenge to develop a simple method for preparing hydrogels with both toughness and controlled drug release performance. In this study, a tannic acid-reinforced poly (sulfobetaine methacrylate) (TAPS) hydrogel was fabricated via free radical polymerization, and the TAPS hydrogel was subjected to a simple electrophoresis process to obtain the hydrogels with a gradient distribution of copper ions. These gradient hydrogels showed tunable mechanical properties by changing the electrophoresis time. When the electrophoresis time reached 15 min, the hydrogel had a tensile strength of 368.14 kPa, a tensile modulus of 16.17 kPa, and a compressive strength of 42.77 MPa. It could be loaded at 50% compressive strain and then unloaded for up to 70 cycles and maintained a constant compressive stress of 1.50 MPa. The controlled release of copper from different sides of the gradient hydrogels was observed. After 6 h of incubation, the hydrogel exhibited a strong bactericidal effect on Gram positive *Staphylococcus aureus* and Gram-negative *Escherichia coli*, with low toxicity to NIH/3T3 fibroblasts. The high toughness, controlled release of copper, and enhanced antimicrobial properties of the gradient hydrogels make them excellent candidates for wound dressings in biomedical applications.

Keywords: gradient hydrogel; electrophoresis; controlled release; toughness; antibacterial property

Citation: Zeng, M.; Huang, Z.; Cen, X.; Zhao, Y.; Xu, F.; Miao, J.; Zhang, Q.; Wang, R. Biomimetic Gradient Hydrogels with High Toughness and Antibacterial Properties. *Gels* **2024**, *10*, 6. https://doi.org/10.3390/gels10010006

Academic Editor: Georgios Bokias

Received: 10 November 2023
Revised: 5 December 2023
Accepted: 15 December 2023
Published: 21 December 2023

1. Introduction

The skin is the largest tissue organ in the body and is vulnerable to external damage [1,2]. When the skin is damaged, wound dressings are needed in order to provide a barrier against secondary damage [3,4]. The ideal wound dressing needs to provide a moist microenvironment on the wound surface, absorb wound exudate, and have sustainable antimicrobial activity [5]. Traditional wound dressings, such as gauze and cotton, tend to adhere to the tissue, are not easy to be removed, and have low antimicrobial activity [6]. As wound dressings, hydrogels have received much attention in recent years as they maintain the wound moist and promote autolytic debridement of necrotic tissue [7,8]. Synthetic hydrogels usually have poor mechanical strength, which greatly limit their application in the biomedical field [9,10]. More recently, the development and application of tough hydrogels has become a research hotspot, and various strategies have been explored to develop tough hydrogels [11,12]. Preparation of tough hydrogels can be done via freeze-assisted salting-out, induced crystallization, and induction via radiation of rays [11]. Using a freezing-assisted salting-out fabrication procedure on poly(vinyl alcohol), multi-length-scale hierarchical hydrogels with high strength and toughness were obtained [13]. In addition, researchers have proposed a method to simulate strong hydrogels with muscle contraction by increasing the hydrogen bonding of the polymer chain dispersion medium [14]. However, most strategies to fabricate tough hydrogels are complicated, expensive, time-consuming, and unfriendly to the environment [15].

Gradient functional materials are non-homogeneous functional composites in which two or more materials with different properties are continuously or quasi-continuously varied in one or more dimensional directions via a special preparation process [16]. The advantages of gradient functional materials include their combination of different properties, modulation of the properties according to the environment [17,18]. In nature, many plants and animals developed biological gradient structures to adapt to the changing living environment [19]. For example, the molecular and mechanical gradients of foot filaments allow mussels to adhere to rocks with irregular surfaces [20], the gradient distribution of porous structure in bones directs cell differentiation and osteogenesis [21], and the osmotic/pH/concentration gradients in some plants endow them with reversible motion capabilities [22]. Gradient structure has been adopted in hydrogel design to improve the mechanical properties [23]. For example, Daniele Vigolo et al. demonstrated that the mechanical properties of soft biocompatible materials can be modified via thermophoresis-induced concentration gradients of aqueous sodium alginate solution in microfluidic channels [24]. Strong nanocomposite hydrogels with a designable gradient network structure and mechanical properties were prepared via a facile site-specific photoirradiation strategy [25].

Homogeneous hydrogels have been widely explored as drug delivery systems [26], but usually suffer from burst drug release due to the large difference in drug concentration between the hydrogel matrix and the external environment [27]. The burst release of drugs may result in toxicity and unnecessary waste of therapeutic drugs [28]. Drug release from a gradient hydrogel could be balanced by controlling the difference in drug concentration in the hydrogel matrix, leading to sustainable drug release at the later stage of release [29]. Recently, gradient hydrogels have been increasingly studied to control drug release due to their special biological gradient structure [30,31]. For example, Samad Ahadian et al. developed a novel application of dielectrophoresis to prepare three-dimensional gelatin methacrylate hydrogels with gradients of micro- and nanoparticles containing drugs. This research confirmed that the release of drugs from hydrogels could be achieved in a gradient manner, inducing a cell viability gradient [32]. Although some strategies have been used to prepare gradient hydrogels, there is still a need to develop a simple method for preparing hydrogels with both toughness and controlled drug release performance [33].

Previously, we obtained tannic acid (TA)-reinforced poly (sulfobetaine methacrylate) (polySBMA) hydrogels (TAPS). The TA molecules act as physical cross-linking agents in the polySBMA and as antioxidants and antibacterial agents, endowing them with biological activity [34]. In this study, gradient hydrogels with high mechanical properties, controlled release of antimicrobial agents, and enhanced antimicrobial properties were fabricated via a simple electrophoresis process of the TAPS hydrogels. The electrical field forced Cu ions to diffuse into the hydrogel matrix and formed a gradient cross-linked structure by chelating with the TA molecules and zwitterionic groups. The mechanical properties, antimicrobial properties, antimicrobial agent release profile, and cytotoxicity of the gradient hydrogels were studied to evaluate the application potential of the hydrogel as an antimicrobial wound dressing.

2. Results and Discussion

2.1. Fabrication of Cu Gradient Hydrogel

First, the TAPS hydrogel was prepared via a straightforward free radical polymerization of SBMA monomers with TA incorporation, as reported in our previous study [34]. The abundant hydrogen bonding and electrostatic interaction between the phenolic hydroxyl groups in TA and the zwitterionic groups in polySBMA formed many physical cross-linking points in the network, which significantly improved the strength and toughness of the hydrogel. Second, the TAPS hydrogel was electrochemically treated to form a gradient structure in the hydrogel (Scheme 1a). The hydrogel was subjected to electrophoresis with a Cu plate as the anode and a graphite plate as the cathode. After the power was connected, the color of the hydrogels near the Cu anode gradually changed from light yellow to reddish brown and then to darker yellow-brown. This phenomenon was induced via an electrochemical reaction of the Cu electrode, where the hydrogel could be considered as an electrolytic cell that transformed electrical energy into chemical energy. The current flowed through the electrolyte, resulting in an oxidation reaction on the Cu anode. The standard electrode potential of Cu ions obtained from electron loss is reported to be 0.34 V [35]. In our study, the potential of the Cu electrode was fixed at 5 V, higher than the value required to convert Cu to Cu^{2+}. Therefore, an oxidation reaction of Cu occurred on the anode to form Cu^{2+} ions. Cu^{2+} ions were driven into the hydrogel matrix under the applied potential and chelated with the phenolic hydroxyl groups of TA molecules in the hydrogel matrix [36] (Scheme 1b). Since the Cu^{2+} ions traveled in the direction from the Cu anode to the graphite cathode [37], they would be complexed with the TA molecules near the Cu plate, then in the central section, and lastly close to the cathode during the electrophoresis process. Therefore, the gradient structure hydrogel cross-linked with Cu ions along the thickness direction (TAPS-CuX) could be obtained. At a fixed voltage of 5 V, the current between the hydrogel decreased from 50–70 mA to around 4 mA within ~1000 s and was maintained at this level afterward. As discussed above, the oxidation of Cu to Cu^{2+} ions occurred due to the electrochemical reaction driven by the applied potential. A proportion of the Cu^{2+} ions that travel in the hydrogel would be chelated by TA molecules. This Cu-TA chelation occurred in a gradient manner along the thickness direction of the hydrogel. In addition, water in the hydrogel was possibly evaporated [38] and electrolyzed [39] as the weight of the TAPS-Cu15 hydrogel decreased by 2.45% after the electrophoresis process. Furthermore, polymer chains and tannic acid molecules in the hydrogel could migrate under the electric field [10]. As a result, the hydrogel near the anode plate first compacted, limiting the transportation of Cu^{2+} ions, and leading to an increase in the resistance of the hydrogel and a decrease in the current (Figure 1a). Since the electrochemical reaction continued to generate Cu^{2+} ions, and some of the chelated Cu^{2+} ions in the hydrogel might escape from the chelating complex under the applied potential, the travel of Cu^{2+} was maintained at a low level, resulting in a low and constant current of ~4 mA within ~1000 s in the electrophoresis process. In sum, the electric charges passing through the hydrogel could be varied by controlling the electrophoretic treatment time (Figure 1b), thus obtaining hydrogels with different gradient cross-linking degrees (i.e., TAPS-Cu0.5, TAPS-Cu1, TAPS-Cu2, TAPS-Cu4, TAPS-Cu7, and TAPS-Cu15). Various parameters and properties of TAPS and TAPS-CuX hydrogels are shown in Table 1.

Table 1. Summary of various parameters of TAPS and TAPS-CuX hydrogels.

Hydrogel Name	Electrophoresis Time (min)	Electric Charge (C)	Tensile Strength (kPa)	Elastic Modulus (kPa)	Compressive Strength (MPa)	Bactericidal Efficacy (%)	
						S. aureus	*E. coli*
TAPS	0	0	100.80 ± 10.62	0.85 ± 0.05	21.00 ± 3.64	/	/
TAPS-Cu0.5	0.5	0.90 ± 0.08	135.54 ± 13.00	1.17 ± 0.16	26.19 ± 1.99	99.66 ± 0.01	92.35 ± 0.87
TAPS-Cu1	1	1.62 ± 0.27	147.20 ± 7.44	1.46 ± 0.14	25.24 ± 3.29	99.68 ± 0.01	94.21 ± 1.24
TAPS-Cu2	2	2.94 ± 0.10	153.35 ± 6.54	1.85 ± 0.08	30.48 ± 0.99	99.81 ± 0.01	95.39 ± 0.73
TAPS-Cu4	4	3.76 ± 0.44	218.28 ± 13.23	3.15 ± 0.31	34.08 ± 2.30	99.77 ± 0.02	99.04 ± 0.09
TAPS-Cu7	7	5.83 ± 0.36	258.59 ± 15.94	5.49 ± 0.44	36.67 ± 2.74	99.99 ± 0	99.39 ± 0.17
TAPS-Cu15	15	7.84 ± 0.80	364.14 ± 34.56	16.1 ± 2.74	42.77 ± 2.52	100 ± 0	99.99 ± 0

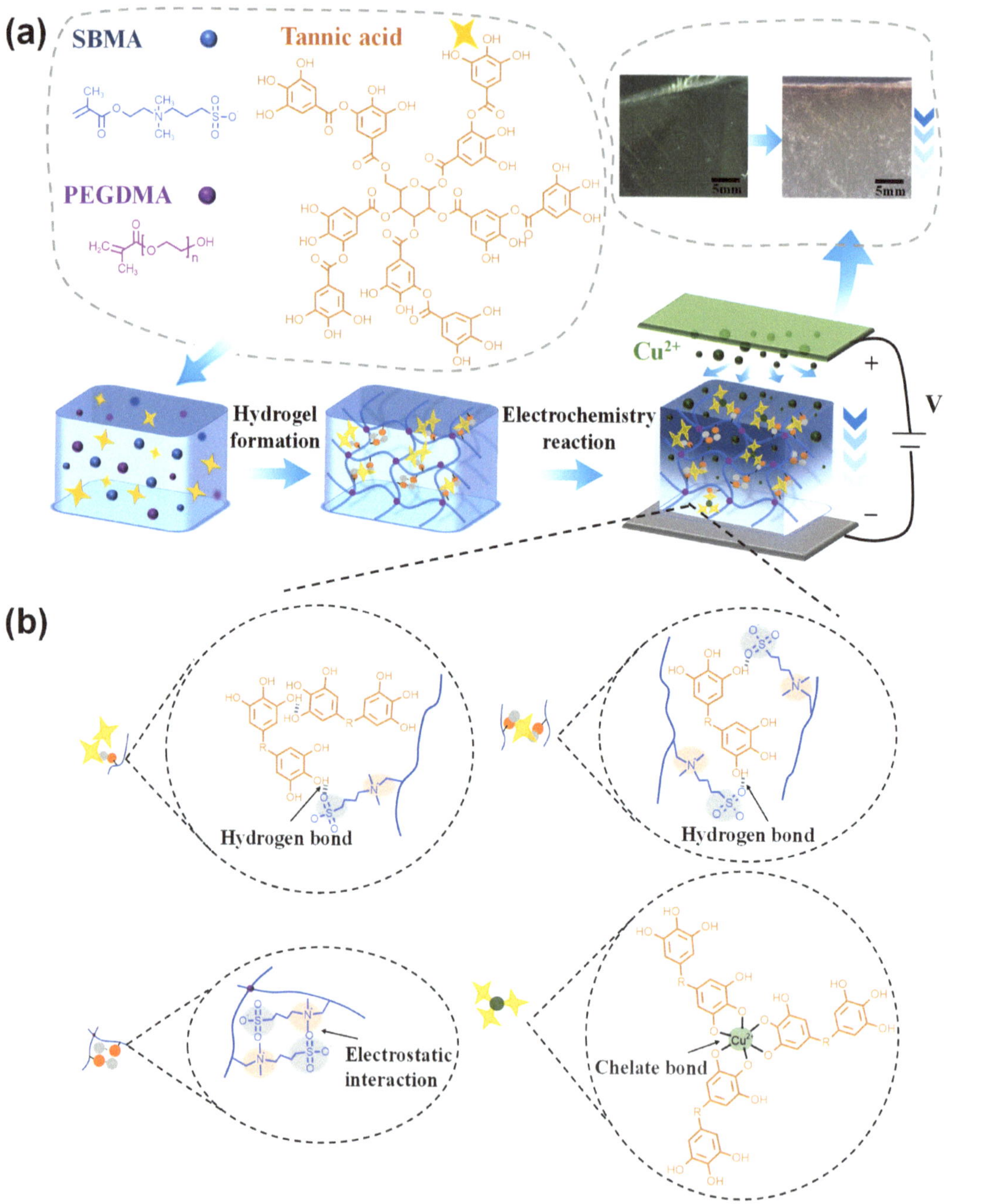

Scheme 1. (**a**) Schematic illustration of the preparation procedure of TAPS-CuX hydrogels and (**b**) the non-covalent interactions in the hydrogel. Scale bar: 5 mm.

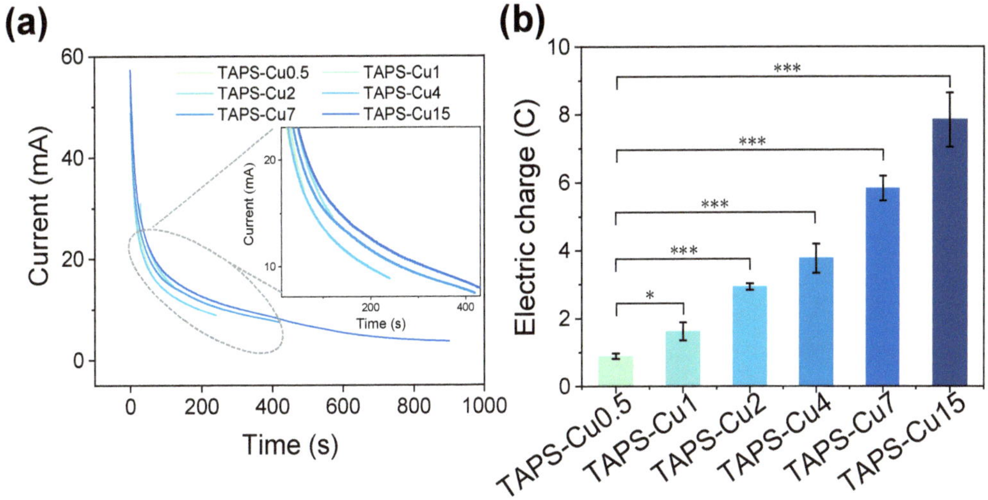

Figure 1. (**a**) Current–time curves and (**b**) different electric charges in the TAPS-CuX hydrogels (n = 3). * $p < 0.05$, *** $p < 0.001$.

2.2. Mechanical Properties

The mechanical properties of the series of TAPS-CuX hydrogels were investigated. As shown in Figure 2a, the tensile strength of the TAPS hydrogel was measured to be 100.80 kPa. The tensile strength of the hydrogels increased with increasing electrophoresis time, and the maximum value was reached at 368.14 kPa for the TAPS-Cu15 hydrogel. The original TAPS hydrogel was weak with a tensile modulus of 0.85 kPa. The electrophoresis treatment reinforced the hydrogel, and the tensile modulus of the hydrogel increased with the increase in the electrophoresis time (Figure 2b). The value of the tensile modulus of the TAPS-Cu0.5 hydrogel was 1.17 kPa and increased to 5.49 kPa of TAPS-Cu7. Furthermore, increasing the electrophoresis time to 15 min obtained a TAPS-Cu15 hydrogel with a tensile modulus of 16.17 kPa. This outcome could be attributed to the fact that there are a high number of metal chelating bonds between Cu ions and tannic acid molecules (Scheme 1b), forming a highly cross-linked structure in the hydrogel, and thereby resulting in stiffening of the hydrogel. Traditional dressings are usually non-stretchable with poor conformity and high modulus compared to skin [41]. The elastic modulus of the TAPS-Cu15 hydrogel falls in the range of human skin (9.5–30.3 kPa) [42], making the TAPS-Cu15 hydrogel less likely to cause tissue contusions when the skin is deformed, and increasing the comfortability and working life of the hydrogel dressing. Therefore, the TAPS-Cu15 hydrogel shows promise as a wound dressing.

Figure 2c shows the compression property of TAPS and TAPS-CuX hydrogels prepared with different electrophoresis times. The compressive strength of the TAPS hydrogel was approximately 20.99 MPa, which is consistent with our previous report [34]. The compressive strength of the hydrogel increased with increasing electrophoresis time and reached a maximum of 42.77 MPa at 15 min of electrophoresis time (that is, the TAPS-Cu15 hydrogel). The compressive strength of the TAPS-Cu15 hydrogel increased approximately 2.04 times compared to that of the TAPS hydrogel (20.99 MPa). It should be noted that the hydrogels were just deformed, not damaged (Figure S1), when compressed at 90% strain. The highest pressure that the skin is subjected to is plantar pressure (maximum of approximately 200 kPa [43]). At 50% compression, the TAPS-Cu15 hydrogel can maintain a pressure of approximately 1.5 MPa, which is sufficient to cope with the in-shoe plantar pressure. Therefore, the TAPS-Cu15 hydrogels were loaded/unloaded for up to 70 cycles at 50% compression strain, and the hydrogels retained excellent integrity, exhibiting a constant compressive stress of 1.50 MPa during the compression cycle (Figure 2d). Therefore, these results confirmed that the tensile and compressive properties of the TAPS hydrogel were enhanced after electrophoresis treatment. This outcome should be attributed to the fact that the Cu ions entered the hydrogel network under the electric field and formed strong metal chelation with the hydroxyl group in the TA molecule, thus enhancing the mechanical properties of the TAPS-CuX hydrogels, and the TAPS-Cu15 was found to be the toughest hydrogel in the study.

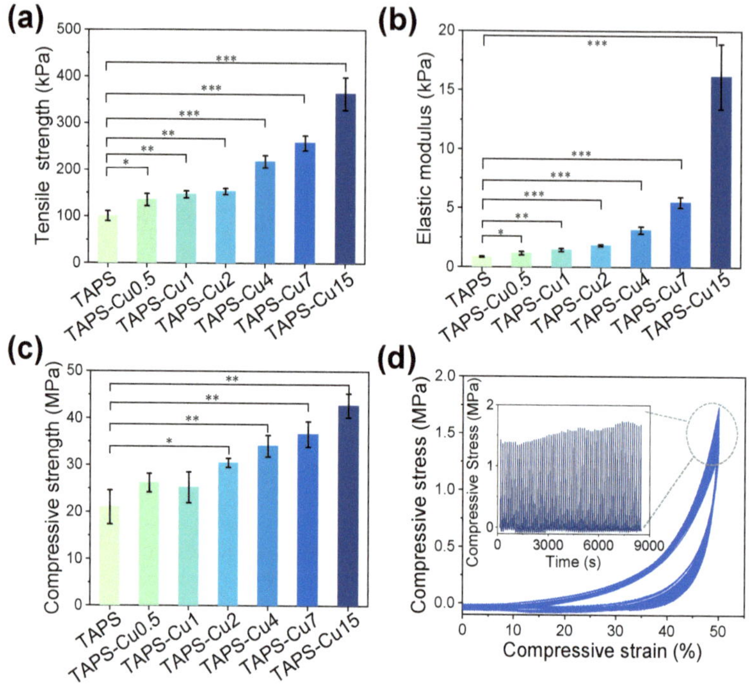

Figure 2. (**a**) Tensile strength and (**b**) tensile elastic modulus of TAPS hydrogels and TAPS-CuX hydrogels (n = 3). * $p < 0.05$, ** $p < 0.01$, *** $p < 0.001$. (**c**) Compressive strength of TAPS and TAPS-CuX hydrogels with the compression strain at 90% (n = 3). * $p < 0.05$, ** $p < 0.01$. (**d**) Compressive stress–strain curves and compressive stress time in 50% compression strain of TAPS-Cu15 hydrogel over 70 cycles.

2.3. Swelling Property

To evaluate the swelling property, the hydrogels were immersed in deionized water to reach the swelling equilibrium (Figure 3a). The swelling ratio of the TAPS hydrogels increased rapidly, ultimately reaching approximately 30% after 1 h. However, the swelling ratio of the TAPS-CuX hydrogels was lower than that of the TAPS hydrogels at 1 h and continued to increase and reached equilibrium after 4 h. The cross-section of the hydrogels after 24 h of swelling was observed using a scanning electron microscope (SEM, regulus 8230, Hitachi, Tokyo, Japan) [44]. The results show that the TAPS-CuX hydrogels exhibited a dense and smooth structure in the cross-section after swelling (Figure 3b). The hydroxyl groups of the TA molecules and the zwitterionic groups of the polySBMA chains in the TAPS hydrogels interact with the water molecules via hydrogen bonding and the solvation effect, causing swelling of the hydrogels. When the TA molecules were chelated with Cu ions, the hydrogen bonding between TA and water molecules was inhibited, resulting in a reduction in the swelling effect of the TAPS-CuX hydrogels.

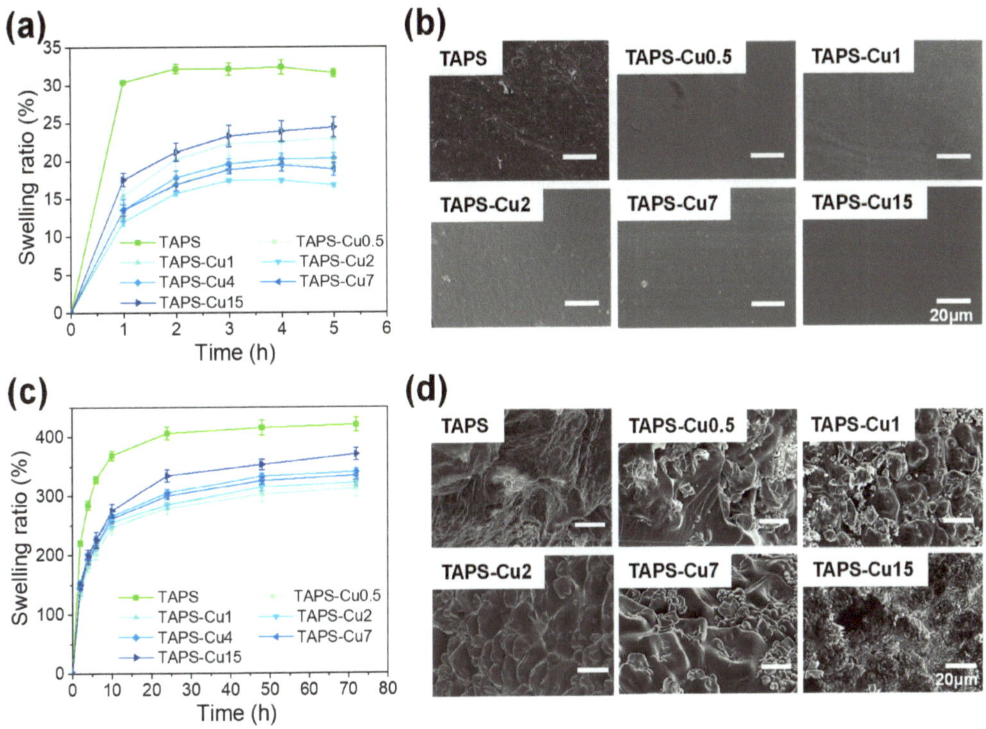

Figure 3. (**a**) Swelling behavior of TAPS hydrogel and TAPS-CuX hydrogels in deionized water for 4 h (n = 3). (**b**) SEM pictures of cross-sections of TAPS and TAPS-CuX hydrogels after 24 h of swelling in deionized water. (**c**) Swelling behaviors of the TAPS hydrogel and TAPS-CuX hydrogels in artificial sweat for 72 h (n = 3). (**d**) SEM pictures of cross-sections of TAPS and TAPS-CuX hydrogels after 24 h of swelling in artificial sweat. Scale bars in (**b**,**d**) were 20 μm.

The swelling property of the hydrogels in artificial sweat was also investigated. As can be seen from Figure 3c, the TAPS hydrogel showed a high swelling ratio of 415.45% in artificial sweat. This is due to the anti-polyelectrolyte effect of zwitterionic polymers, which causes weak interactions between zwitterionic groups in polySBMA chains and results in swelling and extensive hydration in solutions with high ionic strengths [45]. The TAPS-CuX hydrogels showed a significant decrease in equilibrium swelling ratio with the increase in Cu ions in comparison to the TAPS hydrogel. After being soaked in artificial sweat for 24 h, the hydrogel almost reached the equilibrium state, and the equilibrium swelling ratios of the TAPS-CuX hydrogels decreased from 352.86% of TAPS-Cu1 to 307.82% of TAPS-Cu15. The cross-sections of the swelled TAPS and TAPS-CuX hydrogels had distinct wrinkles and build-up of salt deposits in the cross-section (Figure 3d). This is probably due to the shrinkage of the highly swollen hydrogels under the vacuum condition during the freeze-drying process before SEM observation. The swelling results also confirmed that high concentrations of Cu ions in the hydrogel restricted the movement of TA molecules and polySBMA chains via the metal–polyphenol coordination, resulting in an increased degree of cross-linking and a lower swelling ratio of the hydrogel.

2.4. Gradient Cu Cross-Linking Structure and Cu Release from the Hydrogel

As discussed above, the content of Cu in the hydrogel can be controlled by varying the electric charge applied in the electrophoresis process, and a gradient distribution of Cu along the thickness direction of the functionalized hydrogel was achieved. The Cu content on the + side, − side, and middle part of the TAPS-Cu15 hydrogel was quantified using inductively coupled plasma-optical emission spectrometry (ICP-OES, SPECTRO ARCOS, SPECTRO, Kleve, Germany) (Figure 4a). As can be seen, the Cu content was 1452.06 μg/g on the + side (contacting the copper anode in the electrophoresis process). It gradually decreased to 1021.99 μg/g in the middle part, and it further decreased to 900.54 μg/g on the − side of the TAPS-Cu15 hydrogel (contacting with the graphite cathode in the electrophoresis process). The Cu content on the + side was 1.42 and 1.61 times of that in the middle part and the − side, respectively. These results confirmed the gradient distribution of Cu content along the thickness direction of the TAPS-CuX hydrogel (Figure 4b).

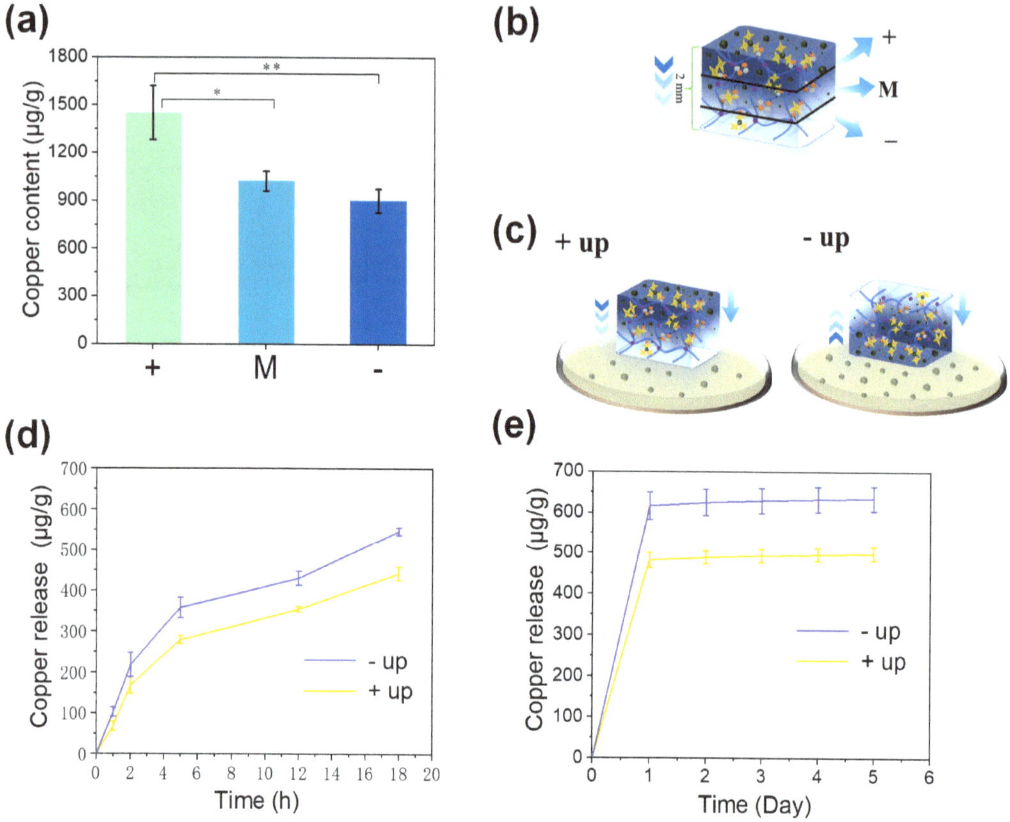

Figure 4. (**a**) Cu content in the + side, the middle part, and the − side of the TAPS-Cu15 hydrogel (n = 3). * $p < 0.05$, ** $p < 0.01$. (**b**) Schematic illustration of the positions of the + side, the middle part and the − side on the TAPS-Cu15 hydrogel. The thickness of the TAPS-Cu15 hydrogel is 2 mm. (**c**) Schematic illustration of the + side and the − side of the TAPS-Cu15 hydrogel contacting the agar gel in the copper release test. (**d**) Copper release from the + side and the − side of the TAPS-Cu15 hydrogel for 18 h (n = 3). (**e**) Copper release from the + side and the − side of the TAPS-Cu15 hydrogel for 5 days (n = 3).

The Cu release profile from either side of the gradient hydrogel was investigated in an agar-contacting experiment to mimic the conditions of the hydrogel dressing applied on skin tissue. In the test, the TAPS-Cu15 hydrogel was placed with either the + side or the − side contacting the agar gel (Figure 4c). The trend of Cu release from both sides of the hydrogels was similar; that is, Cu released rapidly on the first day, and the release rate was reduced and stable on the following 2 to 5 days (in the range of 2–10 μg/g daily). The higher release rate from the hydrogel on the first day could provide more Cu ions to achieve a stronger sterilization capability (Figure 4d), and the continuous release of Cu ions on the following days could provide a sustainable sterilization effect during the application (Figure 4e). Cu release from the − side (with a lower Cu content) on the first day was 482.93 μg/g, which was lower than that from the + side (that is, 616.36 μg/g). The three-dimensional network structure of the hydrogel itself acts as a Cu reservoir and as a retarding system for the diffusion of Cu ions. The Cu ions distributed a gradient along the thickness direction in the hydrogel matrix. The Cu ions near the − side (containing a lower Cu content) could quickly diffuse out to the agar gel, while the Cu ions near the + side (containing a higher Cu content) need to migrate through the hydrogel matrix before they can diffuse out and reach the agar gel. The TA molecules in the hydrogel matrix would chelate the Cu ions and suppress their migration. In this way, the release behavior of the antimicrobial Cu ions could be controlled, and a more sustainable manner of Cu release was achieved. The controllable release of Cu ions from the gradient hydrogel is beneficial in reducing the potential cytotoxicity of high concentrations of Cu ions to the contacting wound bed, thus achieving a prolonged antimicrobial property of the hydrogel dressing. In addition, the higher Cu content on the + side could endow the outer surface of the hydrogel dressing with a higher contact-killing efficacy against the pathogens from the environment, achieving a strong sterilization effect. Moreover, the hydrogel had a dense structure, and there was no particulate matter found, demonstrating the absence of nanoparticles in the hydrogel. Therefore, we think that copper nanoparticles were not formed in the hydrogel (Figure S2).

2.5. Antibacterial Activity

The TAPS hydrogel showed some degree of bactericidal property due to the antibacterial efficacy of TA, as reported in our previous study [34]. It is hypothesized that the antibacterial property could be further enhanced via the introduction of Cu ions in the hydrogel. The antibacterial properties of the hydrogels containing different amounts of Cu ions were evaluated by incubating the hydrogel with Gram-positive *Staphylococcus aureus* (*S. aureus*) and Gram-negative *Escherichia coli* (*E. coli*) (Figure 5). As can be seen, the survival rate of both bacterial species decreased markedly after incubation with the TAPS-CuX hydrogels for 6 h, which indicated that the TAPS-CuX hydrogels had strong antibacterial properties. Specifically, the bactericidal efficacy of the TAPS-Cu0.5 hydrogel against *S. aureus* and *E. coli* was 99.66% and 92.35%, respectively. The bactericidal property of the TAPS-CuX hydrogel increased with increasing Cu content. The TAPS-Cu15 hydrogel almost killed all the bacterial cells of the two strains in the suspension after 6 h of incubation. The TAPS hydrogel is antibacterial due to the addition of TA, possibly because the TA molecule targets peptidoglycans in the cell wall to disrupt bacterial integrity. For the TAPS-CuX hydrogels, the incorporation of Cu would further enhance the antibacterial properties, and its antibacterial capability also depended on the content of Cu in the hydrogel. The positively charged Cu ions released from hydrogels (Figure 4c), reached the negatively charged bacterial cell membrane, adsorbed to the cell membrane via Coulomb gravity, and further penetrated the cell wall, leading to rupture of the cell wall and death of the bacteria [46]. The benefits of using Cu as an antimicrobial agent rather than other metals (such as Ag ions) include the fact that Cu ions are safer compared to other metal antimicrobial agents [47] and that Cu ions are highly stable in light, heat, and aqueous solutions [48]. Furthermore, Cu^{2+} ion is an essential trace element that the body requires to maintain normal hematopoietic function and has an important physiological function in wound healing [49]. According to the literature, the minimum concentration of Cu^{2+} ions with an antibacterial activity of >90% is in the range of 10^{-5} M–10^{-4} M (0.367–3.67 ppm) [50]. The cumulative antimicrobial amount of Cu released from hydrogels over 5 days did not exceed 700 ppm in this study (Figure 4c), which had a good antibacterial effect and low cytotoxicity at this concentration.

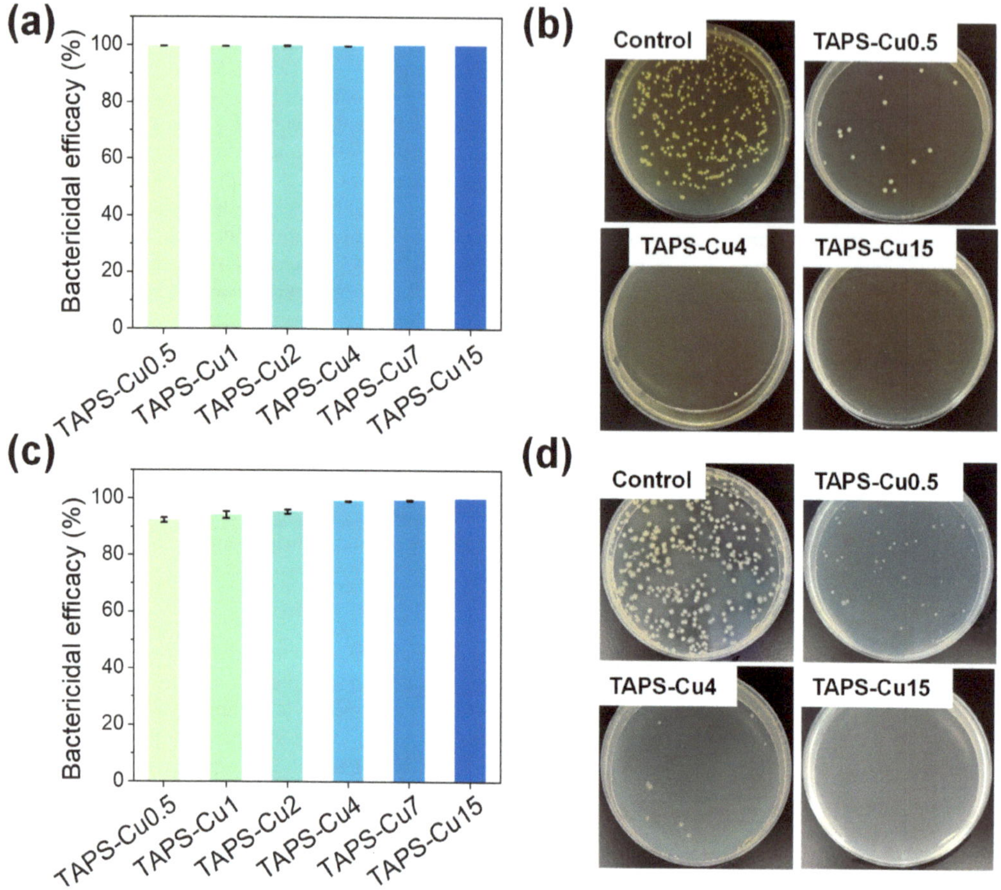

Figure 5. Bactericidal efficacy of TAPS-CuX hydrogels against (**a**,**b**) *S. aureus* and (**c**,**d**) *E. coli* after 6 h of incubation in PBS (n = 3). Digital photographs of the surviving (**b**) *S. aureus* and (**d**) *E. coli* clones in culture plates after 6 h of incubation with TAPS-CuX hydrogel.

2.6. Cytotoxicity Assay

The cytotoxicity test of the hydrogels was performed following the ISO-10993-5 standard with minor modification [51]. As shown in Figure 6, the cell viability of all groups with hydrogel extracts was above 90%, indicating that the TAPS-CuX hydrogels had good biocompatibility. As reported in our previous study, TAPS hydrogels have good biocompatibility [34]. On the other hand, Cu is an essential trace element and serves as a bioactive component in vivo. Cu^{2+} has been reported to promote tube formation in vascular endothelial cells by activating the hypoxia-inducible factor 1 pathway [52] and increasing the expression of angiogenesis-related genes, such as vascular endothelial growth factor [53]. Many studies have shown that Cu^{2+} can promote angiogenesis, cell migration, and collagen deposition and thus be effective in tissue regeneration [54], especially skin tissue regeneration. The good biocompatibility and bioactivity of TAPS and Cu indicate the application potential of the TAPS-CuX hydrogel as wound dressings in the biomedical field.

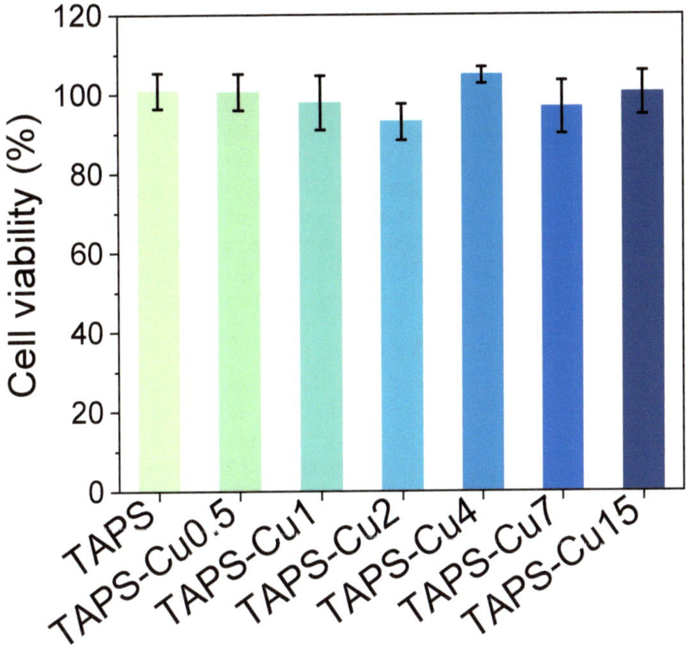

Figure 6. Viability of NIH/3T3 (Catalogue No.: SCSP-515) fibroblast after incubation in extracts of TAPS hydrogel and TAPS-CuX hydrogels for 24 h (n = 6). There was no statistically significant difference in cell viability between each group.

3. Conclusions

In this work, a TA-reinforced polySBMA hydrogel (TAPS) was fabricated via free radical polymerization, and the TAPS hydrogel was subjected to a simple electrophoresis process to obtain TAPS-CuX hydrogels with a gradient distribution of Cu ions. The mechanical properties and swelling properties of the TAPS-CuX hydrogels were controlled and optimized by varying the applied electrical charges in the electrophoresis process. The TAPS-Cu15 hydrogel showed good mechanical properties, a skin-like elastic modulus, and low swelling properties, due to the high and gradient cross-linking degree by Cu chelation with TA molecules. The gradient distribution of Cu in the TAPS-Cu15 hydrogel was verified by ICP-OES. The release profile of Cu from the + and − side of the TAPS-Cu15 hydrogel was differentiated due to the gradient distribution of Cu in the hydrogel matrix. It was demonstrated that the initial release rate of Cu from the − side of the hydrogel (containing a lower Cu content) was lower compared to that from the + side (containing a higher Cu content), and the Cu release in the subsequent days was maintained in a low and sustainable level. The Cu release could thus be controlled to obtain a low cytotoxic effect via contact with the − side and a prolonged release profile. The + side of the hydrogel on the outer surface was expected to have a high bactericidal effect against pathogens from the external environment. The TAPS-Cu15 hydrogels were shown to have good antimicrobial properties against Gram-positive *S. aureus* and Gram-negative *E. coli*, and minimal cytotoxicity to mammalian cells. In conclusion, the biomimetic gradient TAPS-CuX hydrogel with good mechanical properties, antimicrobial capacity, minimal cytotoxicity, and controllable and sustainable release of antimicrobials provided a promising solution for the treatment and healing of infected wounds.

4. Experimental Section

4.1. Materials

Tannic acid, [2-(methacryloyloxy) ethyl] dimethyl-(3-sulfopropyl) ammonium hydroxide (sulfobetaine methacrylate, SBMA, 97%), and ammonium persulfate (APS, 98%) were obtained through Aladdin Chemistry (Shanghai, China). Poly(ethylene glycol) dimethacrylate (PEGDMA, Mn 550 Da) was obtained from Sigma-Aldrich (Shanghai, China). Artificial sweat (Catalogue No.: CF-001) was bought from Chuangfeng Technology (Dongguan, China). *S. aureus* 5622, a strain isolated in patients with skin wound infections, was given by the First Affiliated Hospital of Ningbo University. *E. coli* ATCC 25922 from the American Type Culture Collection. NIH/3T3 fibroblasts cells (Catalogue No.: SCSP-515) were derived from National Collection of Authenticated Cell Cultures, Chinese Academy of Sciences.

4.2. Hydrogel Preparation

To a solution of SBMA monomer (11.16 g) in deionized water (9.8 mL), TA (1.845 g), PEGDMA (1.6 mg), and APS (9 mg) were added at room temperature. After degassing with nitrogen bubbles for 30 min, the mixture was poured into a silicone-coated glass mold (12 cm × 7 cm, thickness of 2 mm) and cured overnight at 60 °C to form a hydrogel with a thickness of 2 mm (denoted as TAPS hydrogel). The as-prepared TAPS hydrogel was sandwiched between a Cu plate as the anode and a graphite plate as the cathode to establish a circuit. A power supply (CHI600E, Chinstruments, Shanghai, China) was used with the applied voltage fixed at 5 V. The series TAPS-CuX hydrogels were obtained after the hydrogel was subjected to electrophoresis for different durations (0.5–15 min), while X represented the electrophoresis duration (in minutes). The hydrogel side connected to the Cu anode during electrophoresis was denoted as the + side, and the side connected to the graphite cathode was denoted as the − side.

4.3. Mechanical Properties and Swelling Behavior

The hydrogels were cut into dumbbell shapes (2 cm in narrow parallel width, 3.5 cm in length, and 2 mm in thickness), and the tensile strength of the hydrogel was measured in a Universal Testing Machine (CMT-1104, SUST, Xi'an, China) with a crosshead speed of 100 mm/min until the hydrogel broke. The force–displacement curve, the tensile strength (axial tensile force at the breakage point divided by the cross-sectional area of the hydrogel, i.e., 0.4 cm^2), and the elastic modulus (the slope in the linear range of the stress–strain curve, i.e., 0% to 2% tensile strain in the stress–strain curve) of the hydrogel were recorded. In the case of the compression test, the hydrogel was cut into discs (10 mm in diameter and 2 mm in thickness), loaded onto the Universal Testing Machine, and compressed at a strain rate of 10% per minute until 90% strain. The compressive strength at 90% compression strain was recorded. In a cyclic compression test, the hydrogel disc was compressed to 50% strain and then relieved at a crosshead rate of 10% strain per minute. The highest compressive stress is recorded in each cycle.

To assess the swelling properties, the hydrogel was cut into discs (10 mm in diameter and 2 mm in thickness) and immersed in 15 mL of deionized water or artificial sweat at 37 °C for a given period. Excess water was removed from the surface of the swollen hydrogel using filter paper. The hydrogels were weighed before and after swelling and the swelling rate of the hydrogels was calculated using Equation (1):

$$S = \frac{W_1 - W_0}{W_0} \times 100\%, \quad (1)$$

where S is the swelling ratio; and W_0 and W_1 are the weight of the hydrogels before and after swelling, respectively. After equilibration in deionized water and artificial sweat for 24 h, the hydrogels were frozen in liquid nitrogen, freeze-dried, and fractured to obtain cross-sections. The morphology of the hydrogel cross-sections was viewed using SEM with a voltage of 10 kV and a working distance of approximately 13 mm.

4.4. Determination of Cu Content and Cu Release from Hydrogel

The TAPS-Cu15 hydrogel discs (10 mm in diameter, 2 mm in thickness) were sectioned using a frozen slicer (NX500, Thermo, Waltham, MA, USA) into slices with a thickness of 0.2 mm. Slices from the + section (0–0.2 mm from the + side), the middle section (1–1.2 mm from the + side), and the − section (1.8–2 mm from the + side) of the hydrogel were collected and dissolved in nitric acid. The amount of copper from the slides was measured by ICP-OES.

For the release experiment, the TAPS-Cu15 hydrogel disk (10 mm in diameter, 2 mm in thickness) was placed on 4 mL of solidified agar gel (20 mm in diameter) in a 12-well plate, in which the + side or the − side of the hydrogel contacting the agar gel. The hydrogel with the agar gel was incubated at 37 °C for 5 days, during which the hydrogel was transferred to a new agar gel daily. The collected agar gel was dissolved in a nitric acid solution (2%), and the content of released Cu was measured by ICP-OES. The amount of Cu released from the hydrogel was calculated according to the standard curve. The cumulative release of Cu was calculated by adding the quantity of Cu released from the hydrogel every day.

4.5. Antibacterial Test

The bacteria were incubated overnight in the appropriate culture broth (tryptic soy broth for *S. aureus* and lysogeny broth for *E. coli*) and shaken at 100 rpm at 37 °C. The bacterial culture was diluted 1000 times with sterilized phosphate-buffered saline (PBS, 10 mM, pH 7.2) to obtain cell suspension (bacterial count 10^5 CFU/mL). Hydrogels (10 mm in diameter, 2 mm in thickness) were placed in 5 mL of the bacterial suspension and cultured with shaking at 100 rpm at 37 °C. A suspension of PBS without hydrogels was used as the control group. The number of viable cells in suspension was counted by plate counting after 6 h of incubation.

4.6. Cytotoxicity Assay

Cytotoxicity tests were performed using NIH/3T3 fibroblasts according to the ISO-10993-5 standard with minor modification [51]. Cells in Dulbecco's modified Eagle's medium (DMEM, Hyclone, containing 10% fetal bovine serum, 100 mg/L streptomycin, and 1.0×10^5 U/L penicillin) were added in a 96-well plate at a concentration of 10^4 cells per well and cultured in DMEM at 37 °C and 5% CO_2 for 24 h. Meanwhile, the hydrogels (500 mg) were disinfected under UV light for 30 min and incubated in sterilized deionized water (8 mL) at 37 °C for 24 h. The resulting extract solution was merged with DMEM at a volume ratio of 1:9. The mixture was used to culture the cells. The cells in the control groups were treated with a mixed deionized water/DMEM solution (1:9 volume ratio), fresh medium, and medium containing zinc diethyldithiocarbamate (10 mg/mL), respectively. After 24 h, cell viability was assessed using the cell counting Kit-8 (CCK-8, TransGen Biotech, Beijing, China) assay according to the manufacturer's protocol. The percentage of absorbance value of the experimental group relative to the negative control group (deionized water/DMEM mixture) was expressed as cell viability.

4.7. Statistical Analysis

All data were presented as the mean ± standard deviation. The difference among different groups was compared, and statistically significant differences were marked with * ($p < 0.05$), ** ($p < 0.01$), *** ($p < 0.001$).

Supplementary Materials: The following supporting information can be downloaded at https://www.mdpi.com/article/10.3390/gels10010006/s1: Figure S1: Photos of TAPS-Cu15 hydrogel (a) before compression and (b) after compression at 90% strain; Figure S2: SEM images of cross-sections of TAPS-Cu15 hydrogel.

Author Contributions: Conceptualization: M.Z. and R.W.; data curation: M.Z., Z.H. and X.C.; investigation: M.Z., Z.H. and X.C.; methodology: M.Z., Y.Z., F.X. and J.M.; resources: Q.Z. and R.W.; supervision: Q.Z. and R.W.; validation: Y.Z., F.X. and J.M.; writing—original draft: M.Z.;

writing—review editing: Q.Z. and R.W.; funding acquisition: R.W. All authors have read and agreed to the published version of the manuscript.

Funding: This research was supported by National Key Research and Development Program of China (2018YFE0119400), Youth Innovation Promotion Association CAS (2021296), Key Research and Development Program of Ningbo (2022Z132), and Ningbo Public Welfare Science and Technology Major Project (2021S106).

Institutional Review Board Statement: Not applicable.

Informed Consent Statement: Not applicable.

Data Availability Statement: All data and materials are available on request from the corresponding author. The data are not publicly available due to ongoing researches using a part of the data.

Conflicts of Interest: The authors declare no conflict of interest.

References

1. Lee, Y.I.; Choi, S.; Roh, W.S.; Lee, J.H.; Kim, T.-G. Cellular Senescence and Inflammaging in the Skin Microenvironment. *Int. J. Mol. Sci.* **2021**, *22*, 3849. [CrossRef] [PubMed]
2. Chen, L.; Shen, Z. Tissue-Resident Memory T Cells and Their Biological Characteristics in the Recurrence of Inflammatory Skin Disorders. *Cell. Mol. Immunol.* **2020**, *17*, 64–75. [CrossRef] [PubMed]
3. Mayandi, V.; Wen Choong, A.C.; Dhand, C.; Lim, F.P.; Aung, T.T.; Sriram, H.; Dwivedi, N.; Periayah, M.H.; Sridhar, S.; Fazil, M.H.U.T.; et al. Multifunctional Antimicrobial Nanofiber Dressings Containing ε-Polylysine for the Eradication of Bacterial Bioburden and Promotion of Wound Healing in Critically Colonized Wounds. *ACS Appl. Mater. Interfaces* **2020**, *12*, 15989–16005. [CrossRef] [PubMed]
4. Li, M.; Pan, G.; Zhang, H.; Guo, B. Hydrogel Adhesives for Generalized Wound Treatment: Design and Applications. *J. Polym. Sci.* **2022**, *60*, 1328–1359. [CrossRef]
5. Zeng, Z.; Zhu, M.; Chen, L.; Zhang, Y.; Lu, T.; Deng, Y.; Ma, W.; Xu, J.; Huang, C.; Xiong, R. Design the Molecule Structures to Achieve Functional Advantages of Hydrogel Wound Dressings: Advances and Strategies. *Compos. Part B Eng.* **2022**, *247*, 11031. [CrossRef]
6. Peng, W.; Li, D.; Dai, K.; Wang, Y.; Song, P.; Li, H.; Tang, P.; Zhang, Z.; Li, Z.; Zhou, Y.; et al. Recent Progress of Collagen, Chitosan, Alginate and Other Hydrogels in Skin Repair and Wound Dressing Applications. *Int. J. Biol. Macromol.* **2022**, *208*, 400–408. [CrossRef] [PubMed]
7. Yao, Y.; Zhang, A.; Yuan, C.; Chen, X.; Liu, Y. Recent Trends on Burn Wound Care: Hydrogel Dressings and Scaffolds. *Biomater. Sci.* **2021**, *9*, 4523–4540. [CrossRef] [PubMed]
8. Zhang, L.; Liu, M.; Zhang, Y.; Pei, R. Recent Progress of Highly Adhesive Hydrogels as Wound Dressings. *Biomacromolecules* **2020**, *21*, 3966–3983. [CrossRef]
9. Bao, Z.; Xian, C.; Yuan, Q.; Liu, G.; Wu, J. Natural Polymer-Based Hydrogels with Enhanced Mechanical Performances: Preparation, Structure, and Property. *Adv. Healthc. Mater.* **2019**, *8*, 1900670. [CrossRef]
10. Fuchs, S.; Shariati, K.; Ma, M. Specialty Tough Hydrogels and Their Biomedical Applications. *Adv. Healthc. Mater.* **2020**, *9*, 1901396. [CrossRef]
11. Zhang, X.; Xiang, J.; Hong, Y.; Shen, L. Recent Advances in Design Strategies of Tough Hydrogels. *Macromol. Rapid Commun.* **2022**, *43*, 2200075. [CrossRef] [PubMed]
12. Fan, H.; Gong, J.P. Fabrication of Bioinspired Hydrogels: Challenges and Opportunities. *Macromolecules* **2020**, *53*, 2769–2782. [CrossRef]
13. Hua, M.; Wu, S.; Ma, Y.; Zhao, Y.; Chen, Z.; Frenkel, I.; Strzalka, J.; Zhou, H.; Zhu, X.; He, X. Strong Tough Hydrogels via the Synergy of Freeze-Casting and Salting Out. *Nature* **2021**, *590*, 594–599. [CrossRef] [PubMed]
14. Deng, W.; Wei, F.; Hu, J. Muscle Contraction-Inspired Tough Hydrogels. *ACS Appl. Mater. Interfaces* **2023**, *15*, 8462–8470. [CrossRef] [PubMed]
15. Wang, Y.J.; Zhang, X.N.; Song, Y.; Zhao, Y.; Chen, L.; Su, F.; Li, L.; Wu, Z.L.; Zheng, Q. Ultrastiff and Tough Supramolecular Hydrogels with a Dense and Robust Hydrogen Bond Network. *Chem. Mater.* **2019**, *31*, 1430–1440. [CrossRef]
16. Naebe, M.; Shirvanimoghaddam, K. Functionally Graded Materials: A Review of Fabrication and Properties. *Appl. Mater. Today* **2016**, *5*, 223–245. [CrossRef]
17. Quan, H.; Kisailus, D.; Meyers, M.A. Hydration-Induced Reversible Deformation of Biological Materials. *Nat. Rev. Mater.* **2020**, *6*, 264–283. [CrossRef]
18. Lavrador, P.; Esteves, M.R.; Gaspar, V.M.; Mano, J.F. Stimuli-Responsive Nanocomposite Hydrogels for Biomedical Applications. *Adv. Funct. Mater.* **2021**, *31*, 2005941. [CrossRef]
19. McCaw, B.A.; Stevenson, T.J.; Lancaster, L.T. Epigenetic Responses to Temperature and Climate. *Integr. Comp. Biol.* **2020**, *60*, 1469–1480. [CrossRef]

20. Wijerathne, B.; Liao, T.; Ostrikov, K.; Sun, Z. Bioinspired Robust Mechanical Properties for Advanced Materials. *Small Struct.* **2022**, *3*, 2100228. [CrossRef]
21. Abbasi, N.; Hamlet, S.; Love, R.M.; Nguyen, N.-T. Porous Scaffolds for Bone Regeneration. *J. Sci. Adv. Mater. Dev.* **2020**, *5*, 1–9. [CrossRef]
22. Must, I.; Sinibaldi, E.; Mazzolai, B. A Variable-Stiffness Tendril-like Soft Robot Based on Reversible Osmotic Actuation. *Nat. Commun.* **2019**, *10*, 344. [CrossRef] [PubMed]
23. Zhao, Q.; Chang, Y.; Yu, Z.; Liang, Y.; Ren, L.; Ren, L. Bionic Intelligent Soft Actuators: High-Strength Gradient Intelligent Hydrogels with Diverse Controllable Deformations and Movements. *J. Mater. Chem. B* **2020**, *8*, 9362–9373. [CrossRef] [PubMed]
24. Vigolo, D.; Ramakrishna, S.N.; deMello, A.J. Facile Tuning of the Mechanical Properties of a Biocompatible Soft Material. *Sci. Rep.* **2019**, *9*, 7125. [CrossRef] [PubMed]
25. Dai, C.F.; Zhang, X.N.; Du, C.; Frank, A.; Schmidt, H.-W.; Zheng, Q.; Wu, Z.L. Photoregulated Gradient Structure and Programmable Mechanical Performances of Tough Hydrogels with a Hydrogen-Bond Network. *ACS Appl. Mater. Interfaces* **2020**, *12*, 53376–53384. [CrossRef]
26. Tong, X.; Pan, W.; Su, T.; Zhang, M.; Dong, W.; Qi, X. Recent Advances in Natural Polymer-Based Drug Delivery Systems. *React. Funct. Polym.* **2020**, *148*, 104501. [CrossRef]
27. Calori, I.R.; Braga, G.; De Jesus, P.D.C.C.; Bi, H.; Tedesco, A.C. Polymer Scaffolds as Drug Delivery Systems. *Eur. Polym. J.* **2020**, *129*, 109621. [CrossRef]
28. Narayanaswamy, R.; Torchilin, V.P. Hydrogels and Their Applications in Targeted Drug Delivery. *Molecules* **2019**, *24*, 603. [CrossRef]
29. Erol, O.; Pantula, A.; Liu, W.; Gracias, D.H. Transformer Hydrogels: A Review. *Adv. Mater. Technol.* **2019**, *4*, 1900043. [CrossRef]
30. Zhang, W.; Zhang, Y.; Zhang, Y.; Dai, Y.; Xia, F.; Zhang, X. Adhesive and Tough Hydrogels: From Structural Design to Applications. *J. Mater. Chem. B* **2021**, *9*, 5954–5966. [CrossRef]
31. Mredha, M.T.I.; Jeon, I. Biomimetic Anisotropic Hydrogels: Advanced Fabrication Strategies, Extraordinary Functionalities, and Broad Applications. *Prog. Mater. Sci.* **2022**, *124*, 100870. [CrossRef]
32. Ahadian, S.; Ramón-Azcón, J.; Estili, M.; Obregón, R.; Shiku, H.; Matsue, T. Facile and Rapid Generation of 3D Chemical Gradients within Hydrogels for High-Throughput Drug Screening Applications. *Biosens. Bioelectron.* **2014**, *59*, 166–173. [CrossRef] [PubMed]
33. Li, R.; Chen, T.; Pan, X. Metal–Organic-Framework-Based Materials for Antimicrobial Applications. *ACS Nano* **2021**, *15*, 3808–3848. [CrossRef] [PubMed]
34. Fang, K.; Gu, Q.; Zeng, M.; Huang, Z.; Qiu, H.; Miao, J.; Fang, Y.; Zhao, Y.; Xiao, Y.; Xu, T.; et al. Tannic Acid-Reinforced Zwitterionic Hydrogels with Multi-Functionalities for Diabetic Wound Treatment. *J. Mater. Chem. B* **2022**, *10*, 4142–4152. [CrossRef] [PubMed]
35. Sanger, M.J.; Greenbowe, T.J. An Analysis of College Chemistry Textbooks As Sources of Misconceptions and Errors in Electrochemistry. *J. Chem. Educ.* **1999**, *76*, 853. [CrossRef]
36. Guo, J.; Ping, Y.; Ejima, H.; Alt, K.; Meissner, M.; Richardson, J.J.; Yan, Y.; Peter, K.; von Elverfeldt, D.; Hagemeyer, C.E.; et al. Engineering Multifunctional Capsules through the Assembly of Metal–Phenolic Networks. *Angew. Chem. Int. Ed.* **2014**, *53*, 5546–5551. [CrossRef]
37. Zhang, Y.; Wang, Y.; Xue, J.; Tang, C. MnO_2-Coated Graphene/Polypyrrole Hybrids for Enhanced Capacitive Deionization Performance of Cu^{2+} Removal. *Ind. Eng. Chem. Res.* **2022**, *61*, 3582–3590. [CrossRef]
38. Fei, J.; Ding, B.; Koh, S.W.; Ge, J.; Wang, X.; Lee, L.; Sun, Z.; Yao, M.; Chen, Y.; Gao, H.; et al. Mechanistic Investigation of Electrostatic Field-Enhanced Water Evaporation. *Adv. Sci.* **2021**, *8*, 2100875. [CrossRef]
39. Liu, H.; Xu, H.; Pan, L.; Zhong, D.; Liu, Y. Porous Electrode Improving Energy Efficiency under Electrode-Normal Magnetic Field in Water Electrolysis. *Int. J. Hydrogen Energy* **2019**, *44*, 22780–22786. [CrossRef]
40. Asoh, T.-A. Electrophoretic Hydrogel Adhesion for Fabrication of Three-Dimensional Materials. *Polym. J.* **2016**, *48*, 1095–1101. [CrossRef]
41. Yang, J.C.; Mun, J.; Kwon, S.Y.; Park, S.; Bao, Z.; Park, S. Electronic Skin: Electronic Skin: Recent Progress and Future Prospects for Skin-Attachable Devices for Health Monitoring, Robotics, and Prosthetics. *Adv. Mater.* **2019**, *31*, 1970337. [CrossRef]
42. Yang, Y.; Wang, L.; Yan, F.; Xiang, X.; Tang, Y.; Zhang, L.; Liu, J.; Qiu, L. Determination of Normal Skin Elasticity by Using Real-time Shear Wave Elastography. *J. Ultrasound Med.* **2018**, *37*, 2507–2516. [CrossRef] [PubMed]
43. Park, J.; Kim, M.; Hong, I.; Kim, T.; Lee, E.; Kim, E.; Ryu, J.-K.; Jo, Y.; Koo, J.; Han, S.; et al. Foot Plantar Pressure Measurement System Using Highly Sensitive Crack-Based Sensor. *Sensors* **2019**, *19*, 5504. [CrossRef] [PubMed]
44. Jin, X.; Wei, C.; Wu, C.; Zhang, W. Gastric Fluid-Induced Double Network Hydrogel with High Swelling Ratio and Long-Term Mechanical Stability. *Composites. Part B* **2022**, *236*, 109816. [CrossRef]
45. Kang, Y.; Xia, Y.; Wang, H.; Zhang, X. 2D Laminar Membranes for Selective Water and Ion Transport. *Adv. Funct. Mater.* **2019**, *29*, 1902014. [CrossRef]
46. Chang, T.; Babu, R.P.; Zhao, W.; Johnson, C.M.; Hedström, P.; Odnevall, I.; Leygraf, C. High-Resolution Microscopical Studies of Contact Killing Mechanisms on Copper-Based Surfaces. *ACS Appl. Mater. Interfaces* **2021**, *13*, 49402–49413. [CrossRef]
47. Dizaj, S.M.; Lotfipour, F.; Barzegar-Jalali, M.; Zarrintan, M.H.; Adibkia, K. Antimicrobial Activity of the Metals and Metal Oxide Nanoparticles. *Biomater. Adv.* **2014**, *44*, 278–284. [CrossRef]
48. Li, Y. Copper Homeostasis: Emerging Target for Cancer Treatment. *IUBMB Life* **2020**, *72*, 1900–1908. [CrossRef]

49. Dai, Y.; Liu, Z.; Bai, Y.; Chen, Z.; Qin, J.; Feng, F. A Novel Highly Fluorescent S, N, O Co-Doped Carbon Dots for Biosensing and Bioimaging of Copper Ions in Live Cells. *RSC Adv.* **2018**, *8*, 42246–42252. [CrossRef]
50. Ning, C.; Wang, X.; Li, L.; Zhu, Y.; Li, M.; Yu, P.; Zhou, L.; Zhou, Z.; Chen, J.; Tan, G.; et al. Concentration Ranges of Antibacterial Cations for Showing the Highest Antibacterial Efficacy but the Least Cytotoxicity against Mammalian Cells: Implications for a New Antibacterial Mechanism. *Chem. Res. Toxicol.* **2015**, *28*, 1815–1822. [CrossRef]
51. *ISO TDC 9 (1697) DTZS*; Biological Evaluation of Medical Devices—Part 5: Tests for In Vitro Cytotoxicity. ISO: Geneva, Switzerland, 2009.
52. Qiao, Y.; Ping, Y.; Zhang, H.; Zhou, B.; Liu, F.; Yu, Y.; Xie, T.; Li, W.; Zhong, D.; Zhang, Y.; et al. Laser-Activatable CuS Nanodots to Treat Multidrug-Resistant Bacteria and Release Copper Ion to Accelerate Healing of Infected Chronic Nonhealing Wounds. *ACS Appl. Mater. Interfaces* **2019**, *11*, 3809–3822. [CrossRef] [PubMed]
53. Liu, L.; Zhang, H.; Peng, L.; Wang, D.; Zhang, Y.; Yan, B.; Xie, J.; Xing, S.; Peng, F.; Liu, X. A Copper-Metal Organic Framework Enhances the Photothermal and Chemodynamic Properties of Polydopamine for Melanoma Therapy. *Acta Biomater.* **2023**, *158*, 660–672. [CrossRef] [PubMed]
54. Xiao, J.; Zhu, Y.; Huddleston, S.; Li, P.; Xiao, B.; Farha, O.K.; Ameer, G.A. Copper Metal–Organic Framework Nanoparticles Stabilized with Folic Acid Improve Wound Healing in Diabetes. *ACS Nano* **2018**, *12*, 1023–1032. [CrossRef] [PubMed]

gels
MDPI

Article

Chitosan–Polyethylene Glycol Inspired Polyelectrolyte Complex Hydrogel Templates Favoring NEO-Tissue Formation for Cardiac Tissue Engineering

Angelo Keklikian [1], Natan Roberto de Barros [2], Ahmad Rashad [2], Yiqing Chen [3], Jinrui Tan [3], Ruoyu Sheng [3], Dongwei Sun [3], Huinan Liu [3] and Finosh G. Thankam [1,*]

1 Department of Translational Research, College of Osteopathic Medicine of the Pacific and Graduate College of Biomedical Sciences, Western University of Health Sciences, Pomona, CA 91766, USA; akeklo28@gmail.com
2 Terasaki Institute for Biomedical Innovation (TIBI), Los Angeles, CA 90064, USA; nbarros@terasaki.org (N.R.d.B.); ahmad.elsebahy@terasaki.org (A.R.)
3 Department of Bioengineering, University of California, Riverside, CA 92521, USA; yiqing.chen@ucr.edu (Y.C.); ruoyu.sheng@email.ucr.edu (R.S.); dongwei.sun@email.ucr.edu (D.S.); huinanliu@engr.ucr.edu (H.L.)
* Correspondence: fthankam@westernu.edu; Tel.: +1-909-469-7042

Abstract: Neo-tissue formation and host tissue regeneration determine the success of cardiac tissue engineering where functional hydrogel scaffolds act as cardiac (extracellular matrix) ECM mimic. Translationally, the hydrogel templates promoting neo-cardiac tissue formation are currently limited; however, they are highly demanding in cardiac tissue engineering. The current study focused on the development of a panel of four chitosan-based polyelectrolyte hydrogels as cardiac scaffolds facilitating neo-cardiac tissue formation to promote cardiac regeneration. Chitosan-PEG (CP), gelatin-chitosan-PEG (GCP), hyaluronic acid-chitosan-PEG (HACP), and combined CP (CoCP) polyelectrolyte hydrogels were engineered by solvent casting and assessed for physiochemical, thermal, electrical, biodegradable, mechanical, and biological properties. The CP, GCP, HACP, and CoCP hydrogels exhibited excellent porosity (4.24 ± 0.18, 13.089 ± 1.13, 12.53 ± 1.30 and 15.88 ± 1.10 for CP, GCP, HACP and CoCP, respectively), water profile, mechanical strength, and amphiphilicity suitable for cardiac tissue engineering. The hydrogels were hemocompatible as evident from the negligible hemolysis and RBC aggregation and increased adsorption of plasma albumin. The hydrogels were cytocompatible as evident from the increased viability by MTT (>94% for all the four hydrogels) assay and direct contact assay. Also, the hydrogels supported the adhesion, growth, spreading, and proliferation of H9c2 cells as unveiled by rhodamine staining. The hydrogels promoted neo-tissue formation that was proven using rat and swine myocardial tissue explant culture. Compared to GCP and CoCP, CP and HACP were superior owing to the cell viability, hemocompatibility, and conductance, resulting in the highest degree of cytoskeletal organization and neo-tissue formation. The physiochemical and biological performance of these hydrogels supported neo-cardiac tissue formation. Overall, the CP, GCP, HACP, and CoCP hydrogel systems promise novel translational opportunities in regenerative cardiology.

Keywords: neo-tissue formation; cardiac tissue engineering; polyelectrolyte hydrogels; chitosan; regenerative cardiology

Citation: Keklikian, A.; de Barros, N.R.; Rashad, A.; Chen, Y.; Tan, J.; Sheng, R.; Sun, D.; Liu, H.; Thankam, F.G. Chitosan–Polyethylene Glycol Inspired Polyelectrolyte Complex Hydrogel Templates Favoring NEO-Tissue Formation for Cardiac Tissue Engineering. *Gels* **2024**, *10*, 46. https://doi.org/10.3390/gels10010046

Academic Editor: Xin Zhao

Received: 11 December 2023
Revised: 22 December 2023
Accepted: 27 December 2023
Published: 8 January 2024

1. Introduction

Neo-tissue formation and host tissue regeneration are the major checkpoints in determining the success of cardiac tissue engineering which is driven by the chemistry and functional performance of scaffolds employed. This process is decided by the biocompatibility, physiochemical features and the contact guidance provided by the hydrogel scaffolds for promoting the proliferation and signaling by regenerative cell phenotypes [1]. Hence, neo-tissue formation is driven by the cardiac-mimetic features of the hydrogel templates employed, allowing for the seeded/host cell migration and accelerating the deposition of the ECM components and angiogenesis [2–4]. Unfortunately, despite the greater demand, the capability of hydrogel templates in promoting neo-cardiac tissue formation is currently limited. Interestingly, polyelectrolyte complex hydrogels promise tissue engineering applications owing to their tunable features, electrical conductivity and structural similarities with mammalian ECM [5]. However, to the best of our knowledge, the polyelectrolyte complex hydrogels that supports neo-cardiac tissue formation is currently unavailable.

Chitosan-based hydrogels demonstrates myocardial regeneration as evidenced by the improved myocardial wall integrity, decreased infarct expansion and left ventricular (LV) remodeling, and retention of contractile function in pre-clinical myocardial infarction (MI) models [6–8]. Additionally, the inherent antimicrobial and antioxidant properties and the ability to form spontaneous polyelectrolyte complex impart additional benefits to chitosan for cardiac applications [9]. Likewise, hydrogels based on polyethylene glycol (PEG), a FDA approved synthetic polymer, have been praised for improving the cardiac function in MI models owing to their exceptional biocompatibility and mechanical strength [10,11]. Additionally, the ECM-derived protein gelatin and cardiac ECM component hyaluronic acid (HA) offer significant benefits for cardiac tissue engineering by providing biological adhesion sites and homely environment to the seeded cells and the subsequent neo-tissue formation [12–15]. Interestingly, the ample positive charge density in chitosan and the anionic functional groups in gelatin, HA and PEG provides enhanced opportunities for forming polyelectrolyte complex hydrogels. However, such templates supporting neo-tissue formation for cardiac applications are currently unavailable. Given this background, we hypothesize that chitosan-based polyelectrolyte complex hydrogels containing the native cardiac ECM biomolecules gelatin and HA promote neo-cardiac tissue formation. Hence, the current study aims to synthesize novel polyelectrolyte hydrogels from chitosan, PEG, gelatin, and HA, characterize the properties and performance, and evaluate their potential in neo-tissue formation for cardiac tissue engineering applications. The hydrogels were tested for physiochemical, mechanical and thermal characteristics and biological performance to assess their ability for neo-tissue formation.

2. Results and Discussion

CP blend in the aqueous phase was casted, allowing for the formation of the parent polyelectrolyte hybrid CP hydrogel. The CP hydrogels were interpenetrated with the biomolecules, gelatin and HA for improved biocompatibility and biological performance. The addition of gelatin to the CP blend resulted in GCP hydrogel. Similarly, CP was blended with hyaluronic acid to form HACP hydrogel. CoCP hydrogel was synthesized by incorporating both gelatin and HA to the CP blend. All four preparations, CO, GCP, HACP and CoCP, were casted at 55 °C to prevent the denaturation of biological domains in the gelatin and HA components of the hydrogels, thereby ensuring the maximum biological responses. Also, the four hydrogels exhibited flexibility of handling using tweezers without breaking.

2.1. *Physiochemical Characterization*

2.1.1. ATR-IR Spectroscopy

AT-IR spectroscopy was used to determine the surface functional groups on CP, GCP, HACP, and CoCP hydrogels (Figure 1A–D). A broad peak at about 3200 cm^{-1} denoted the presence of -OH groups with a peak area of 5281.1 ± 1427.7 for CP, 4479.6 ± 1547.6 for GCP, 3664.9 ± 1010.4 for HACP, and 4492.2 ± 639.2 for CoCP. The difference in the -OH peak intensity was statistically not significant among the groups, thereby suggesting similar level of surface -OH groups contributed by chitosan, PEG, gelatin, and HA. C-H stretch was demonstrated by a peak around 2800 cm^{-1} with a peak area of 1142.7 ± 291.9 for CP, 1338.2 ± 26.7 for GCP, 2537.3 ± 557.8 for HACP, and 1847.2 ± 378.0 for CoCP. HACP displayed significantly higher intensity for the C-H stretch when compared to GCP

(p = 0.0062) and CoCP (p = 0.0007). However, the difference in the C-H peak intensity was statistically not significant among the other groups, thus suggesting similar surface chemistry for C-H. The peak around 1550 cm^{-1} demonstrated N-H bending with a peak area of 618.3 ± 281.4 for CP, 527.2 ± 322.6 for GCP, 365.1 ± 182.5 for HACP, and 608.8 ± 134.2 for CoCP, thereby suggesting the formation of polyelectrolyte complex [16]. Also, the difference in the N-H bending peak intensity was statistically not significant among the groups suggesting similar surface chemistry for N-H, revealing the presence of chitosan, gelatin, and HA. C-H bending of methyl groups was characterized by a peak at about 1440 cm^{-1} with a peak area of 294.7 ± 37.8 for CP, 1264.0 ± 79.3 for GCP, 386.3 ± 54.0 for HACP, and 204.6 ± 32.3 for CoCP. Interestingly, GCP displayed significantly increased C-H bending of methyl groups relative to CP (p = 0.0006), HACP (p = 0.0023), and CoCP (p = 0.0003). However, the difference in the C-H bending peak intensity was statistically not significant among the other groups, thus suggesting similar surface chemistry for C-H. Lastly, C-O stretching of vinyl ethers was characterized by a peak around 1400 cm^{-1} with a peak area of 262.2 ± 163.6 for CP, 192.2 ± 110.2 for GCP, 308.7 ± 129.1 for HACP, and 473.7 ± 238.8 for CoCP. Also, the difference in the C-O stretching peak intensity was statistically not significant among the groups, thereby suggesting the presence of chitosan and PEG fragments on the surface of all the four hydrogels. Overall, the IR spectrum demonstrated similar surface chemistry in all four hydrogels (Figure 1A–D).

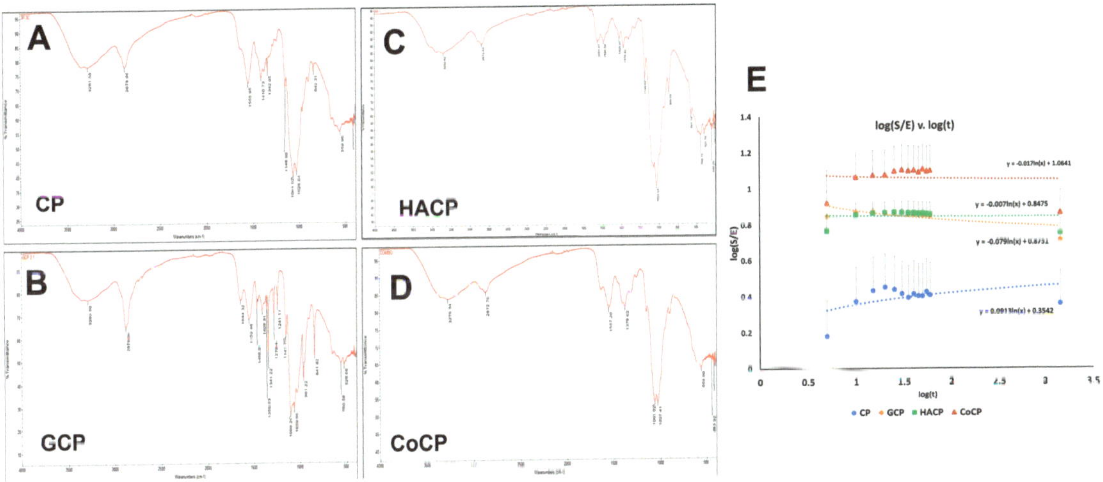

Figure 1. ATR-IR spectrum for CP (**A**), GCP (**B**), HACP (**C**), and CoCP (**D**) hydrogels; the X-axis shows % transmittance, and the Y-axis shows wavenumber (cm^{-1}). (**E**) Determination of diffusional exponent (n) from the slope and swelling constant (k) from the Y-intercept.

2.1.2. Contact Angle

The ACA and RCA for CP hydrogels were 29.38 ± 2.57° and 31.83 ± 2.42°, respectively. GCP displayed an ACA of 43.70 ± 4.60° and 46.55 ± 5.99°. HACP hydrogels exhibited 44.40 ± 3.90° and 42.38 ± 7.13° for ACA and RCA, respectively. ACA for CoCP hydrogels was 47.21 ± 7.20°, and RCA was 49.05 ± 5.39 °. Overall, the contact angle measurements of all the four hydrogels were in the amphiphilic range. Furthermore, the difference between ACA and RCA of CP (p = 0.1594), GCP (p = 0.4233), HACP (p = 0.6359), and CoCP (p = 0.6294) hydrogels were not statistically significant, thus suggesting the absence of phase transitions and migration of functional moieties in the hydrogels upon interacting with the aqueous environment (Table 1).

Table 1. Characteristic features of CP, GCP, HACP and CoCP hydrogels.

Parameters	CP	GCP	HACP	CoCP
ACA (°) (n = 10)	29.38 ± 2.57	43.70 ± 4.60	44.401 ± 3.90	47.21 ± 7.20
RCA (°) (n = 10)	31.83 ± 2.42	46.55 ± 5.99	42.38 ± 7.10	49.05 ± 5.37
Swelling Ratio (S) (n = 10)	2.47 ± 0.12	4.48 ± 0.66	5.51 ± 1.14	9.27 ± 2.46
Equilibrium Swelling Ratio (E) (n = 10)	0.71 ± 0.01	0.79 ± 0.02	0.80 ± 0.03	0.81 ± 0.05
% Swelling (%) (n = 10)	247.3 ± 11.8	447.8 ± 65.62	551.2 ± 113.6	926.8 ± 245.9
Equilibrium Water Content (EWC) (%) (n = 10)	70.92 ± 0.94	79.12 ± 2.43	79.56 ± 3.30	80.65 ± 4.54
Total Water Absorption Sites (TWAS) (n = 10)	$1.129 \times 10^{21} \pm 3.67 \times 10^{19}$	$1.520 \times 10^{21} \pm 2.85 \times 10^{20}$	$1.679 \times 10^{21} \pm 2.20 \times 10^{20}$	$1.433 \times 10^{21} \pm 4.53 \times 10^{20}$
Diffusional Exponent (n) (n = 10)	0.0913	−0.079	−0.007	−0.017
Diffusion Constant (k) (n = 10)	0.3542	0.8751	0.8475	1.0641
Pore Length (μm) (n > 20)	4.241 ± 0.181	13.089 ± 1.130	12.532 ± 1.301	15.881 ± 1.104
Aspect Ratio (n > 20)	1.256 ± 0.054	1.746 ± 0.165	1.728 ± 0.106	1.461 ± 0.047
Enthalpy of Melting Endotherm (J/g)	237.3	122.0	188.2	277.0
Freezing Water Content (%)	69.58	34.28	52.93	71.92
Non-freezing Water Content (%)	1.34	44.91	26.63	8.74
Conductance (μS/cm) (n = 5)	0.28 ± 0.01	0.69 ± 0.07	2.60 ± 1.23	0.44 ± 0.02
Young's Modulus (kPa) (n = 6)	1141.0 ± 241.1	179.4 ± 51.5	368.4 ± 95.4	131.7 ± 16.7
Tensile Stress at Failure (kPa) (n = 6)	401.9 ± 54.2	39.22 ± 2.12	96.41 ± 18.81	47.22 ± 5.71
Load at Failure (N) (n = 6)	5.63 ± 0.75	0.94 ± 0.03	1.74 ± 0.31	1.34 ± 0.12
Direct Contact—MTT assay (%Viability) (n = 5)	71.88 ± 3.26	82.84 ± 2.82	92.44 ± 0.79	74.09 ± 1.43
Test on Extract—MTT assay (% Cell) (n = 4)	96.1 ± 5.2	98.8 ± 10.4	114.4 ± 11.1	94.9 ± 3.2
% Hemolysis (n = 3)	0.66 ± 0.02	0.76 ± 0.02	0.27 ± 0.08	0.63 ± 0.04
Absorption of Total Plasma Protein (μg/μL) (n = 5)	1.67 ± 0.51	5.48 ± 0.26	2.78 ± 0.26	8.19 ± 1.8

2.1.3. Water Profile

The EWC for CP, GCP, HACP, and CoCP was 70.92 ± 0.94%, 79.12 ± 2.43%, 79.56 ± 3.30, and 80.65 ± 4.54%, respectively. Similarly, the % swelling was 247.3 ± 11.8%, 447.8 ± 65.62%, 551.2 ± 113.6%, and 926.8 ± 245.9% for CP, GCP, HACP, and CoCP, respectively. The values for E and S are displayed in Table 1, which were used to calculate the diffusional exponent (n). The respective diffusional exponent and diffusion constant (k) were 0.0913 and 0.3542 for CP, −0.079 and 0.8751 for GCP, −0.007 and 0.8475 for HACP, and −0.017 and 1.0641 for CoCP. Linear regressions for log(S/E) vs. log(t) are displayed in Figure 1E. The alterations in E, EWC, S and %S were statistically not significant, thereby suggesting similar water profile for CP, GCP, HACP, and CoCP hydrogels. Also, the water profiling displayed the superabsorbent nature of GCP, HACP, and CoCP hydrogels. GCP ($p < 0.0001$), HACP ($p < 0.0001$), and CoCP ($p < 0.0001$) hydrogels displayed significantly increased TWAS compared to the CP hydrogels; however, all the hydrogels exhibited a TWAS in the order of 10^{21} (Table 1).

2.1.4. Cross-Sectional Pore Morphometry

SEM analysis revealed the cross-sectional morphometry of CP, GCP, HACP, and CoCP hydrogels, showing excellent porosity with interconnectivity (Figure 2A–D). Moreover, the pore sizes for the hydrogels were heterogenous and the CP hydrogels displayed significantly decreased pore lengths compared to GCP ($p < 0.0001$), HACP ($p < 0.0001$), and CoCP ($p < 0.0001$) (Table 1) (Figure 2E). Similarly, CP displayed a significantly lower aspect ratio than GCP ($p = 0.0023$) and HACP ($p = 0.0036$); however, the overall trend was toward a oval pore shape (Table 1) (Figure 2F).

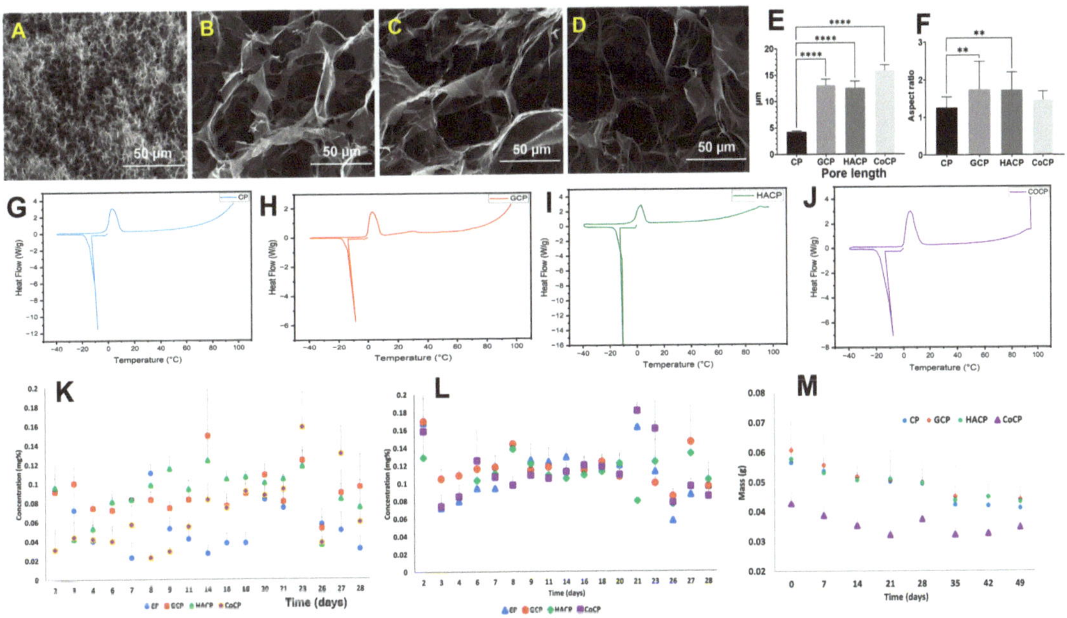

Figure 2. Representative SEM images demonstrating cross-sectional morphology of CP (**A**), GCP (**B**), HACP (**C**), and CoCP (**D**) hydrogels. Bar diagram depicting hydrogel pore length (**E**) and pore aspect ratio (**F**). DSC thermograms depicting the melting and crystallization of water in CP (**G**), GCP (**H**), HACP (**I**), and CoCP (**J**) hydrogels. Long-term dye dissociation of Methylene Blue (**K**) and Trypan Blue (**L**) from CP, GCP, HACP, and CoCP hydrogels. (**M**) Alterations in the weight of CP, GCP, HACP, and CoCP hydrogels upon aging in PBS. (** $p < 0.01$, **** $p < 0.0001$ and unlabeled parameters are nonsignificant).

2.1.5. Water Transition Status

The DSC thermograms for CP, GCP, HACP and CoCP hydrogels displayed corresponding endothermic and exothermic peaks, reflecting the melting of frozen water and crystallization of freezing water, respectively (Figure 2G–J). CP hydrogel contained 69.58% freezable water and 1.34% non-freezing bound water; GCP hydrogel contained 34.28% freezable water and 44.91% non-freezable water; HACP contained 52.93% freezable water and 26.63% non-freezable water; and CoCP hydrogel contained 71.92% freezable water and 8.74 non-freezing bound water (Table 1) (Figure 2G–J).

2.2. Release Profile

CP, GCP, HACP, and CoCP hydrogels demonstrated the ability to absorb and release hydrophobic and hydrophilic dyes. Overall, the release profiles for CP, GCP, HACP, and CoCP displayed a constant release rate of hydrophobic and hydrophilic dyes (Figure 2K,L). Specifically, GCP, HACP, and CoCP hydrogels displayed consistent release profile compared to CP hydrogel.

2.3. Conductance

CP hydrogels exhibited a conductance of 0.28 ± 0.01 μS/cm, while the conductance for GCP hydrogels was 0.69 ± 0.07 μS/cm, HACP hydrogels was 2.60 ± 1.23 μS/cm, and CoCP hydrogels was 0.44 ± 0.02 μS/cm (Table 1). HACP displayed significantly increased conductance compared to CP (p = 0.0001), GCP (p = 0.0008), and CoCP (p = 0.0002).

2.4. Biodegradation

The CP, GCP, HACP, and CoCP hydrogels were biodegradable, as evidenced by the progressive weight loss upon aging in PBS. The degradation profiles of CP, GCP, HACP, and CoCP hydrogels proceeded as surface erosion in a regulated manner (Figure 2M). Also, the slight drop in pH for CP (−0.03 ± 0.01), GCP (−0.03 ± 0.01), HACP (−0.01 ± 0.00), and CoCP (−0.04 ± 0.01) hydrogels were evident, which suggest very mild acidic degradation products.

2.5. Mechanical Characterization

Using UTM, Young's modulus, maximum tensile stress, and load at failure were calculated for the CP, GCP, HACP, and CoCP hydrogels. CP hydrogels exhibited 1141.0 ± 24.1 kPa Young's modulus, and 401.9 ± 54.2 kPa tensile stress with 5.62 ± 0.75 N load at failure. Similarly, GCP displayed 179.4 ± 51.5 kPa Young's modulus, 39.22 ± 2.12 kPa tensile stress, and 0.94 ± 0.03 N load at the failure. The Young's modulus for HACP hydrogel was 368.4 ± 95.4 kPa, tensile stress was 96.41 ± 18.81 kPa, and load at the failure was 1.74 ± 0.31 N. For CoCP hydrogel, Young's modulus was 131.7 ± 6.7 kPa, tensile stress was 47.22 ± 5.71 kPa, and load at the failure was 1.34 ± 0.12 N (Table 1). The variations in modulus, tensile stress and load at the failure were statistically not significant among CP, GCP, HACP, and CoCP hydrogels.

2.6. Cytocompatibility

Direct contact assay using CP, GCP, HACP, and CoCP hydrogels revealed the absence of changes in the characteristic morphology of H9c2, demonstrating that all four hydrogels were non-cytotoxic and supported the cell proliferation upon contact (Figure 3A–E). Direct contact viability assay displayed >70% viability of H9c2 cells where the respective viability was 71.88 ± 3.26 ($p < 0.0001$), 82.84 ± 2.82 (p = 0.0002), 92.44 ± 0.79 (p = 0.1215) and 74.09 ± 1.43 ($p < 0.0001$) for CP, GCP, HACP, and CoCP hydrogels (Table 1). Also, the viability assay using the hydrogel extracts for CP hydrogel was 96.1 ± 5.2%, for GCP was 98.8 ± 10.4%, for HACP was 114.4 ± 11.1%, and for CoCP was 94.9 ± 3.2% (Table 1) (Figure 3F). The alteration in cell viability on extracts was statistically not significant among the four hydrogels and the control.

2.7. Hemocompatibility

Hemolysis was 0.66 ± 0.02%, 0.76 ± 0.02%, 0.27 ± 0.08%, and 0.63 ± 0.04% for CP, GCP, HACP, and CoCP hydrogels, respectively (Table 1) (Figure 3G). HACP hydrogel exhibited significantly decreased levels of hemolytic potential compared to CP (p = 0.0046), GCP (p = 0.0004), and CoCP (p = 0.0113) hydrogels. Also, the CP, GCP, HACP, and CoCP hydrogels evoked minimal RBC aggregations, rouleaux formations, and changes in characteristic RBC morphology, thus revealing their hemocompatibility (Figure 3H–L). The total serum protein adsorption was 1.67 ± 0.51 μg/μL, 5.48 ± 0.26 μg/μL, 2.78 ± 0.26 μg/μL, 8.19 ± 1.80 μg/μL for CP, GCP, HACP, and CoCP hydrogels, respectively (Table 1) (Figure 3M). CoCP displayed significantly increased adsorption of plasma proteins compared to the control (p = 0.0060), CP (p = 0.0062) and HACP (p = 0.0031); however, the increase was statistically not significant for GCP (p = 0.2677). SDS-PAGE analysis revealed that albumin adsorption was prevalent in all the four hydrogels compared to the control (Figure 3N). Alterations in the relative amounts of protein and albumin adsorption were similar among the four hydrogel groups, suggesting similar protein and albumin adsorption capacities.

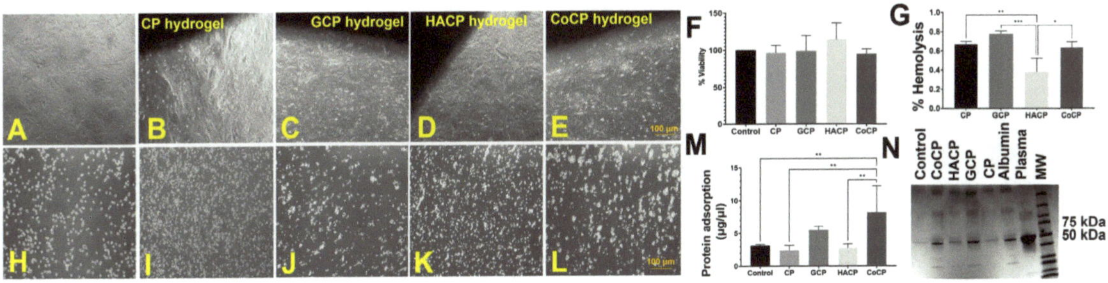

Figure 3. (**A**) Direct contact assay control with no hydrogels. Representative images displaying a lack of morphological changes in H9c2 cardiomyoblasts when in direct contact with CP (**B**), GCP (**C**), HACP (**D**), and CoCP (**E**) hydrogels. (**F**) Cell viability following the growth of H9c2 cells in hydrogel extracts. (**G**) Hemolytic potential of the hydrogels. Representative images of RBC aggregation assay showing negative control (**H**), CP (**I**), GCP (**J**), HACP (**K**), and CoCP (**L**) hydrogels. Quantification of total plasma protein adsorption by the hydrogels (**M**) and the SDS-PAGE (**N**) showing the adsorption of albumin (* $p < 0.05$, ** $p < 0.01$, *** $p < 0.001$, unlabeled parameters are not significant).

2.8. Biological Performance

The CP, GCP, HACP, and CoCP hydrogels facilitated H9c2 cardiomyoblasts adhesion to the porous network, migration within interstices, and proliferation as observed from the rhodamine staining. The H9c2 cells formed 3D clusters within the hydrogels, highlighting their cytocompatibility and ability to support the 3D growth of cells (Figure 4) (Supplementary Videos S1–S5). Neo-tissue formation was confirmed using the explant culture from the left ventricle of rat and swine models by growing onto the CP, GCP, HACP, and CoCP hydrogels after seven days. Rhodamine phalloidin staining displayed cytoskeletal organization and neo-tissue formation in both rat (Figure 5) (Supplementary Videos S6–S9) and swine (Figure 6) (Supplementary Videos S10–S13) cultures where the neo-tissue adhered to the hydrogels. The cells migrated onto the hydrogel network and into the interstices. The rat neo-tissue migration was ~70 µm for all the hydrogels. In contrast, the swine neo-tissues penetrated ~70 µm, 130 µm, 150 µm, and 330 µm for CP, GCP, HACP, and CoCP hydrogels, respectively, as evidenced by the Z-stack analysis. Even though neo-tissue formation was evidenced in the CP, GCP, HACP, and CoCP hydrogels, CP, HACP hydrogels were superior to both GCP and CoCP.

2.9. Discussion

The natural polysaccharide chitosan has been hailed for cardiac tissue engineering applications owing to its biocompatibility and inherent antimicrobial capacity [17]. Being a polysaccharide, chitosan is superior in myocardial healing and does not provoke aggressive immune reactions [17]. Chitosan retains its crystallinity, rigidity, and stability in the physiological milieu. Additionally, the ample -NH_3^+ groups in chitosan interact with anionic polymers, such as gelatin, HA, and PEG, forming polyelectrolyte complexes [18]. Chitosan complexes with biocompatible polymers improves the cardiogenic and biomechanical properties [17,19]. Interestingly, the PEG, upon spontaneously receiving cations including H+, form supra-polyelectrolyte aggregate imparting additional stability and biological features [20]. Hence, the polyelectrolyte complex chemistry of CP hydrogel provides stability and degradability to the hydrogels apart from the inherent biological features of chitosan and PEG. Additionally, the extensive hydrogen bonding stabilize the CP hydrogels and provides opportunities for further complexation with functional groups of complementarily/oppositely charged density. The collagen-derived biomolecule, gelatin, imparts natural adhesion cues for the cell survival and performance [21,22] which inspired us to exploit the biocompatibility of gelatin by modifying CP hydrogel by incorporating gelatin to engineer GCP hydrogels. Similarly, the non-immunogenic cardiac ECM component HA has been successfully exploited in cardiac tissue engineering owing to its active involvement in the biological properties including wound healing, morphogenesis, angiogenesis, and ECM organization [14,23]. Hence, the incorporation of HA in the parent CP hydrogel offers superior benefits in HACP hydrogels and the synergistic effects of gelatin and HA have been exhibited by CoCP.

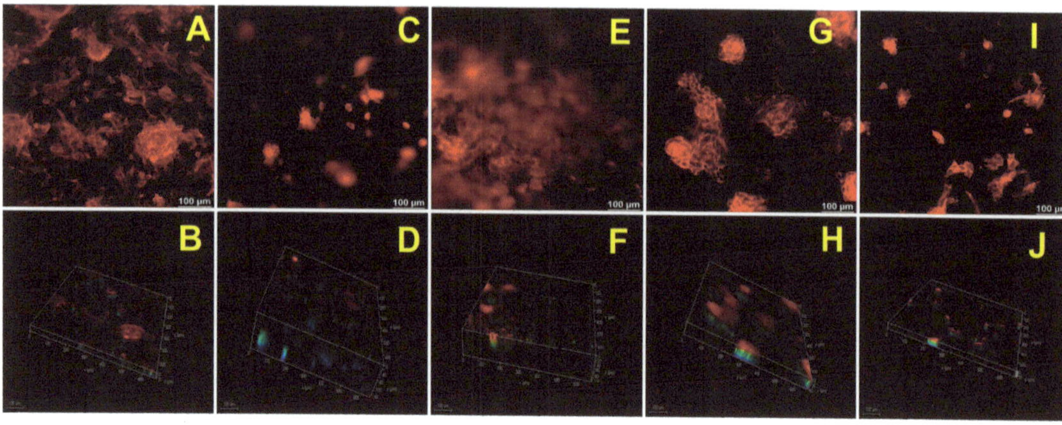

Figure 4. Rhodamine phalloidin-stained H9c2 cells seeded atop control (**A**), CP hydrogel (**C**), GCP hydrogel (**E**), HACP hydrogel (**G**), and CoCP hydrogel (**I**). Twenty consecutive Z-stack images and 3D rendering showing control (**B**), CP hydrogel (**D**), GCP hydrogel (**F**), HACP hydrogel (**H**), and CoCP hydrogel (**J**). Images were acquired at 20× magnification; scale bar: 100 µm.

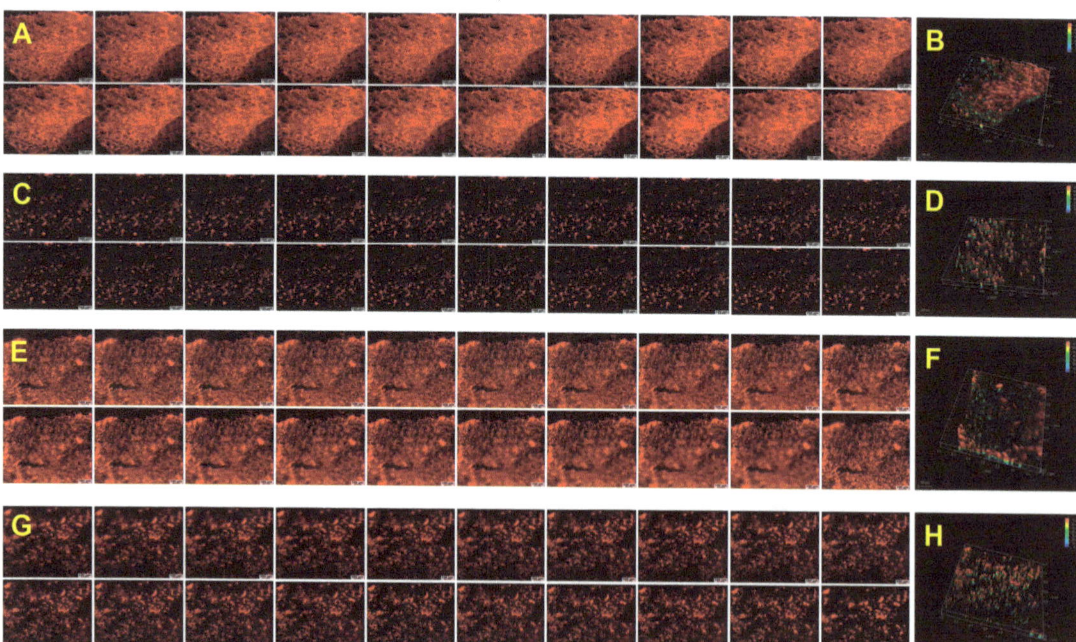

Figure 5. Neo-tissue formation by rat LV explants: 20 consecutive z-stack images and 3D rendering images of rhodamine phalloidin for CP hydrogel (**A**,**B**), GCP hydrogel (**C**,**D**), HACP hydrogel (**E**,**F**), and CoCP hydrogel (**G**,**H**), respectively, after a week of seeding. Images were acquired at 10× magnification; scale bar: 100 µm.

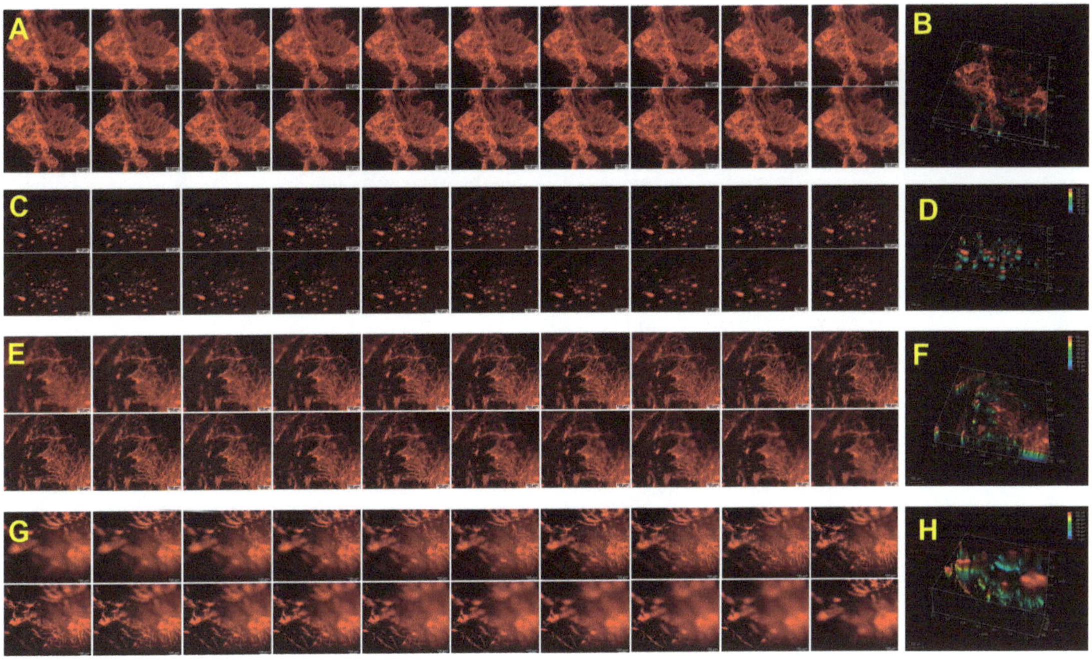

Figure 6. Neo-tissue formation by swine LV explants: 20 consecutive z-stack images and 3D rendering images of rhodamine phalloidin for CP hydrogel (**A**,**B**), GCP hydrogel (**C**,**D**), HACP hydrogel (**E**,**F**), and CoCP hydrogel (**G**,**H**), respectively, after a week of seeding. Images were acquired at 10× magnification; scale bar: 100 µm.

The ATR-IR analysis revealed the surface decoration of diverse functional groups and chitosan, PEG, HA, and gelatin fragments in the respective hydrogels. The peaks for -OH groups and C-H stretches revealed the formation of CP-based hydrogels consistent with prior reports [24,25]. The N-H bending reflects the polyelectrolyte formation in all four hydrogels [16], which was greater for CP and CoCP, followed by GCP and HACP, as evidenced by the area under the peaks. Also, the IR data suggests the presence of hydrophobic and hydrophilic functional groups/moieties distributed on the surface of CP, GCP, HACP, and CoCP hydrogels. The increased surface -OH groups also facilitate hydrogen bonding, resulting in enhanced water kinetics, cellular adhesion, and biocompatibility [26]. Overall, the surface functional moieties/domains in the CP-based hydrogels favor cell adhesion and survival and guide the seeded cells to deposit ECM, where the surface chemistry allows the integration of neo-ECM with the native cardiac ECM, thereby promoting regeneration [27].

Amphiphilicity promotes biocompatibility, cell migration and protein adsorption in cardiac tissue engineering hydrogels [28,29]. Also, the similar range of ACA and RCA reflects a consistent and biomimetic surface for the cells and biomolecules to interact [30]. Evidently, HA- and gelatin-incorporated hydrogels exhibit a contact angle around 42° [31] as displayed by GCP, HACP and CoCP hydrogels. Additionally, the surfaces with contact angles between 50° and 80° support maximal spreading, growth, and proliferation of multiple cell types, including cardiac progenitor cells and cardiomyocytes (CMs) [32]. Plasma proteins, being the first line of biomolecules encountering the implanted biomaterials and the amphiphilic surfaces, provide cues for interacting with plasma proteins, exhibiting the binding opportunities for both hydrophilic and hydrophobic domains [28,33]. Hence, the amphiphilicity of CP, GCP, HACP, and CoCP hydrogels benefits their proposed cardiac tissue engineering applications.

The water-holding capacity of hydrogels directly impacts biocompatibility, permeability, protein adsorption, and hemocompatibility [34]. Interestingly, superabsorbent hydrogels are characterized by their superior water-holding capacities (EWC > 80%) [24]. When comparing GCP, HACP, and CoCP hydrogels with CP hydrogel, the EWC ~80% in the former hydrogels highlights their superabsorbent nature contributed by the superior water binding capacity of gelatin and HA. Additionally, the increased EWC facilitates the transport and diffusion of oxygen nutrients, metabolites, and exhausts to and from the hydrogel interstices, thus promoting the survival and existence of seeded/recruited cells [35]. A higher EWC and swelling correlate with improved biocompatibility and immunocompatibility [36]. Also, the swelling in hydrogels is a multi-step process involving the diffusion of water molecules, which is a function of the diffusional exponent (n). Interestingly, $n < 0.5$ drives swelling-controlled Fickian diffusion, and $n > 0.5$ favors Super Case II Transport (SCIIT) diffusion, which is controlled by network relaxation in hydrogels [1,5]. Our findings revealed that all four, CP, GCP, HACP, and CoCP, hydrogels favor Fickian diffusion, which is crucial for the integration with the native myocardium and for controlling the trafficking of nutrients, biomolecules, and exhausts [37,38]. Additionally, all four hydrogels exhibited a superior TWAS in the order of 10^{21}, suggesting their potential to absorb ample water within a relatively short interval, which is responsible for their exceptional water-holding capability [24]. Overall, all four hydrogels bear excellent water profile, which supports cell adhesion, growth, and proliferation and enhances biocompatibility.

Porosity and pore size are critical for the performance of tissue engineering hydrogels as the porosity guides the trafficking of biomolecules, biocompatibility, water kinetics, biodegradation, and cell homing and migration [24,31,35]. Interestingly, a previously prepared chitosan-collagen co-polymer hydrogel displayed pore length similar to GCP, HACP, and CoCP hydrogels, suggesting the impact of chitosan chemistry in inducing porosity [39]. Furthermore, the superabsorbent hydrogels are characterized by superior water content, which in turn, increases porosity. Ideally, cardiac tissue engineering hydrogels require the pore length in the range of 10–100 μm, where pore heterogeneity with small and large pores favor myocardial regeneration [16]. Notably, the larger pores (>80 μm) drive cell migration, angiogenesis, and trafficking of nutrients, metabolites, signaling molecules, and metabolic exhaust, whereas smaller pores (5–20 μm) determine neo-vascularization and cell adhesion [40,41]. Additionally, the porosity and pore size guide the adhered cells to secrete and deposit ECM onto the interstices of the hydrogel, which promote neo-tissue formation [7,8]. Furthermore, imbibing the hydrogels in the physiological medium the pore opening occurs, thereby increasing the room for cellular performance potentiating cardiac regeneration.

Freezing free water and freezing bound water together constitute the total freezable water content (W_f). The freezable and non-freezing bound water are essential for successful cardiac tissue engineering [25]. Generally, the distribution of hydrophobic/hydrophilic moieties, crosslinking density, and chemical structure play a critical role in the water content and water transition status of hydrogels. Freezable water plays a critical role in trafficking, biocompatibility, adsorption of proteins, and anti-thrombogenicity. In contrast, non-freezable water contributes to pore morphology and mechanical strength [42]. Hence, the key to successful cardiac applications is to balance both water types within the hydrogels. Myocardium is a highly energy-demanding tissue, requiring constant transport of nutrients, metabolites, waste, and ample mechanical strength; thus, hydrogels with appreciable water profile are required [24]. Interestingly, the GCP and HACP hydrogels displayed a balance between freezable and non-freezable water. In contrast, the CP and CoCP hydrogels exhibited minimal non-freezing bound water, suggesting fewer hydrogen bond formation opportunities than the other two counterparts.

The ideal tissue engineering hydrogels support the absorption and release of loaded molecules and cells [24]. We attempted to assess the release profile of the hydrogel using the hydrophilic dye methylene blue and the hydrophobic dye Trypan blue to represent the hydrophilic and hydrophobic biological molecules. Both the dyes exhibited similar release trend supported by the amphiphilic nature of CP, GCP, HACP, and CoCP hydrogels, suggesting the ability of the hydrogels to release loaded molecules, nutrients, and cells. Hence, the CP, GCP, HACP, and CoCP hydrogels establish a proper tissue-hydrogel interface with the native myocardium by facilitating the interactions and transport of biomolecules for cardiac regeneration [43].

Cardiac hydrogels promote the conductance of electrical impulses as in the native myocardium, which is ~1 mS/cm. Practically, the conductive hydrogels inhibit the decoupling of CMs and promote the synchronization of distinct clusters [44] thereby significantly limiting the infarct expansion and CM necrosis due to decoupling [45,46]. The CP, GCP, HACP, and CoCP hydrogels conduct electrical signals. However, the electrical conductivity of the hydrogels was lower than the native myocardium, warranting further modifications, including the addition of conductive polymers, fillers, or ions. Additionally, the biodegradation mechanism by surface erosion displays a relatively constant rate of weight loss, sustaining the release of loaded cells and molecules in a regulated manner. Hence, the stability of CP, GCP, HACP, and CoCP hydrogels in the simulated physiological fluid regulates bulk degradation providing opportunities for protein adsorption and cellular adhesion, proliferation and integration with the host myocardium [24]. Also, the biodegradation of CP, GCP, HACP, and CoCP hydrogels comply with cardiac applications.

The elastic properties imparted by the constant contraction and relaxation cycles of the myocardium warranted to be complied by the mechanical strength of tissue engineering hydrogels to provide ample mechanical support to the damaged areas. An ideal cardiac hydrogel requires the Young's modulus value to be greater than that of the native myocardium, which in turn has been calculated to be between 20 and 500 kPa. Also, the hydrogels with sufficient mechanical strength significantly inhibit infarct expansion, formation of collagenous scar, and necrosis of the surviving myocardium, and also, cardiac hydrogels capable of providing sufficient mechanical support to the encapsulated cells intensively promote ECM deposition and neo-tissue formation. The GCP, HACP, and CoCP hydrogels exhibited mechanical properties, complying with the native myocardium. In contrast, CP hydrogels, with the lowest concentration of natural materials, exceeded the limit. The increased hydrophilicity imparted by gelatin and HA in GCP, HACP, and CoCP hydrogels possibly reduced the mechanical properties compared to CP hydrogels, representing a strong polyelectrolyte complex. Additionally, the decreased porous nature of CP hydrogels suggests the increased mechanical properties compared to the other three. Translationally, the degradation of hydrogels results in a subsequent decrease in mechanical strength, where the hydrogels with Young's modulus greater than that of the native tissue exhibit maximal regenerative benefits in cardiac tissue engineering. Overall, all four hydrogels displayed mechanical strength sufficient to support the seeded cells and the diseased myocardium, suggesting ample cross-linking density and regenerative potential.

CP, GCP, HACP, and CoCP hydrogels displayed excellent cytocompatibility with H9c2 cells, as evidenced by the direct contact and MTT viability assays, revealing the non-toxic nature of hydrogel degradation products and leachates. Generally, the hydrogels exhibiting more than 90% cell viability in the test on extract assay have been considered cytocompatible. Interestingly, the hydrogel extracts promoted the H9c2 cell viability over the 90% threshold, suggesting that the degradation products and leachates are non-toxic to the growth and proliferation of cells [42]. Furthermore, the non-toxic nature of the hydrogels is the key to cell adherence, growth, migration, and proliferation [47]. Hence, CP, GCP, HACP, and CoCP hydrogels promise the long-term viability of cells facilitating myocardial regeneration [48]. These findings suggest the non-cytotoxic nature CP, GCP, HACP, and CoCP hydrogels reflecting excellent cytocompatibility and promise in cardiac applications.

The ideal tissue engineering scaffold exhibits minimal hemolytic potential as excessive hemolysis results in immune reactions, anemia, jaundice, and even renal failure and alters the host circulatory system. Interestingly, the hemolytic potential of chitosan-based hydrogels has been reported to be less than 3% [49]. The hemocompatibility of CP, GCP, HACP, and CoCP hydrogels was less than 1%, promising cardiac applications without adverse reactions with the host circulatory system. Additionally, the absence of RBC aggregation by CP, GCP, HACP, and CoCP hydrogels suggests unaltered blood rheology and rouleaux formation upon contact with the circulatory system [24]. These results were consistent with previous reports on PVA-based hydrogels [50]. Furthermore, the interaction of plasma proteins, especially albumin and hydrogel implants, is crucial in determining hemocompatibility. Notably, the amphiphilicity of the hydrogel surface drives plasma protein adsorption, especially that of serum albumin following implantation [24]. Also, the increased freezable water plays a critical role in protein adsorption, where the hydrogels with superior freezing water content display higher protein adsorption [51]. For instance, albumin adsorption onto hydrogel surfaces significantly enhances cell adhesion. Binding to integrin molecules present on cells and to the hydrogel surface facilitates ECM deposition onto hydrogels. Thus, the adsorption of proteins directly enhances the ability of hydrogels to bind to cells and facilitate the formation of neo-tissue [52]. Moreover, the passivation effect of serum albumin enhances biocompatibility, cell adhesion, and hemocompatibility by preventing thrombogenesis, immune reactions, and inflammation [42]. Interestingly, CP, GCP, HACP, and CoCP hydrogels exhibited negligible hemolysis and RBC aggregation and could adsorb serum albumin, reflecting enhanced biocompatibility, hemocompatibility, and anti-thrombogenicity.

A major determinant in the success of cardiac hydrogels is their ability to support the adhesion, growth, and proliferation of cells, resulting in ECM deposition and subsequent integration with the native myocardium. The H9c2 cells grown on CP, GCP, HACP, and CoCP hydrogels displayed the attachment, cytoskeletal spreading, and penetration onto the interstices, suggesting the healthy performance of the cells. The H9c2 cells successfully formed colonies and existed as clusters on CP, GCP, HACP, and CoCP hydrogels, which agrees with previously validated reports [53] suggesting the potential of these hydrogels in initiating tissue formation. Interestingly, all the four hydrogels were capable of neo-tissue formation as tested with rat LV explants which was confirmed with swine myocardial tissue. However, CP and HACP hydrogels were superior to GCP and CoCP. The possible mechanism for the reduced capacity of gelatin-containing hydrogels to support tissue formation is the inherently short degradation time, extreme hydrophilicity, and the involvement of charged domains in poly electrolyte formation limiting the bioavailability of functional cues for cell adhesion [54,55]. Additionally, the cells migrating from the explants and adhering to the surface of hydrogels further accelerate the degradation by breaking hydrogen bonds and activating MMPs prevalent in gelatin-based hydrogels. Nonetheless, the optimal architectural, physiochemical, and biological properties of CP, GCP, HACP, and CoCP hydrogels support neo-tissue organization ex vivo. Translationally, neo-tissue formation reflects the potential of the hydrogels to integrate with the surviving myocardium and to mediate the anterograde and retrograde migration of seeded/host cells post-implantation, which in turn is critical for successful regeneration.

The overall findings from this study demonstrate that the CP, GCP, HACP, and CoCP hydrogels exhibited adequate physiochemical and biological properties, thus benefiting translational cardiac tissue engineering applications. However, further in vivo optimizations in normal and ischemic animal models are warranted to assess the clinical application of these hydrogels. Additionally, the information regarding the cellular/ECM protein deposition/binding to the hydrogels, immuno/histological/biochemical dissection of neo-tissues, and the surgical strategies for implantation warrant careful consideration for translating these hydrogels in clinical arena. Nevertheless, the findings from the present study extend promise for CP, GCP, HACP, and CoCP hydrogel systems in translational regenerative cardiology.

3. Conclusions

A panel of four polyelectrolyte complex hydrogels, CP, GCP, HACP, and CoCP, were successfully synthesized. The hydrogels displayed favorable characteristics, such as porosity, water profile and transition status, mechanical strength, hemocompatibility, biocompatibility, enhanced protein adsorption, cell adhesion, growth, proliferation, and neo- tissue formation. Compared to GCP and CoCP, CP and HACP hydrogels were superior owing to cell viability, hemocompatibility, and conductance, resulting in the highest degree of cytoskeletal organization and neo-tissue formation. The excellent physiochemical and biological performance of these hydrogels supported neo-cardiac tissue formation. Overall, the CP, GCP, HACP, and CoCP hydrogel systems promise novel translational opportunities in regenerative cardiology.

4. Materials and Methods

4.1. Materials

All the chemicals and reagents solutions used for the experiments were of synthetic or analytical grade. Chitosan (medium molecular weight, Cat#448877), glacial acetic acid (Cat#695092), PEG (MW 35,000, Cat#81310), gelatin (gelatin from bovine skin, Cat#G-9382), HA sodium from *Streptococcus equi* (Cat#53747), and dimethyl sulfoxide (Cat#41639) were purchased from Sigma-Aldrich, St. Louis, MO, USA. Dulbecco's modified Eagle's medium (DMEM) (Cat#30-2002) was obtained from ATCC, MTT (Cat#M6494), Rhodamine/phalloidin (Cat#R415) was purchased from Invitrogen, 10% neutral buffered formalin (Cat#16004-128) was procured from VWR Labs Private Limited and Pierce Bicinchoninic Acid (BCA) Protein Assay Kit (Cat#23225) was purchased from Thermo Fisher Scientific, Waltham, MA, USA.

4.2. Preparation of Chitosan-PEG-Based Polyelectrolyte Hydrogels

Briefly, 3% chitosan in 1% acetic acid and 5% gelatin and 0.1% HA in water were prepared and stored individually at room temperature in airtight containers and were used within 2–3 days for the experiments. The 3% PEG in 3% chitosan solution constituted the parent Chitosan-PEG (CP) blend, which was warmed at 42 °C with constant stirring at 400 rpm for 30 min and casted at 55 °C for 48 h, thus constituting the CP hydrogel. A panel of four hydrogels were prepared from the parent CP blend by incorporating gelatin, and HA in optimal concentrations. Next, 1% gelatin solution was combined with CP blend at 1:2 v/w ratio, warmed to 42 °C with constant stirring and casted at 55 °C for 48 h to form the gelatin-chitosan-PEG (GCP) hydrogel. HA-chitosan-PEG (HACP) hydrogel was prepared by casting 0.1% HA with CP blend at 1:10 v/w ratio at 55 °C for 48 h. Finally, 1% gelatin and 0.1% HA were combined (Co) in a 4:1 ratio by stirring for 30 min and mixed with CP blend 1:2 v/w ratio, heated to 42 °C and casted at 55 °C for 48 h to form the gelatin-HA chitosan-PEG (CoCP) hydrogel. The hydrogel sheets were then submerged in distilled water, lyophilized, UV sterilized, and stored in aseptic airtight containers for further characterizations and biological studies.

4.3. Physiochemical Characterization

4.3.1. Attenuated Total Reflection Infrared Spectroscopy (ATR-IR)

The surface functional groups of lyophilized CP, GCP, HACP, and CoCP hydrogels (n = 5, freeze-dried discs) were determined by AT-IR spectral analysis using the IR spectrometer (NICOLET iS50, Thermo Fisher Scientific, Jacksonville, FL, USA). The significant peaks were examined, and the area under the curves was calculated using SPA software associated with the instrument and compared among the groups [56].

4.3.2. Contact Angle

Water-swollen CP, GCP, HACP, and CoCP hydrogels (n = 10) of dimension 1 cm × 3 cm with known weight and thickness were used for contact angle analysis. The contact angle was measured using the Wilhelmy method in the KSV Sigma 701 Tensiometer). The advancing contact angle (ACA) and receding contact angle (RCA) in water were automatically recorded by the software associated with the instrument [57].

4.3.3. Water Profiling

Freeze-dried CP, GCP, HACP, and CoCP hydrogels (n = 10) were soaked in distilled water and swelling ratio (S), % swelling (%S), equilibrium swelling ratio (E), equilibrium water content (EWC) and total water absorption sites (TWAS) were calculated following our previous protocols [57,58]. The swelling constant and the diffusional exponent were calculated from the intercept and slope of the log(S/E) v. log(time) graph using the equations $y = 0.0913 \ln(x) + 0.3542$, $y = -0.079 \ln(x) + 0.8751$, $y = -0.007 \ln(x) + 0.8475$, and $y = -0.017 \ln(x) + 1.0641$ for CP, GCP, HACP, and CoCP hydrogels, respectively.

4.3.4. Scanning Electron Microscopy (SEM)

The microarchitecture and morphology of the cross-sectional area of freeze-dried hydrogels were characterized using scanning electron microscopy (Nova NanoSEM 450; Fei Co., Hillsboro, OR, USA). Samples were sputter-coated with a Pd/Pt layer for 60 s at 20 mA using a sputter coater (108 Auto Sputter Coater; Ted Pella, Inc., Redding, CA, USA) before being imaged. An accelerating voltage of 20 kV and a working distance of 5 mm were used, along with a spot size of 3.0. The images were processed using ImageJ(1.46) software to quantify pore size and aspect ratios of the pores, following our previously validated protocols [59,60].

4.3.5. Thermal Evaluation

The thermal properties of water swollen hydrogels were assessed using differential scanning calorimetry (DSC214; Netzsch, Germany) to determine water transition states following previous protocols. Water-swollen hydrogels were cooled from 0 °C to −40 °C, and then heated to 100 °C at a rate of 5 K/min in a nitrogen (N2) atmosphere. The heating and cooling curves were recorded. The freezing water content (W_f), freezing free water (W_{ff}), freezing bound water (slightly structured) (W_{fb}), and non-freezing bound water (W_{nb}) were calculated following our previously published reports [57,60].

4.4. Release Kinetics

CP, GCP, HACP, and CoCP hydrogels (n = 4) of known dry weights were submerged in the hydrophobic dye trypan blue (5 mg%), and the hydrophilic dye methylene blue (5 mg%), for 24 h. The unbound dyes were washed in PBS, and the burst release profiles of the dyes from each hydrogel were observed every hour for 8 h, followed by the sustained release profiles once in 2 days for 28 days by extracting PBS. The concentration of released dyes was calculated from the absorption values measured using a plate reader. A standard curve was plotted, and the percentage release of the dyes was determined from our already established protocols [56].

4.5. Electrical Conductivity

Water-swollen CP, GCP, HACP, and CoCP hydrogel discs (n = 5) were placed on a non-conductive surface, and resistance was assessed using a multimeter. Conductance was then calculated from the resistance using the following equation, where G is conductance, and R is resistance.

$$G = \frac{1}{R}$$

4.6. Biodegradation

Degradation of the CP, GCP, HACP, and CoCP hydrogels (n = 5) was assessed by aging in PBS (pH = 7.33) at 37 °C. Weight loss (difference wet weight/initial wet weight × 100) was assessed weekly for seven weeks and the changes in pH were measured weekly to determine the acidity/basicity of degradation products [61].

4.7. Mechanical Characterization

The mechanical strength of the water swollen CP, GCP, HACP, and CoCP hydrogels was (n = 6) was measured in the universal testing machine (Instron 5943 dual-column testing system, Norwood, MA, USA) equipped with a 10 N load cell (Instron). The tensile strength was determined using a 0.005 N preload with a 10 mm/min crosshead speed. Tensile stress at failure, load at failure, and Young's modulus were calculated automatically by the program associated with the instrument [62].

4.8. Cytocompatibility

4.8.1. Cell Culture and Maintenance

Rat cardiomyoblasts, H9c2 cells (Cat# CLR1446; ATCC), were used for cytocompatibility assessment, and the cells were maintained in high glucose DMEM with 10% FBS under standard culture conditions (5% CO_2, 37 °C, and antibiotics). H9c2 from passages 2–6 was used for the studies.

4.8.2. Direct Contact Assay

Lyophilized hydrogels (n = 5) were swollen in DMEM with 10% FBS for 24 h, placed atop a sub-confluent layer of H9c2 cells and incubated for 24 h. The cells at the interface of the hydrogels were imaged to assess the changes in morphological alterations using a phase-contrast microscope (Olympus CKx41). The overall survival of the H9c2 cells was quantified by MTT cell viability assay as previously validated [63]. H9c2 cells cultured under the same conditions without hydrogels served as control.

4.8.3. Test on Extract

DMEM with 10% FBS was used to extract the lyophilized CP, GCP, HACP, and CoCP hydrogels (n = 4) for 48 h and to culture H9c2 cells. The cell viability was quantified after three days using the MTT assay. H9c2 cells cultured under the same conditions without hydrogels served as control.

4.9. Hemocompatibility

4.9.1. Hemolysis Assay Red Blood Cell Aggregation

Lyophilized CP, GCP, HACP, and CoCP hydrogels (n = 3) were soaked in 0.9% saline solution for 48 h. Briefly, 5 mL blood for the hemocompatibility was collected from Yucatan micro swine (from a different study) immediately prior to sacrifice (*Sus scorfa*, Sinclair BioResources) with approval from the IACUC at Western University of Health Sciences (R21IACUC012). The plasma was separated, and the RBCs were washed and diluted 10 times with a 0.9% saline solution. Then, 0.5 mL diluted RBC was mixed gently with 0.5 mL saline extract of each hydrogel, incubated at 37 °C for 30 min, centrifuged, and the absorbance of the supernatant was read at 540 nm. RBC in 0.5 mL 0.9% saline solution served as the negative control, and 0.5 mL RBC in 0.5 mL deionized water served as the positive control. The percentage of hemolysis was calculated using previously reported protocols [56]. The smears of the above samples were examined under a brightfield microscope (Olympus CKx41) to assess the morphology of red blood cells and rouleau formations [56].

4.9.2. Protein Adsorption

Lyophilized CP, GCP, HACP, and CoCP hydrogels (n = 5) were incubated in 10% plasma isolated from Yucatán micro swine (*Sus scorfa*, Sinclair Bioresources) at 37 °C overnight. The loosely bound proteins adsorbed onto the hydrogels were washed with sterile-filtered PBS, and the adsorbed proteins were extracted in 500 µL PBS by vigorous vortexing and centrifuged to remove the debris. The extracted protein content was quantified using the BCA assay [16]. Also, the level of serum albumin adsorbed on CP, GCP, HACP, and CoCP hydrogels was assessed using sodium dodecyl sulfate-polyacrylamide gel electrophoresis (SDS-PAGE) using bovine serum albumin (BSA) as a reference. Relative protein concentration was determined by calculating band intensity using ImageJ software [56].

4.10. Biological Performance

4.10.1. Cell Spreading and Penetration

H9c2 cells were seeded onto the CP, GCP, HACP, and CoCP DMEM-swollen hydrogels ($n = 3$) and allowed to grow under standard culture conditions for six days. The hydrogels were fixed in formalin, blocked with 1% horse-blocking serum (HBS), stained with rhodamine phalloidin for 2 h, and imaged using a fluorescent microscope (Leica Thunder, Wetzlar, Germany). Z-stack images were captured, 3D rendering was performed, and Z-stack depth was measured for each sample to assess the penetration [56]. Cells grown on glass coverslips were used as controls.

4.10.2. Ex Vivo Explant Culture and Neo-Tissue Formation

Left ventricular tissue sections (~50 mg weight) were harvested from cadaver rats and swine in serum-free DMEM and were allowed to grow on CP, GCP, HACP, and CoCP hydrogels ($n = 3$) equilibrated in DMEM with 20% FBS. After 1 h of attachment, additional media was added without disturbing the tissue and allowed to grow for one week. Then, the hydrogels were fixed after removing the tissues, and rhodamine staining was performed to observe the neo-tissue organization in the hydrogels, as mentioned above.

4.10.3. Statistical Analysis

The results of all experiments were expressed as mean ± SEM, and all the experiments were run in at least three or more replicates, excluding SEM and DSC. ImageJ software was used for pore measurements, DSC area calculations, and SDS-PAGE band quantification. The statistical significance for all the experimental data was determined by one-way ANOVA with a two-stage linear step-up procedure with Tukey's multiple comparisons. However, the statistical significance for contact angle measurement was determined by an unpaired "t" test for comparing two groups. Statistical significance of $p < 0.05$ was set for all experiments, and GraphPad Prism 9.4.1 (681) program was employed for the analysis.

Supplementary Materials: The following supporting information can be downloaded at: https://www.mdpi.com/article/10.3390/gels10010046/s1; Video S1: Z-stack for the rhodamine phalloidin staining for cell spreading and penetration in 2D control. Video S2: Z-stack for the rhodamine phalloidin staining for cell spreading and penetration in CP hydrogel. Video S3: Z-stack for the rhodamine phalloidin staining for cell spreading and penetration in GCP hydrogel. Video S4: Z-stack for the rhodamine phalloidin staining for cell spreading and penetration in HACP hydrogel. Video S5: Z-stack for the rhodamine phalloidin staining for cell spreading and penetration in CoCP hydrogel. Video S6: Z-stack for the rhodamine phalloidin staining for neo-tissue formation by rat LV explants in CP hydrogel. Video S7: Z-stack for the rhodamine phalloidin staining for neo-tissue formation by rat LV explants in GCP hydrogel. Video S8: Z-stack for the rhodamine phalloidin staining for neo-tissue formation by rat LV explants in HACP hydrogel. Video S9: Z-stack for the rhodamine phalloidin staining for neo-tissue formation by rat LV explants in CoCP hydrogel. Video S10: Z-stack for the rhodamine phalloidin staining for neo-tissue formation by swine LV explants in CP hydrogel. Video S11: Z-stack for the rhodamine phalloidin staining for neo-tissue formation by swine LV explants in GCP hydrogel. Video S12: Z-stack for the rhodamine phalloidin staining for neo-tissue formation by swine LV explants in HACP hydrogel. Video S13: Z-stack for the rhodamine phalloidin staining for neo-tissue formation by swine LV explants in CoCP hydrogel.

Author Contributions: Conceptualization—A.K., N.R.d.B. and F.G.T.; methodology—A.K., N.R.d.B., A.R., Y.C., J.T., R.S., D.S., H.L. and F.G.T.; software—A.K., N.R.d.B., A.R., J.T. and F.G.T.; validation—N.R.d.B., H.L. and F.G.T.; formal analysis—A.K., N.R.d.B., A.R., Y.C., J.T., R.S., D.S., H.L. and F.G.T.; investigation—A.K., N.R.d.B., A.R., Y.C., J.T., R.S., D.S., H.L. and F.G.T.; resources—F.G.T., N.R.d.B. and H.L.; data curation—A.K., A.R., Y.C., J.T., R.S. and D.S.; writing—original draft preparation—F.G.T. and A.K.; writing—review and editing—A.K., N.R.d.B., A.R., Y.C., J.T., R.S., D.S., H.L. and F.G.T.; supervision—F.G.T., N.R.d.B. and H.L.; project administration—F.G.T.; funding acquisition—F.G.T. All authors have read and agreed to the published version of the manuscript.

Funding: The research work of F.T. is supported by the startup funds from WU.

Institutional Review Board Statement: Not applicable.

Informed Consent Statement: Not applicable.

Data Availability Statement: All data and materials are available on request from the corresponding author. The data are not publicly available due to ongoing research using a part of the data.

Conflicts of Interest: The authors declare no conflicts of interest.

References

1. Best, C.; Onwuka, E.; Pepper, V.; Sams, M.; Breuer, J.; Breuer, C. Cardiovascular Tissue Engineering: Preclinical Validation to Bedside Application. *Physiology* **2016**, *31*, 7–15. [CrossRef] [PubMed]
2. Wissing, T.B.; Bonito, V.; Bouten, C.V.C.; Smits, A.I.P.M. Biomaterial-Driven in Situ Cardiovascular Tissue Engineering—A Multi-Disciplinary Perspective. *npj Regen. Med.* **2017**, *2*, 18. [CrossRef] [PubMed]
3. Sodian, R.; Hoerstrup, S.P.; Sperling, J.S.; Daebritz, S.; Martin, D.P.; Moran, A.M.; Kim, B.S.; Schoen, F.J.; Vacanti, J.P.; Mayer, J.E. Early In Vivo Experience with Tissue-Engineered Trileaflet Heart Valves. *Circulation* **2000**, *102*, Iii-22–Iii-29. [CrossRef]
4. Bouten, C.V.C.; Cheng, C.; Vermue, I.M.; Gawlitta, D.; Passier, R. Cardiovascular Tissue Engineering and Regeneration: A Plead for Further Knowledge Convergence. *Tissue Eng. Part A* **2022**, *28*, 525–541. [CrossRef] [PubMed]
5. Wang, C.-G.; Surat'man, N.E.B.; Chang, J.J.; Ong, Z.L.; Li, B.; Fan, X.; Loh, X.J.; Li, Z. Polyelectrolyte Hydrogels for Tissue Engineering and Regenerative Medicine. *Chem. Asian J.* **2022**, *17*, e202200604. [CrossRef] [PubMed]
6. Henning, R.J.; Khan, A.; Jimenez, E. Chitosan Hydrogels Significantly Limit Left Ventricular Infarction and Remodeling and Preserve Myocardial Contractility. *J. Surg. Res.* **2016**, *201*, 490–497. [CrossRef]
7. Deng, B.; Shen, L.; Wu, Y.; Shen, Y.; Ding, X.; Lu, S.; Jia, J.; Qian, J.; Ge, J. Delivery of Alginate-Chitosan Hydrogel Promotes Endogenous Repair and Preserves Cardiac Function in Rats with Myocardial Infarction. *J. Biomed. Mater. Res. A* **2015**, *103*, 907–918. [CrossRef]
8. Fu, B.; Wang, X.; Chen, Z.; Jiang, N.; Guo, Z.; Zhang, Y.; Zhang, S.; Liu, X.; Liu, L. Improved Myocardial Performance in Infarcted Rat Heart by Injection of Disulfide-Cross-Linked Chitosan Hydrogels Loaded with Basic Fibroblast Growth Factor. *J. Mater. Chem. B* **2022**, *10*, 656–665. [CrossRef]
9. Li, J.; Shu, Y.; Hao, T.; Wang, Y.; Qian, Y.; Duan, C.; Sun, H.; Lin, Q.; Wang, C. A Chitosan–Glutathione Based Injectable Hydrogel for Suppression of Oxidative Stress Damage in Cardiomyocytes. *Biomaterials* **2013**, *34*, 9071–9081. [CrossRef]
10. Dobner, S.; Bezuidenhout, D.; Govender, P.; Zilla, P.; Davies, N. A Synthetic Non-Degradable Polyethylene Glycol Hydrogel Retards Adverse Post-Infarct Left Ventricular Remodelling. *J. Card. Fail.* **2009**, *15*, 629–636. [CrossRef]
11. Chow, A.; Stuckey, D.J.; Kidher, E.; Rocco, M.; Jabbour, R.J.; Mansfield, C.A.; Darzi, A.; Harding, S.E.; Stevens, M.M.; Athanasiou, T. Human Induced Pluripotent Stem Cell-Derived Cardiomyocyte Encapsulating Bioactive Hydrogels Improve Rat Heart Function Post Myocardial Infarction. *Stem Cell Rep.* **2017**, *9*, 1415–1422. [CrossRef] [PubMed]
12. Li, Z.; Masumoto, H.; Jo, J.-I.; Yamazaki, K.; Ikeda, T.; Tabata, Y.; Minatoya, K. Sustained Release of Basic Fibroblast Growth Factor Using Gelatin Hydrogel Improved Left Ventricular Function through the Alteration of Collagen Subtype in a Rat Chronic Myocardial Infarction Model. *Gen. Thorac. Cardiovasc. Surg.* **2018**, *66*, 641–647. [CrossRef] [PubMed]
13. Ptaszek, L.M.; Lara, R.P.; Sani, E.S.; Xiao, C.; Roh, J.; Yu, X.; Ledesma, P.A.; Yu, C.H.; Annabi, N.; Ruskin, J.N. Gelatin Methacryloyl Bioadhesive Improves Survival and Reduces Scar Burden in a Mouse Model of Myocardial Infarction. *J. Am. Heart Assoc.* **2020**, *9*, e014199. [CrossRef] [PubMed]
14. Abdalla, S.; Makhoul, G.; Duong, M.; Chiu, R.C.J.; Cecere, R. Hyaluronic Acid-Based Hydrogel Induces Neovascularization and Improves Cardiac Function in a Rat Model of Myocardial Infarction. *Interact. Cardiovasc. Thorac. Surg.* **2013**, *17*, 767–772. [CrossRef]
15. Ifkovits, J.L.; Tous, E.; Minakawa, M.; Morita, M.; Robb, J.D.; Koomalsingh, K.J.; Gorman, J.H.; Gorman, R.C.; Burdick, J.A. Injectable Hydrogel Properties Influence Infarct Expansion and Extent of Postinfarction Left Ventricular Remodeling in an Ovine Model. *Proc. Natl. Acad. Sci. USA* **2010**, *107*, 11507–11512. [CrossRef]
16. Martins, A.F.; Pereira, A.G.B.; Fajardo, A.R.; Rubira, A.F.; Muniz, E.C. Characterization of Polyelectrolytes Complexes Based on N,N,N-Trimethyl Chitosan/Heparin Prepared at Different pH Conditions. *Carbohydr. Polym.* **2011**, *86*, 1266–1272. [CrossRef]
17. Acosta, B.B.; Advincula, R.C.; Grande-Tovar, C.D. Chitosan-Based Scaffolds for the Treatment of Myocardial Infarction: A Systematic Review. *Molecules* **2023**, *28*, 1920. [CrossRef]
18. Buranachai, T.; Praphairaksit, N.; Muangsin, N. Chitosan/Polyethylene Glycol Beads Crosslinked with Tripolyphosphate and Glutaraldehyde for Gastrointestinal Drug Delivery. *AAPS PharmSciTech* **2010**, *11*, 1128–1137. [CrossRef]
19. Kim, Y.; Zharkinbekov, Z.; Raziyeva, K.; Tabyldiyeva, L.; Berikova, K.; Zhumagul, D.; Temirkhanova, K.; Saparov, A. Chitosan-Based Biomaterials for Tissue Regeneration. *Pharmaceutics* **2023**, *15*, 807. [CrossRef]
20. Cao, N.; Zhao, Y.; Chen, H.; Huang, J.; Yu, M.; Bao, Y.; Wang, D.; Cui, S. Poly(Ethylene Glycol) Becomes a Supra-Polyelectrolyte by Capturing Hydronium Ions in Water. *Macromolecules* **2022**, *55*, 4656–4664. [CrossRef]
21. McCain, M.L.; Agarwal, A.; Nesmith, H.W.; Nesmith, A.P.; Parker, K.K. Micromolded Gelatin Hydrogels for Extended Culture of Engineered Cardiac Tissues. *Biomaterials* **2014**, *35*, 5462–5471. [CrossRef] [PubMed]

22. Zhang, F.; Zhang, N.; Meng, H.-X.; Liu, H.-X.; Lu, Y.-Q.; Liu, C.-M.; Zhang, Z.-M.; Qu, K.-Y.; Huang, N.-P. Easy Applied Gelatin-Based Hydrogel System for Long-Term Functional Cardiomyocyte Culture and Myocardium Formation. *ACS Biomater. Sci. Eng.* **2019**, *5*, 3022–3031. [CrossRef] [PubMed]
23. Chang, C.Y.; Chan, A.T.; Armstrong, P.A.; Luo, H.-C.; Higuchi, T.; Strehin, I.A.; Vakrou, S.; Lin, X.; Brown, S.N.; O'Rourke, B.; et al. Hyaluronic Acid-Human Blood Hydrogels for Stem Cell Transplantation. *Biomaterials* **2012**, *33*, 8026–8033. [CrossRef] [PubMed]
24. Masood, N.; Ahmed, R.; Tariq, M.; Ahmed, Z.; Masoud, M.S.; Ali, I.; Asghar, R.; Andleeb, A.; Hasan, A. Silver Nanoparticle Impregnated Chitosan-PEG Hydrogel Enhances Wound Healing in Diabetes Induced Rabbits. *Int. J. Pharm.* **2019**, *559*, 23–36. [CrossRef] [PubMed]
25. Joo, H.; Park, J.; Sutthiwanjampa, C.; Kim, H.; Bae, T.; Kim, W.; Choi, J.; Kim, M.; Kang, S.; Park, H. Surface Coating with Hyaluronic Acid-Gelatin-Crosslinked Hydrogel on Gelatin-Conjugated Poly(Dimethylsiloxane) for Implantable Medical Device-Induced Fibrosis. *Pharmaceutics* **2021**, *13*, 269. [CrossRef] [PubMed]
26. Basara, G.; Ozcebe, S.G.; Ellis, B.W.; Zorlutuna, P. Tunable Human Myocardium Derived Decellularized Extracellular Matrix for 3D Bioprinting and Cardiac Tissue Engineering. *Gels* **2021**, *7*, 70. [CrossRef]
27. Geckil, H.; Xu, F.; Zhang, X.; Moon, S.; Demirci, U. Engineering Hydrogels as Extracellular Matrix Mimics. *Nanomedicine* **2010**, *5*, 469–484. [CrossRef]
28. Thankam, F.G.; Muthu, J. Infiltration and sustenance of viability of cells by amphiphilic biosynthetic biodegradable hydrogels. *J. Mater. Sci. Mater. Med.* **2014**, *25*, 1953–1965. [CrossRef]
29. Thankam, F.G.; Muthu, J. Biosynthetic hydrogels—Studies on chemical and physical characteristics on long-term cellular response for tissue engineering. *J. Biomed. Mater. Res.* **2014**, *107*, 2238–2247. [CrossRef]
30. Shoichet, M.S. Polymer Scaffolds for Biomaterials Applications. *Macromolecules* **2010**, *43*, 581–591. [CrossRef]
31. Sharifisistani, M.; Khanmohammadi, M.; Badali, E.; Ghasemi, P.; Hassanzadeh, S.; Bahiraie, N.; Lotfibakhshaiesh, N.; Ai, J. Hyaluronic Acid/Gelatin Microcapsule Functionalized with Carbon Nanotube through Laccase-Catalyzed Crosslinking for Fabrication of Cardiac Microtissue. *J. Biomed. Mater. Res. A* **2022**, *110*, 1866–1880. [CrossRef] [PubMed]
32. Vogler, E.A. Water and the Acute Biological Response to Surfaces. *J. Biomater. Sci. Polym. Ed.* **1999**, *10*, 1015–1045. [CrossRef] [PubMed]
33. Nuttelman, C.R.; Henry, S.M.; Anseth, K.S. Synthesis and Characterization of Photocrosslinkable, Degradable Poly(Vinyl Alcohol)-Based Tissue Engineering Scaffolds. *Biomaterials* **2002**, *23*, 3617–3626. [CrossRef] [PubMed]
34. Aswathy, S.H.; Narendrakumar, U.; Manjubala, I. Commercial Hydrogels for Biomedical Applications. *Heliyon* **2020**, *6*, e03719. [CrossRef] [PubMed]
35. Camci-Unal, G.; Annabi, N.; Dokmeci, M.R.; Liao, R.; Khademhosseini, A. Hydrogels for Cardiac Tissue Engineering. *NPG Asia Mater.* **2014**, *6*, e99. [CrossRef]
36. Sievers, J.; Sperlich, K.; Stahnke, T.; Kreiner, C.; Eickner, T.; Martin, H.; Guthoff, R.F.; Schünemann, M.; Bohn, S.; Stachs, O. Determination of Hydrogel Swelling Factors by Two Established and a Novel Non-Contact Continuous Method. *J. Appl. Polym. Sci.* **2021**, *138*, 50326. [CrossRef]
37. Jacques, C.H.M.; Hopfenberg, H.B.; Stannett, V. Super Case II Transport of Organic Vapors in Glassy Polymers. In *Permeability of Plastic Films and Coatings: To Gases, Vapors, and Liquids*; Hopfenberg, H.B., Ed.; Polymer Science and Technology; Springer: Boston, MA, USA, 1974; pp. 73–86. ISBN 978-1-4684-2877-3.
38. Das, R.; Dasgupta, C.; Karmakar, S. Time Scales of Fickian Diffusion and the Lifetime of Dynamic Heterogeneity. *Front. Phys.* **2020**, *8*, 210. [CrossRef]
39. Si, R.; Gao, C.; Guo, R.; Lin, C.; Li, J.; Guo, W. Human Mesenchymal Stem Cells Encapsulated-Coacervated Photoluminescent Nanodots Layered Bioactive Chitosan/Collagen Hydrogel Matrices to Indorse Cardiac Healing after Acute Myocardial Infarction. *J. Photochem. Photobiol. B* **2020**, *206*, 111789. [CrossRef]
40. Wu, T.; Liu, W. Functional Hydrogels for the Treatment of Myocardial Infarction. *NPG Asia Mater.* **2022**, *14*, 9. [CrossRef]
41. Wang, R.M.; Christman, K.L. Decellularized Myocardial Matrix Hydrogels: In Basic Research and Preclinical Studies. *Adv. Drug Deliv. Rev.* **2016**, *96*, 77–82. [CrossRef]
42. Li, Z.; Fan, Z.; Xu, Y.; Niu, H.; Xie, X.; Liu, Z.; Guan, J. Thermosensitive and Highly Flexible Hydrogels Capable of Stimulating Cardiac Differentiation of Cardiosphere-Derived Cells under Static and Dynamic Mechanical Training Conditions. *ACS Appl. Mater. Interfaces* **2016**, *8*, 15948–15957. [CrossRef] [PubMed]
43. Qu, Y.; Tang, J.; Liu, L.; Song, L.; Chen, S.; Gao, Y. α-Tocopherol Liposome Loaded Chitosan Hydrogel to Suppress Oxidative Stress Injury in Cardiomyocytes. *Int. J. Biol. Macromol.* **2019**, *125*, 1192–1202. [CrossRef] [PubMed]
44. Mihic, A.; Cui, Z.; Wu, J.; Vlacic, G.; Miyagi, Y.; Li, S.-H.; Lu, S.; Sung, H.-W.; Weisel, R.D.; Li, R.-K. A Conductive Polymer Hydrogel Supports Cell Electrical Signaling and Improves Cardiac Function After Implantation into Myocardial Infarct. *Circulation* **2015**, *132*, 772–784. [CrossRef] [PubMed]
45. Zhang, L.; Li, T.; Yu, Y.; Shi, K.; Bei, Z.; Qian, Y.; Qian, Z. An Injectable Conductive Hydrogel Restores Electrical Transmission at Myocardial Infarct Site to Preserve Cardiac Function and Enhance Repair. *Bioact. Mater.* **2022**, *20*, 339–354. [CrossRef] [PubMed]
46. Mousavi, A.; Mashayekhan, S.; Baheiraei, N.; Pourjavadi, A. Biohybrid Oxidized Alginate/Myocardial Extracellular Matrix Injectable Hydrogels with Improved Electromechanical Properties for Cardiac Tissue Engineering. *Int. J. Biol. Macromol.* **2021**, *180*, 692–708. [CrossRef] [PubMed]
47. Jiang, L.; Chen, D.; Wang, Z.; Zhang, Z.; Xia, Y.; Xue, H.; Liu, Y. Preparation of an Electrically Conductive Graphene Oxide/Chitosan Scaffold for Cardiac Tissue Engineering. *Appl. Biochem. Biotechnol.* **2019**, *188*, 952–964. [CrossRef] [PubMed]

48. Chacon, E.; Acosta, D.; Lemasters, J.J. 9—Primary Cultures of Cardiac Myocytes as In Vitro Models for Pharmacological and Toxicological Assessments. In *In Vitro Methods in Pharmaceutical Research*; Castell, J.V., Gómez-Lechón, M.J., Eds.; Academic Press: San Diego, CA, USA, 1997; pp. 209–223. ISBN 978-0-12-163390-5.
49. Hao, Y.; Zhao, W.; Zhang, H.; Zheng, W.; Zhou, Q. Carboxymethyl Chitosan-Based Hydrogels Containing Fibroblast Growth Factors for Triggering Diabetic Wound Healing. *Carbohydr. Polym.* **2022**, *287*, 119336. [CrossRef]
50. Semenova, M.V.; Osadchenko, S.V.; Mezhuev, Y.O.; Shtil'man, M.I.; Semenova, I.N. Synthesis of Hemocompatible Materials Based on Branched Polyvinyl Alcohol. *Russ. J. Appl. Chem.* **2016**, *89*, 1286–1291. [CrossRef]
51. Amdursky, N.; Mazo, M.M.; Thomas, M.R.; Humphrey, E.J.; Puetzer, J.L.; St-Pierre, J.-P.; Skaalure, S.C.; Richardson, R.M.; Terracciano, C.M.; Stevens, M.M. Elastic Serum-Albumin Based Hydrogels: Mechanism of Formation and Application in Cardiac Tissue Engineering. *J. Mater. Chem. B* **2018**, *6*, 5604–5612. [CrossRef]
52. Kuten Pella, O.; Hornyák, I.; Horváthy, D.; Fodor, E.; Nehrer, S.; Lacza, Z. Albumin as a Biomaterial and Therapeutic Agent in Regenerative Medicine. *Int. J. Mol. Sci.* **2022**, *23*, 10557. [CrossRef]
53. De Oliveira, A.C.; Sabino, R.M.; Souza, P.R.; Muniz, E.C.; Popat, K.C.; Kipper, M.J.; Zola, R.S.; Martins, A.F. Chitosan/Gellan Gum Ratio Content into Blends Modulates the Scaffolding Capacity of Hydrogels on Bone Mesenchymal Stem Cells. *Mater. Sci. Eng. C* **2020**, *106*, 110258. [CrossRef] [PubMed]
54. Tondera, C.; Hauser, S.; Krüger-Genge, A.; Jung, F.; Neffe, A.T.; Lendlein, A.; Klopfleisch, R.; Steinbach, J.; Neuber, C.; Pietzsch, J. Gelatin-Based Hydrogel Degradation and Tissue Interaction In Vivo: Insights from Multimodal Preclinical Imaging in Immunocompetent Nude Mice. *Theranostics* **2016**, *6*, 2114–2128. [CrossRef] [PubMed]
55. Boucard, E.; Vidal, L.; Coulon, F.; Mota, C.; Hascoët, J.-Y.; Halary, F. The Degradation of Gelatin/Alginate/Fibrin Hydrogels Is Cell Type Dependent and Can Be Modulated by Targeting Fibrinolysis. *Front. Bioeng. Biotechnol.* **2022**, *10*, 920929. [CrossRef] [PubMed]
56. Sedighim, S.; Chen, Y.; Xu, C.; Mohindra, R.; Liu, H.; Agrawal, D.K.; Thankam, F.G. Carboxymethyl Cellulose–Alginate Interpenetrating Hydroxy Ethyl Methacrylate Crosslinked Polyvinyl Alcohol Reinforced Hybrid Hydrogel Templates with Improved Biological Performance for Cardiac Tissue Engineering. *Biotechnol. Bioeng.* **2023**, *120*, 819–835. [CrossRef] [PubMed]
57. Finosh, G.T.; Jayabalan, M.; Vandana, S.; Raghu, K.G. Hybrid Alginate-Polyester Bimodal Network Hydrogel for Tissue Engineering—Influence of Structured Water on Long-Term Cellular Growth. *Colloids Surf. B Biointerfaces* **2015**, *135*, 855–864. [CrossRef] [PubMed]
58. Thankam, F.G.; Diaz, C.; Chandra, I.; Link, J.; Newton, J.; Dilisio, M.F.; Agrawal, D.K. Hybrid Interpenetrating Hydrogel Network Favoring the Bidirectional Migration of Tenocytes for Rotator Cuff Tendon Regeneration. *J. Biomed. Mater. Res. B Appl. Biomater.* **2022**, *110*, 467–477. [CrossRef]
59. Thankam, F.G.; Muthu, J.; Sankar, V.; Gopal, R.K. Growth and Survival of Cells in Biosynthetic Poly Vinyl Alcohol–Alginate IPN Hydrogels for Cardiac Applications. *Colloids Surf. B Biointerfaces* **2013**, *107*, 137–145. [CrossRef]
60. Thankam, F.G.; Muthu, J. Alginate–polyester comacromer based hydrogels as physiochemically and biologically favorable entities for cardiac tissue engineering. *J. Colloid Interface Sci.* **2015**, *457*, 52–61. [CrossRef]
61. Thankam, F.G.; Muthu, J. Influence of Plasma Protein–Hydrogel Interaction Moderated by Absorption of Water on Long-Term Cell Viability in Amphiphilic Biosynthetic Hydrogels. *RSC Adv.* **2013**, *3*, 24509–24520. [CrossRef]
62. Finosh, G.T.; Jayabalan, M. Hybrid Amphiphilic Bimodal Hydrogels Having Mechanical and Biological Recognition Characteristics for Cardiac Tissue Engineering. *RSC Adv.* **2015**, *5*, 38183–38201. [CrossRef]
63. Thankam, F.G.; Muthu, J. Alginate Based Hybrid Copolymer Hydrogels—Influence of Pore Morphology on Cell–Material Interaction. *Carbohydr. Polym.* **2014**, *112*, 235–244. [CrossRef] [PubMed]

gels

MDPI

Article

Agar Graft Modification with Acrylic and Methacrylic Acid for the Preparation of pH-Sensitive Nanogels for 5-Fluorouracil Delivery

Ivelina Ivanova [1], Marta Slavkova [1,*], Teodora Popova [1], Borislav Tzankov [1], Denitsa Stefanova [2], Virginia Tzankova [2], Diana Tzankova [3], Ivanka Spassova [4], Daniela Kovacheva [4] and Christina Voycheva [1,*]

1. Faculty of Pharmacy, Department Pharmaceutical Technology and Biopharmacy, Medical University—Sofia, 1000 Sofia, Bulgaria; ivelinavasileva23@gmail.com (I.I.); tpopova@pharmfac.mu-sofia.bg (T.P.); btzankov@pharmfac.mu-sofia.bg (B.T.)
2. Faculty of Pharmacy, Department of Pharmacology, Pharmacotherapy and Toxicology, Medical University—Sofia, 1000 Sofia, Bulgaria; denitsa.stefanova@pharmfac.mu-sofia.bg (D.S.); vtzankova@pharmfac.mu-sofia.bg (V.T.)
3. Faculty of Pharmacy, Department of Pharmaceutical Chemistry, Medical University—Sofia, 1000 Sofia, Bulgaria; d.tsankova@pharmfac.mu-sofia.bg
4. Institute of General and Inorganic Chemistry, Bulgarian Academy of Sciences, 1113 Sofia, Bulgaria; ispasova@svr.igic.bas.bg (I.S.); didka@svr.igic.bas.bg (D.K.)

* Correspondence: mslavkova@pharmfac.mu-sofia.bg (M.S.); hvoycheva@pharmfac.mu-sofia.bg (C.V.)

Abstract: Agar, a naturally occurring polysaccharide, has been modified by grafting it with acrylic (AcA) and methacrylic (McA) acid monomers, resulting in acrylic or methacrylic acid grafted polymer (AA-g-AcA or AA-g-McA) with pH-sensitive swelling behavior. Different ratios between agar, monomers, and initiator were applied. The synthesized grades of both new polymer series were characterized using FTIR spectroscopy, nmR, TGA, DSC, and XRD to ascertain the intended grafting. The percentage of grafting (% G), grafting efficiency (% GE), and % conversion (% C) were calculated, and models with optimal characteristics were further characterized. The swelling behavior of the newly synthesized polymers was studied over time and in solutions with different pH. These polymers were subsequently crosslinked with varying amounts of glutaraldehyde to obtain 5-fluorouracil-loaded nanogels. The optimal ratios of polymer, drug, and crosslinker resulted in nearly 80% loading efficiency. The performed physicochemical characterization (TEM and DLS) showed spherical nanogels with nanometer sizes (105.7–250 nm), negative zeta potentials, and narrow size distributions. According to FTIR analysis, 5-fluorouracil was physically incorporated. The swelling and release behavior of the prepared nanogels was pH-sensitive, favoring the delivery of the chemotherapeutic to tumor cells. The biocompatibility of the proposed nanocarrier was proven using an in vitro hemolysis assay.

Keywords: pH-sensitive polymers; grafted agar; nanogels; 5-Fluorouracil; pH-sensitive delivery

Citation: Ivanova, I.; Slavkova, M.; Popova, T.; Tzankov, B.; Stefanova, D.; Tzankova, V.; Tzankova, D.; Spassova, I.; Kovacheva, D.; Voycheva, C. Agar Graft Modification with Acrylic and Methacrylic Acid for the Preparation of pH-Sensitive Nanogels for 5-Fluorouracil Delivery. *Gels* **2024**, *10*, 165. https://doi.org/10.3390/gels10030165

Academic Editors: Daihua Fu and Jieyu Zhang

Received: 11 January 2024
Revised: 30 January 2024
Accepted: 1 February 2024
Published: 23 February 2024

1. Introduction

Nanotechnology is a rapidly developing and promising field in pharmaceutical science due to the versatile properties and applications it may provide over conventional drug delivery strategies [1]. The advantages of nanoparticles are their potential to pass through the smallest capillary vessels; avoidance of rapid clearance by phagocytes, thereby prolonging their stay in the bloodstream; penetration into cells and tissue to arrive at target organs; reduced toxicity of loaded drugs based on their controlled release [2]. Nanogels represent one of the many known nanocarriers. They are defined as nanoscale three-dimensional networks of hydrogel polymers with the ability to absorb significant amounts of water [3–5]. They simultaneously possess the properties of nanoparticles and hydrogels, which endows them with more advantages [6]. Their nanometer range can improve the enhanced permeation and retention effect (EPR) and improve the delivery of chemotherapeutics [5].

Based on the method of their preparation, nanogels can be classified chemically (covalently) or physically crosslinked [4]. The level of crosslinking can finely tune the drug release rate. The polymer can be obtained or polymerized from monomers together with the crosslinking step [7]. By utilizing newly synthesized polymers, new properties can be attributed to the nanogels, such as stimuli-sensitive delivery. The development of stimuli-sensitive multi-particulate formulation is a promising platform for sustained release and drug targeting [8,9]. The stimuli to which the used polymers respond—pH, temperature, light, magnetic field, enzymes, ionic strength, ultrasound, and redox change—depend on their properties [10]. One of the triggers used for smart delivery is the pH or proton concentration because of anatomical and pathological differences in the human body. It is well known that the gastrointestinal tract has pH-varying media. The stomach pH is between 1 and 3, in the duodenum 4.8–5.2, small intestines 6.8, and colon 7–7.5 [11]. In chemotherapy, the pH-dependent release is significantly exploited due to the difference between the physiological pH of 7.4 for the blood and normal cells and the neoplastic cells. The latter are characterized by lower extracellular pH (pH $\approx$ 4–5.8) [12,13] and 4.5–5.0 in the lysosomes [5]. Thus, the choice of polymer for the nanogel preparation can determine its properties. If the swelling/deswelling properties of the polymer are pH-dependent, then the drug release can be controlled for the desired place or cells [14]. Both natural and synthetic polymers can be used for the subsequent crosslinking. Some examples of natural polymers, such as agar, sodium alginate, chitosan, and dextran belong to the polysaccharide class. They are an object of biomedical scientific interest because of their biocompatibility, biodegradability, and enhanced adhesion with biological tissues such as epithelia and mucous membranes [15]. They all contain various functional groups, allowing grafting or crosslinking with synthetic polymers. This chemical modification could lead to stimuli-sensitive behavior [16]. Graft polymers are usually obtained through the use of initiators or gamma ray, UV, or microwave radiation in order to create a free radical site on the polymer chain. Monomeric units are attached similarly to a side chain to a polymer backbone to form a branched polymer with desirable properties non-intrinsic for the base polymer [17]. In regard to the pH-sensitive delivery, specific polymers which change their physico-chemical properties based on the change in the pH can be utilized. They polyelectrolytes with ionizable groups whose solubility in aqueous of solutions is changeable by environmental pH. Typical monomers and polymers used for pH-sensitive systems include two types—anionic and cationic. Examples of anionic monomers are acrylic acid, methacrylic acid, propionic acid, ethylenesulfonic acid, styrenesulfonic acid, and cationic monomers include acrylamide, aminoethyl methacrylate, N,N′-dimethylaminomethylacrylamide, N,N′-dimethylaminoethyl methacrylate, N,N′-dimethylaminopropyl methacrylate, N,N′-diethylaminoethyl methacrylate, diallyldimethylammonium chlorid [18]. Acrylic and methacrylic acid have been employed to graft modification of agar for various purposes including waste water treatment, antimicrobial properties, or drug delivery [19–22]. Agar is a natural unbranched hydrophilic polysaccharide. One of its interesting properties is the capacity to form gels even at low concentration. It is biocompatible, inexpensive and can be easily chemically modified [23].

As a model drug in the current study, 5-Fluorouracil (5-FU) (5-fluoro-2,4-pyrimidinedione) was chosen. This drug is an antimetabolite used to treat colorectal, stomach, breast, pancreas, ovary, liver, and other solid tumors [24–26]. It is a pyrimidine analog and has good water solubility, mainly administered intravenously [26,27]. The major drawback of 5-FU clinical usage is the development of drug resistance. Moreover, the drug is associated with high toxicity. It can be manifested with several side effects, such as gastrointestinal, hematological, neural, dermatological, myelosuppression, and cardiotoxicity [28,29]. Its inclusion in different nanoparticles could affect its efficacy by increasing circulation time, reducing side effects, and improving its therapeutic index.

Even though the increasing use of nanoparticles in pharmaceutical technology provides opportunities for innovative approaches in therapy and diagnosis, the toxicity risk related to the new nanostructured drug delivery systems needs to be well established.

The hemolytic assay serves as a primary screening tool for in vitro biocompatibility assessments of drugs and new drug delivery systems. Hemolysis refers to the breakdown of the erythrocyte membrane, leading to the release of hemoglobin into the plasma. Various factors contribute to hemolysis, such as immunologic responses, antigen–antibody reactions, mechanical injury, specific infections, hereditary and acquired cell membrane disorders, G6PD deficiency, hemoglobinopathies (e.g., sickle cell diseases, thalassemia), and certain chemotherapeutic agents [30]. This process causes anemia and poses a substantial limitation for the direct use of chemotherapeutic drugs [31].

The aim of the current study is to investigate the possibility of synthesizing agar-grafted pH-sensitive polymers with the help of acrylic or methacrylic acid and further utilize them as a polymer for nanogel preparation. The prepared nanogel could serve as a carrier for a model chemotherapeutic agent. Thus, it could be more efficiently and safely delivered to the target tumor cells where the pH-sensitivity of the nanogel could guarantee its release.

2. Results and Discussion

2.1. Preparation of Agar Agar-g-Polyacrylic Acid (AA-g-AcA) and Agar Agar-g-Polymethacrylic Acid (AA-g-McA) Polymers

Ag-g-AcA and AA-g-McA were synthesized by graft polymerization using cerium ammonium nitrate (CAN) as a free radical initiator in an inert nitrogen atmosphere. By direct oxidation, CAN generates free radicals on the agar backbone [17]. Graft copolymers are generated in the presence of these active free radicals and the presence of acrylic or methacrylic acid.

A series of polymers were synthesized, varying the ratio between agar, acid monomers, and the initiator. The investigated parameters, together with the sample coding, are presented in Table 1.

Table 1. Polymer, monomer, and initiator ratios for the model grafted polymers. Characteristics of the grafted polymers ($\overline{M\eta}$)– viscosity average molecular weight; G%—grafting; GE%—grafting efficiency (see Equations (1) and (2)); η_{intr} 35 °C—intrinsic viscosity at 35 °C).

Code	Wt. of Agar (g)	Wt. of AcA (g)	Wt of McA (g)	Wt of CAN (g)	$\overline{M\eta}$	Grafting (G%)	Grafting Efficiency (GE%)	$[\eta_{intr}]$ dL/g
Agar (AA)	-	-	-		175,455	-	-	2.89
A1	1	10	-	0.1	1,011,263	709.0	70.9	10.2
A2	1	10	-	0.2	1,840,781	982.0	98.2	15.7
A3	1	10	-	0.3	1,094,816	764.0	76.4	10.8
A4	1	5	-	0.2	461,687	169.6	33.9	5.8
A5	1	15	-	0.2	1,209,053	779.7	51.9	11.6
M1	1	-	10	0.1	889,499	638.3	63.8	9.3
M2	1	-	10	0.2	1,711,811	966.6	96.6	14.9
M3	1	-	10	0.3	943,076	663.5	66.3	9.7
M4	1	-	5	0.2	418,065	234.2	46.8	5.4
M5	1	-	15	0.2	997,519	668.6	44.5	10.1

2.2. Characterization of Newly Synthesized Polymers

The effects of different ratios on the %G and %GE were investigated. Initially, with increasing monomer concentration, an increase in %G and %GE was observed, but after the optimal value was achieved, they decreased. The initial increase in grafting is probably due to the greater availability of grafting sites on the monomer. After that, the monomer remains in excess. With an excess of monomer, the formation of more homopolymer is observed compared with the grafted polymer. Additionally, upon homopolymer formation, the viscosity in the reaction medium increases, which creates an obstacle to the movement of free radicals to the active sites, resulting in lower %G and %GE.

The lower concentration of the initiator probably generates fewer free radicals to attack the AA backbone, resulting in lower values of %G and %GE. At the highest concentration, ceric ions increase their participation in the termination of growing grafted chains, and the parameters' values decrease.

All obtained AA-g-AcA and AA-g-McA polymers show greater intrinsic viscosity values than agar at 35 °C. The obtained intrinsic viscosity of the agar was in agreement with previously determined values [32]. Therefore, the method could be used to calculate the viscosity of the average molecular weight. The increase in hydrodynamic volume measured by [η] correlates with higher molecular weight [33]. With the grafting of more AcA or McA units to the AA backbone, the $\overline{M\eta}$ of grafted polymer increases. After achieving the optimum ratio between polymer, monomer, and initiator the molecular weight decreases. These results are consistent with the literature [32].

Thus, the highest G% and GE% are obtained for A2 and M2 polymers, respectively (Table 1). These models have been further characterized. From now on, in the text, they will be referred to as AA-g-AcA and AA-g-McA for clarity.

The swelling behavior over time in deionized water and its dependence on pH were investigated for the two selected polymers. The results are depicted in Figure 1.

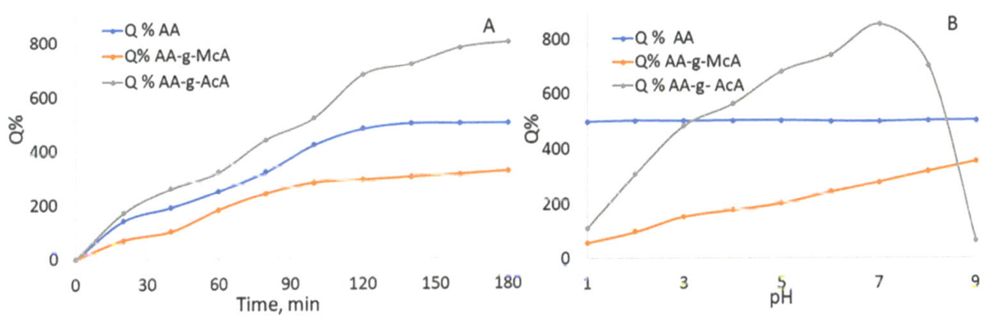

Figure 1. Effect: (**A**) of time on Q% (**B**) of pH on Q% of AA, AA-g-McA, AA-g-AcA.

All of the tested polymers showed swelling upon contact with an aqueous medium (Figure 1A). This is expected because of their ability to absorb and retain water. The effect of pH on their swelling behavior is shown in Figure 1B. Q% values for AA are independent of pH because AA macromolecules do not contain weakly acidic or weakly basic groups. AA-g-AcA2 shows expressed pH-dependent swelling behavior. The presence of COOH groups in the weak polyacid and the increase in pH affect Q%. The ionization of carboxylate groups at pH values around 7 results in electrostatic repulsion and enhanced swelling. A further increase in the pH is associated with a screening effect of the present Na^+ ions in the medium. Similar findings were reported for other polysaccharide hydrogels grafted with acrylic acid [34]. Dependence on pH is observed for the AA-g-McA polymer but to a lower extent. This is probably attributed to the presence of a hydrophobic methyl group in its structure as opposed to the AA-g-AcA polymer. Similar results of Q% about grafting with methacrylic acid were found in the literature [35–37]. Rapid deswelling in the case of AA-g-McA could be expected as well, but at higher pH values [38]. In the present study, the swelling behavior was investigated only in the physiologically relevant pH range, as the prepared hydrogel is intended for biomedical applications.

Moreover, the absolute value of Q% of AA-g-McA is lower than that of pure AA. That is probably due to the hydrophobic interactions between McA and AA, which further stabilize the structure of the resulting hydrogel after swelling. Therefore, the less stable AA-g-AcA swells to a greater extent at comparable pH values. In addition, the AA-g-AcA hydrogel loses its mechanical strength at pH = 8, which is expressed in forming a dispersion of insoluble particles.

Further, 1H nmR spectroscopy was applied to confirm the grafting of AA with acrylic and methacrylic acids using deuterated DMSO as solvent. Pure deuterated DMSO shows no peaks in 1H nmR spectroscopy and is commonly used as an nmR solvent. However, the commercially available samples are not 100% pure, and a residual DMSO-d5 1H nmR signal is observed at 2.50 ppm [39]. The 1H nmR spectrum of agar shows peaks at δ = G1′-5.22, A1-5.08 and agarose skeleton (G2–G6) and (A2′–A6′)-4.83–3.96 ppm. The 1H nmR spectrum of acrylic acid consists of three quadruplets centered at 6.52 ppm (cis proton), 6.14 ppm (germinal proton), and 5.96 ppm (trans proton) and proton of the carboxylic group at 12.0 ppm [40]. When the acrylic or methacrylic acid was grafted on AA, the peak at 3.96 ppm showed decreased signal strength as monomer moieties replaced the hydroxyl group. The AcA carboxylic group's proton peak shifted to 12.16 ppm and 12.22 ppm for the McA carboxylic group.

Compared with the spectrum of AA, new resonances for AA-g-AcA and AA-g-McA new proton signals appeared at 1.5 ppm, 1.7 ppm, and 2.2 ppm for –OH and methylene protons of AcA, respectively, as well as 0.9 ppm and 1.7 and 2.2 ppm, because of the presence of –OH, methyl and methylene groups in McA respectively. These results demonstrated the formation of AA-g-AcA and AA-g-McA.

FTIR spectra of AA, AcA, McA, AA-g-AcA, and AA-g-McA were recorded to prove the successful grafting of AcA and McA. The results are presented in Figure 2A,B. The peak at 3361 cm^{-1} corresponds to stretching vibrations of hydroxyl groups on the spectrum of AA (Figure 2A,B). At 1033 cm^{-1}, stretching vibrations of CH_2OH are found in the same spectrum. The signal at 2896 cm^{-1} is related to C–H stretching vibration, and the one at 1634 cm^{-1} is associated with C=O stretching vibrations. The stretching for the C–O bond was observed at 1047 cm^{-1} [41]. Figure 2A,B show the FTIR spectrum of AcA and McA, respectively. The stretching vibration of –OH from the acid group is detected at 3100 cm^{-1}, while at 1431 cm^{-1}, the O–H bending vibration of the same group for AcA is at 3361 cm^{-1}, and McA is at 1425 cm^{-1}. The bending movement of terminal $=CH_2$ is detected at 1294 cm^{-1} [42] for AcA and 1296 cm^{-1} for McA. At 2988 cm^{-1}, the C–H stretching of the allyl group for AcA and at 2986 cm^{-1} for McA. The carbonyl group is discovered at 1695 cm^{-1} for AcA and 1689 cm^{-1} for McA, and the vinyl group's stretching vibration signal at 1634 cm^{-1} for AcA and 1632 cm^{-1} for McA guarantees the absence of homopolymerization for both monomers. McA shows a peak at 1375 cm^{-1} due to CH_3 bending.

The difference between the monomer and the grafted polymer spectra is the absence of a signal at 1634 cm^{-1} for the vinyl group of AcA and a signal at 1632 cm^{-1} for McA. In addition, the absence of the stretching vibrations found in the AA spectrum of CH_2OH at 1033 cm^{-1} in spectra of grafted polymers shows the successful grafting of monomer units onto the AA backbone. The thermal behavior of AA, AA-g-AcA, and AA-gMcA were compared using DSC analysis at a heating rate of 10 °C/min to distinguish thermal transitions. Agar composition exhibited two glass transitions (Tg). One was observed at 26 °C by agarose and another at 63 °C by agaropectin. The exothermal signal noticed at 255 °C is associated with its thermal decomposition (Figure 3). AA-g-AcA shows Tg at 24 °C, as well as two endothermic peaks at 211 °C and the other at 278 °C while AA-g-McA shows Tg at 23 °C and endothermic peaks at 210 °C and the other at 276 °C. The signals noticed at 23 and 24 °C are Tg belonging to AA on the grafted polymer. The 211 °C and 210 °C peaks were new and corresponded to the AA-g-AcA and AA-g-McA melting points. They are assigned by the formation of the new polymeric structure by the combination of AA and AcA or McA. The grafted polymers' thermal profiles demonstrated the presence of agar on their structure and the obtention of new polymeric material. Additionally, the AA-g-AcA and AA-g-McA exhibited an enhancement in thermal resistance compared with agar since no exothermic signals were observed.

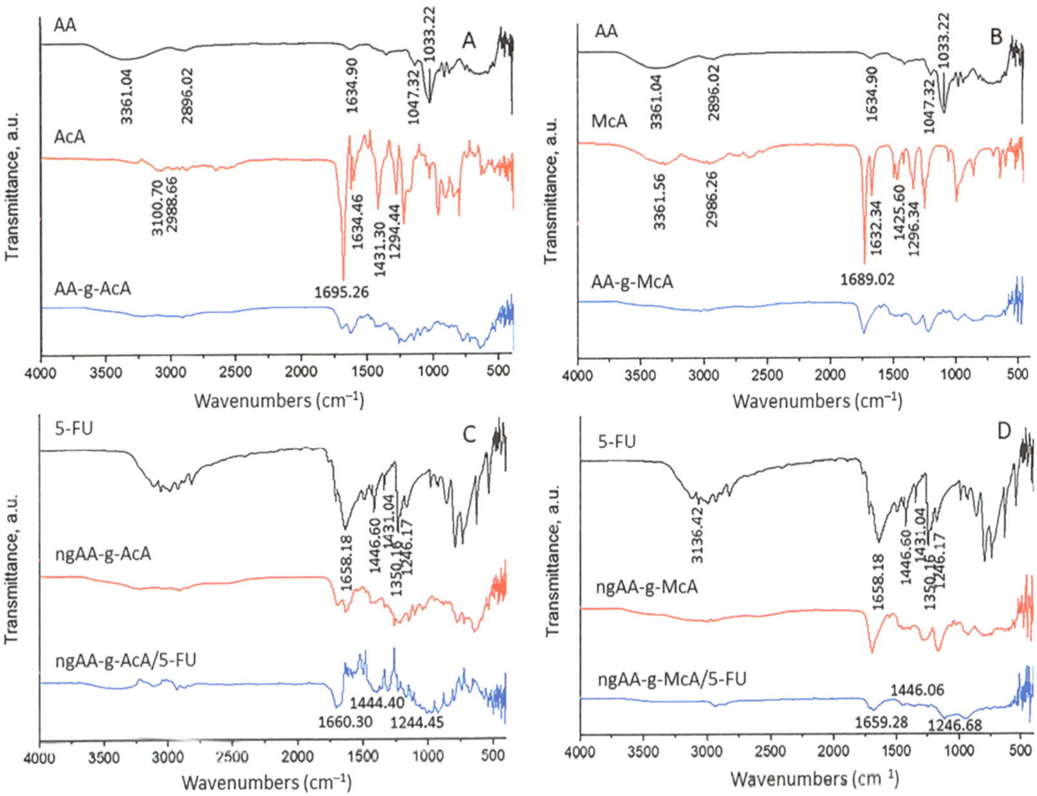

Figure 2. FTIR spectra of (**A**) AA (black line), AcA (red line), AA-g-AcA (blue line) and (**B**) AA (black line), McA (red line), AA-g-McA (blue line), (**C**) 5-Fluorouracil (black line), ngAA-g-AcA (red line), ngAA-g-AcA/5-FU, (**D**) 5-Fluorouracil (black line), ngAA-g-McA (red line), ngAA-g-McA/5-FU.

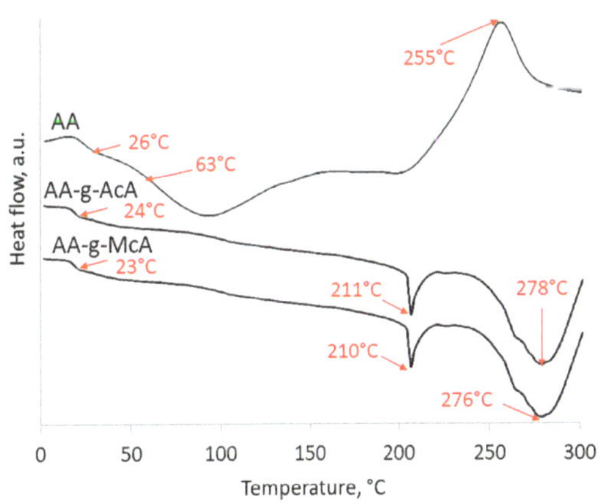

Figure 3. DSC profiles of AA, AA-g-AcA, and AA-g-McA.

Additionally, the thermal stability of the initial AA and the grafted polymers was studied using thermogravimetric analysis. Figure 4 shows the TG and DTG curves for the thermal decomposition of AA and the grafted polymers AA-g-AcA and AA-g-McA in Ar. The TG curve for AA comprises two thermal steps: first small one up to ~120 °C, due to loss of physically absorbed water, which is about 3% of the mass, and second large step at 250–350 °C with a mass loss of ~50%, due to the degradation of agar [43,44]. The DTG curve of AA revealed the main decomposition peaks are centered at 85 °C and 300 °C.

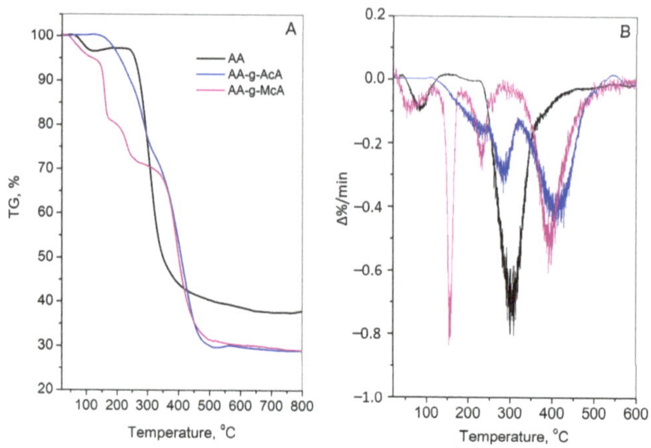

Figure 4. TGA (**A**) and DTG (**B**) curves of initial AA and the grafted polymers AA-g-AcA (A4) and AA-g-McA (M4) in Ar.

The final mass loss is 62% up to 800 °C. At the thermogram of AA-g-AcA, a gradual mass loss (22%) up to about 300 °C due to the loss of free water and dehydration of the polymer backbone is observed. The sharp step registered from 300 to 500 °C with a mass loss of 50% should be attributed to de-polymerization and degradation of the AA and AcA compounds. The DTG maxima are found to be ~290 °C and ~10 °C. The final mass loss is about 71% up to 800 °C. Four well-established thermal stages are visible in the AA-g-McA TG curve. The first one resembles those of AA with a mass loss of 5% up to 125 °C due to loss of physically absorbed water. The second stage at temperatures 130–200 °C is usually interpreted as water release due to a reaction between carboxylic groups of the McA forming polymethacrylic anhydride [45,46]. The next step, with 9% mass loss at 200–300 °C, could be due to de-polymerization, and the final stage, 310–520 °C is associated with the degradation of the residual fragments to form carbonaceous material [46,47]. The corresponding maxima of DTG peaks of degradation are registered at 75 °C, 160 °C, 230 °C and 400 °C. In the case of AA-g-McA, the final mass loss is 71%. Thermal behavior data for the samples, according to a study [48], are placed in Table 2.

Table 2. Thermal behavior data.

Sample	Decomposition Stage	Temperature Range (°C) [a]	DTG Peak (°C) [b]	Mass Loss [a], %	Residue, %
AA	1	~44–130	85	3	97
	2	~225–450	305	56	41
AA-g-AcA (A4)	1	~115–250	230	10	90
	2	~250–320	290	14	73
	3	~320–520	410	44	29
AA-g-McA (M4)	1	~40–125	75	5	95
	2	~130–200	160	13	82
	3	~200–250	230	9	73
	4	~300–520	400	44	29

[a]—From TG. [b]—From DTG.

X-ray powder diffraction was also applied for the polymer characterization. Figure 5 represents X-ray powder diffraction patterns (PXRD) of initial compounds—agar (AA), Acrylic acid (AcA), and Methacrylic acid (McA) as well as that of the prepared grafted polymers—(AA-g-AcA) and (AA-g-McA). The PXRD pattern of agar (AA) consists of four amorphous peaks. The first one is at 13.5°2θ and appears as a shoulder of the second peak at 18.6°2θ, which is the most intensive peak of the diffractogram. The intensity ratio between these two peaks is a measure of the degree of hydration and crystallinity of the agar structure [49]. Two more peaks with lower intensities are visible at 29.9°2θ and 42.7°2θ. The patterns of acrylic and methacrylic acids are similar and consist of broad humps at about 22.6°2θ and 24.1°2θ, respectively.

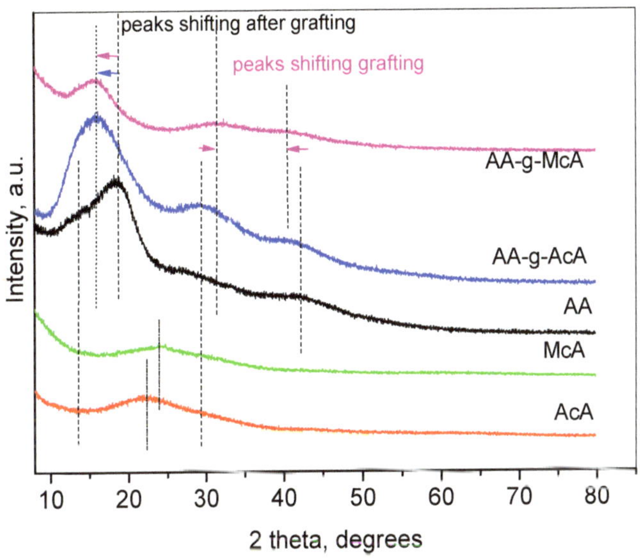

Figure 5. X-ray powder diffraction patterns of the initial compounds and the grafted polymers.

The patterns of the grafted polymers AA-g-AcA and AA-g-McA are also similar, showing three more or less intensive peaks. The transformation of the aar peaks from a clear doublet to a single peak in the range of 13–18° 2θ after the interaction with the acrylic and methacrylic acids indicates the successful formation of the grafted polymers. For AA-g-AcA, the first one is at 15.9° 2θ, the next peaks are at 29.6° 2θ and 41.1° 2θ while for AA-g-McA, the first one appears at 15.9° 2θ also but the second and third peaks could be found at 31.2° 2θ and 40.6° 2θ. Compared with the pure agar pattern, the shift of the three peaks appears more pronounced for the sample with methacrylic acid, which may be an indication that the latter modifies the chain structure of agar to a higher extent [38].

2.3. *Preparation of Nanogels—Parent and 5-FU Loaded*

The newly prepared graft polymers were subsequently applied in the preparation of nanogels by crosslinking with glutaraldehyde (GA). The preparation followed an established protocol with modifications [50]. GA is the crosslinker most frequently used to prepare polymeric hydrogel beads, microspheres, and nanogels [51]. The ratio between polymer, monomer, and crosslinker, as well as drug loading efficiency and swelling, were investigated, and the results are summarized in Table 3. The resulting nanogels were named as follows: parent nanogels AA-g-AcA (ngAA-g-AcA), AA-g-McA (ngAA-g-McA), loaded nanogels AA-g-McA (ngAA-g-McA/5-FU), and AA-g-McA (ngAA-g-McA/5-FU).

Table 3. Formulation parameters and characterization of the AA-g-AcA and AA-g-McA nanogels, (EE%—encapsulation efficiency, %).

Polymer	Drug/Polymer Ratio (*w*/*w*)	GA (mL)	EE%
AA-g-AcA	1/1	2	20.81 ± 2.80
	1/1	3	32.66 ± 1.52
	1/1	4	39.12 ± 2.72
AA-g-AcA	1/2	2	35.25 ± 1.42
	1/2	3	56.86 ± 0.03
	1/2	4	62.44 ± 2.40
AA-g-AcA	1/3	2	53.33 ± 2.17
	1/3	3	72.90 ± 1.69
	1/3	4	78.42 ± 1.19
AA-g-McA	1/1	2	19.20 ± 3.20
	1/1	3	30.30 ± 2.15
	1/1	4	40.25 ± 1.35
AA-g-McA	1/2	2	33.15 ± 2.25
	1/2	3	55.60 ± 1.14
	1/2	4	60.98 ± 2.05
AA-g-McA	1/3	2	54.54 ± 2.65
	1/3	3	74.20 ± 0.54
	1/3	4	79.12 ± 2.16

2.4. Characterization of Nanogels

Nanogels were prepared in different drug/polymer ratios and crosslinker amounts, and these data are presented in Table 3. Increasing the amount of polymer in the drug/polymer ratio results in the retention of a greater amount of 5-FU in the nanogels. It is most likely due to increased hydrogen bonds that the drug and the polymer create between themselves because of the larger number of hydroxyl groups on the polymer's side. The drug loading efficiency decreases with an increase in the amount of drug used during the preparation of the nanoparticles. Maximum encapsulation efficiency (EE%) was found at a drug/polymer ratio of 1/3 for both polymers. The amount of 5-FU above the maximum encapsulation efficiency may be washed after the lavation of the obtained particles because of the drug's good water solubility. At the same time, increasing the amount of GA leads to an increase in the EE% value for both polymers. It is most likely due to the tighter crosslinking, which allows a greater amount of 5-FU to be retained. Based on the results of the drug loading efficiency calculations, the experiments continued with the nanoparticles prepared with a drug/polymer ratio of 1/3 and a 4 mL crosslinker.

FTIR spectra of 5-FU, ngAA-g-AcA, ngAA-g-McA, ngAA-g-AcA/5-FU, and ngAA-g-McA/5-FU were recorded to prove the loading of 5-FU. The spectra can be seen on Figure 2C,D. The spectrum of 5-FU shows characteristic picks at 3136 cm^{-1} due to N-H stretching, C=O stretching at 1658 cm^{-1}, C=C stretching at 1446 cm^{-1}, C-F stretching at 1431 cm^{-1}, C-N stretching at 1246 cm^{-1}, and vibration of the pyrimidine ring at 1350 cm^{-1} [52,53].

It was observed that there were no changes in these main peaks in the IR spectra of ngAA-g-AcA/5-FU and ngAA-g-McA/5-FU, which assumes the physical incorporation of 5-FU in both types of nanoparticles.

Further, the size, shape, and structure of ngAA-g-AcA and ngAA-g-AcA/5-FU as well as ngAA-g-McA and ngAA-g-McA/5-FU were studied using TEM (Figure 6). As seen, empty nanoparticles show spherical shape, nanoscale size, matrix structure, and narrow distribution. The same shape, structure, and distribution are observed for the loaded nanoparticles. The increase in size after loading is visible.

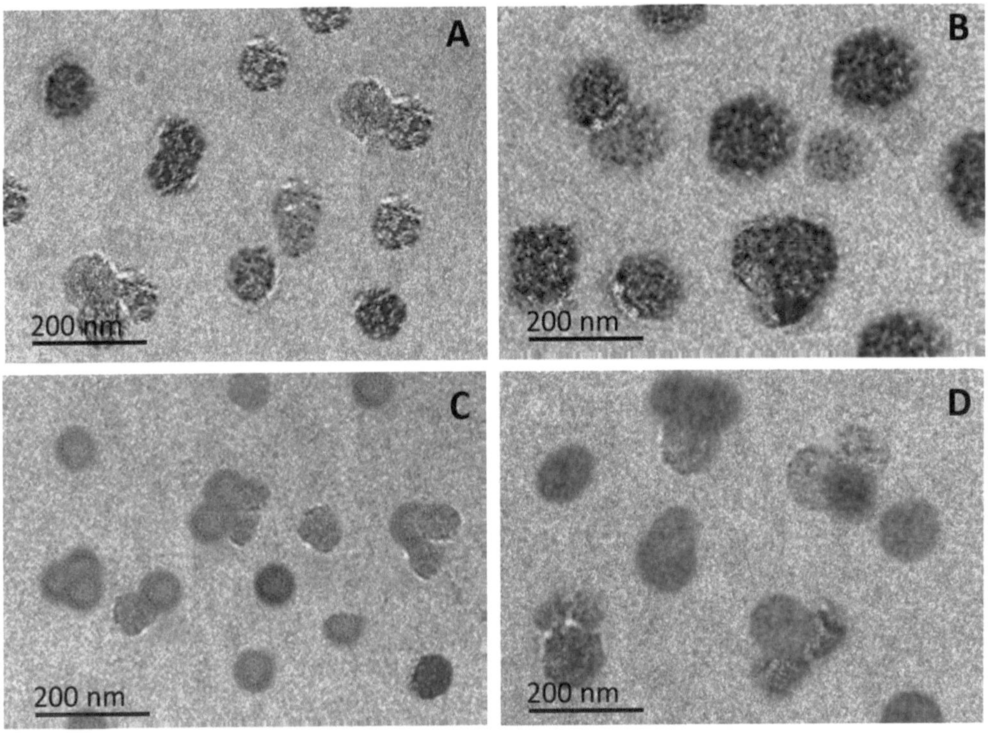

Figure 6. Transmission electron micrographs of empty ngAA-g-AcA (**A**), 5-FU loaded ngAA-g-AcA (**B**), empty ngAA-g-McA (**C**), and 5-FU loaded ngAA-g-McA (**D**) nanoparticles.

The obtained TEM data are in accordance with the DLS analysis. The particle sizes are in the nano range for the obtained nanogels (Table 4). The nanogels' diameter for both polymers increased with the loading of the drug. Probably, the entrapment of the drug in the free volume areas of the polymer particles prevents the additional shrinking. The AA-g-McA nanogels show a smaller size compared with the AA-g-AcA. Most likely, the higher viscosity of the polymer with attached AcA units forms larger droplets when the polymer solution is sprayed into the crosslinker solution, resulting in a larger particle size. This fact is also confirmed by the results obtained from the swelling test. ngAA-g-AcA swells to a greater extent compared with ngAA-g-McA (Table 4). Similar results regarding the nanogels' swelling behavior have been reported before [54].

Table 4. Average size, polydispersity index (PDI), and zeta potential results from DLS analysis, as well as the equilibrium swelling degree (ESD) in different media.

Parameter	ngAA-g-AcA Empty	ngAA-g-AcA/5-FU	ngAA-g-McA Empty	ngAA-g-McA/5-FU
Size, nm	124.7 ± 7.3	250.0 ± 8.4	105.7 ± 3.3	211.8 ± 4.2
PDI	0.38	0.27	0.39	0.28
Zeta potential, mV	−20.1 ± 3.7	−33.6 ± 3.6	−18.8 ± 2.4	−31.8 ± 3.1
ESD %, pH 5.0	168.2 ± 7.2	-	124.7 ± 6.5	-
ESD %, pH 7.4	465.5 ± 3.8	-	354.9 ± 4.9	-

The PDI value shows the physical stability of the empty nanogels and narrow limits in size distribution. Upon 5-FU loading, the polydispersity index for both polymer-based nanogels decreases, most probably due to a change in the zeta potential value. The 5-FU molecule (pKa = 8.0) is negatively charged in distilled water [55]. After loading, empty polymer particles accumulate the drug's negative charge, reflecting on the zeta potential value. This hypothesis is supported by FTIR data showing the physical entrapment of the drug within the nanogels.

The pH-dependent release of 5-FU was observed by performing an in vitro dissolution test in pH 7.4 and 5.0 buffer media at 37 °C. The results are presented in Figure 7.

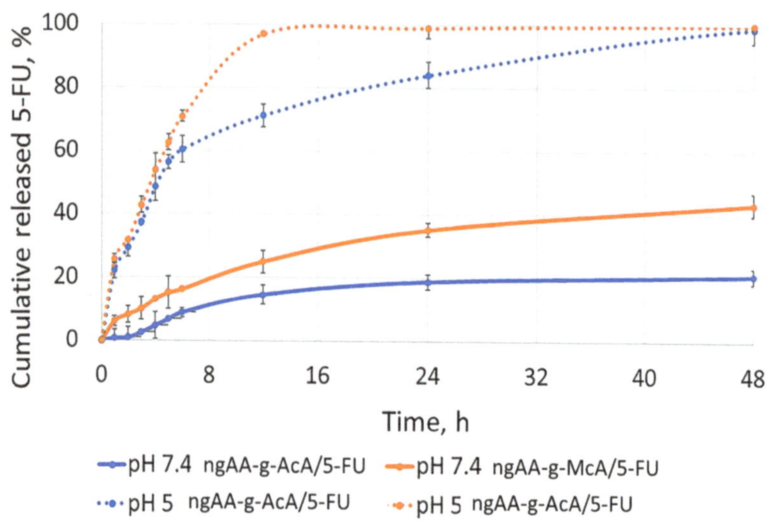

Figure 7. Release profiles of 5-Fluorouracil from npAA-g-AcA and npAA-g-AcA in a buffer medium with a pH of 7.4 and pH of 5.0 at temperature 37 °C, 100 rpm; n = 3 ± SD.

At pH 7.4, the nanogels obtained from both polymers showed a sustained release. For ngAA-g-AcA, it was about 20% in 48 h, and for ngAA-g-McA, it was about 50%. This delay is most likely due to the swelling of the polymer under these conditions. Swelling significantly slows the release of the drug due to the additional crosslinking of the polymer that occurs during the process of preparation of the nanoparticles. The obtained gel layer, by the swelling of the polymer, slows down the release of 5-FU.

At pH 5.0 (Figure 7), ngAA-g-AcA/5-FU and ngAA-g-McA/5-FU show increased and faster release. Such pH conditions are associated with the tumor cells because of their extensive metabolism [16]. The nanogels at the simulated tumor conditions (pH 5.0) show an initial burst release for the first 6 h. A gradual, controlled release was observed over time, and almost 100% of the released 5-FU from ngAA-g-McA was reached at 12 h and 100% from ngAA-g-AcA at 24 h. The faster release is probably connected with polymers' swelling behavior, which is used for nanogel preparation and physical entrapment of 5-FU in them. At pH 5.0, both polymers swell to a lower extent, which allows faster drug release.

The slower release of 5-FU at physiological pH (7.4) is considered to be an advantage as it may reduce drug loss until it reaches the target tumor tissue and is a prerequisite for lowering systemic toxicity. In the bloodstream, as the particles pass from a pH 7.4 region to a pH 5.0 region, the release would be promoted due to the contraction of the polymer chains and the release of API located in the voids of the swollen polymer at pH 7.4. This is in accordance with equilibrium swelling data presented in Table 3. Similar results have been reported in other studies [56].

Nanosized drug delivery systems are designed for different routes of administration (e.g., oral or parentheral use), but in most cases, these systems would have direct contact with circulating blood. The hemocompatibility of nanoparticulate systems, as a part of their preliminary biocompatibility evaluation, includes an assessment of their hemolytic activity as an important safety marker. In this study, the hemolytic activity of the tested compounds was compared with Triton X-100 (20%) in erythrocyte samples (see Figure 8). After 1 h incubation, Triton X-100 led to complete hemolysis (100%). The empty nanogels (ngAA-g-McA and ngAA-g-AcA) exhibited no hemolytic effects (Figure 8A,B). Free 5-FU caused weak hemolysis (2.94%) at the highest concentration (30 µM) (Figure 8C). At the same concentration (30 µM), 5-FU loaded in ngAA-g-McA/5-FU and ngAA-g-AcA/5-FU nanoparticles induced lower levels of hemolysis (1.45% and 1.73%, respectively), compared with the free drug, which is however below the threshold level of 5% according to ISO 10993-4 [57]. Our results are in accordance with the findings of Belman-Flores et al. (2020), who found a good hemocompatibility of pH-sensitive hydrogel nanoparticles based on N-isopropyl acrylamide (NIPAM) and methacrylic acid (MAA) [58]. In another interesting study performed by Zhang et al. (2019), the membrane integrity of poly(N-isopropylacrylamide) (PNIPAM)—treated red blood cells showed that hemolysis was lower than 1%, thus proving that PNIPAM did not significantly impaired red blood cells membrane integrity at a concentration of up to 10 mg/mL [59]. Moreover, data from our study support data from other studies indicating that incorporating 5-FU in different drug delivery systems could be a successful approach to reducing its hemolytic potential [31,60].

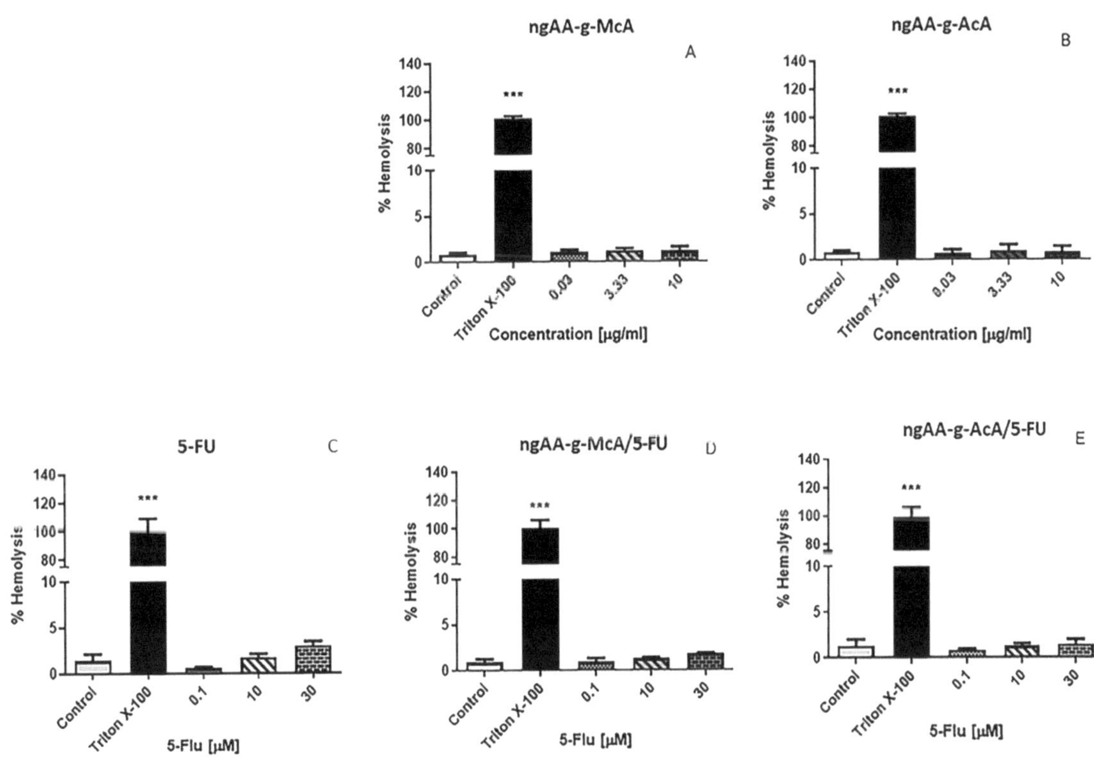

Figure 8. Hemolytic effects of (**A**) AA-g-McA, (**B**) AA-g-AcA, (**C**) free 5-FU, (**D**) 5-FU loaded in AA-g-McA, (**E**) 5-FU loaded in AA-g-AcA on human erythrocytes. Data are presented as means ± SD from triplicate assays (n = 3). Statistical comparisons were made against the untreated controls using one-way ANOVA followed by Dunnett's post-test *** $p < 0.001$ vs. control.

3. Conclusions

In the present study, the grafting of acrylic or methacrylic acid onto agar was successfully carried out by free radical polymerization using cerium ammonium nitrate as an initiator. The optimal polymers were further utilized to prepare 5-FU loaded nanogels by crosslinking with glutaraldehyde. The results demonstrated that the drug release depends on the pH of the medium, being about 40% and 20% for ngAA-g-AcA/5-FU and ngAA-g-McA/5-FU, respectively, in pH 5.0. On the other hand, at pH 7.4, the 5-FU release is more pronounced and complete within 8 h from ngAA-g-McA/5-FU and within 48 h from ngAA-g-AcA/5-FU. Therefore, it can be expected that ngAA-g-AcA/-FU could provide release of the chemotherapeutic predominantly at the target site over a prolonged period of time. Data from biocompatibility characterization showed that both synthesized nanogels were hemocompatible with red blood cells; moreover, the loading of 5-FU in ngAA-g-McA and ngAA-g-AcA leads to a decrease in the hemolitic potential of the chemotherapeutic drug.

4. Materials and Methods

4.1. Materials

Agar-agar powder (AA), Acrylic acid (AcA), Methacrylic acid (McA), Ammonium ceric nitrate (CAN), Nitric acid, and glutaraldehyde were purchased from Sigma-Aldrich (St. Louis, MO, USA). Hydrochloric acid, Potassium dihydrogen phosphate, and Disodiumhydrogen phosphate dihydrate were purchased from Merck (Darmstadt, Germany). Hydroquinone was purchased from TCI Europe (Zwijndrecht, Belgium). Deionized water was prepared in the laboratory.

4.2. Methods

4.2.1. The Preparation of Agar Agar-g-Polyacrylic Acid (AA-g-AcA) and Agar Agar-g-Polymethacrylic Acid (AA-g-McA)

Agar was dissolved in distilled water at 90 °C in a three-neck round bottom flask with a gas inlet system and a condenser. Then, acrylic or methacrylic acid was added to the agar solution and stirred. Nitrogen gas was poured for 60 min before the solution of CAN in water, acidified with concentrated HNO_3, was added, and the flask was closed. The ratio between polymer, monomer, and initiator was varied, and it is shown in Table 1. Then, the reaction was continued under constant stirring in a thermostatic paraffin bath for 6 h at a constant temperature of 40 °C. Finally, the grafting procedure was terminated upon adding a saturated hydroquinone solution. Next, a separating funnel splits the resulting polymer AA-g-AcA (A1-5) or AA-g-McA (M1-5) from the homopolymer. Next, the graft copolymer hydrogel was precipitated over acetone, separated by centrifugation, dried to constant weight, and ground for further use. The proposed synthesis mechanism of AA-g-AcA and AA-g-McA is represented in Figure 9.

The nanogels ngAA-g-AcA and ngAA-g-McA parent containing 5-FU were prepared as per a modified protocol [50]. Briefly, the polymer is placed in distilled water or a solution of 5-FU (0.1 mg/20 mL) and stirred to form a homogenous solution for 2 h. Then, it was added by spraying into stirred water containing glutaraldehyde (crosslinker) and HNO_3 (catalyst) using a spraying device under ultrasound for 4 min (0.04 wats). The formed nanogels were then removed from the crosslinking solution by centrifugation and washed with water repeatedly to remove the glutaraldehyde and acid residue. Due to the known toxic effect of GA, its total amount was kept as low as possible during the crosslinking procedure, and the free GA was further eliminated. Finally, the nanogels were completely dried under vacuum at 40 °C. The varying ratios between polymer, monomer, and crosslinker are shown in Table 3. The resulting nanogels were named as follows: parent nanogels AA-g-AcA (ngAA-g-AcA), AA-g-McA (ngAA-g-McA), loaded nanoparticles AA-g-McA (ngAA-g-McA/5-FU), and AA-g-McA (ngAA-g-McA/5-FU).

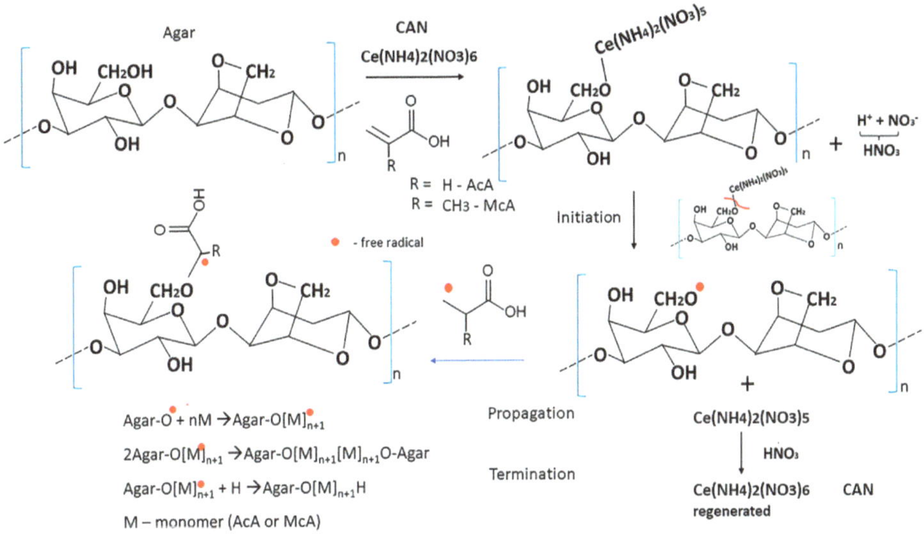

Figure 9. Schematic representation of mechanism for preparation of Ag-g-AcA and AA-g-McA in an inert atmosphere.4.2.2. Preparation of nanogels—parent and 5-FU loaded.

4.2.2. The Percentage of Grafting (% G) and Grafting Efficiency (% GE) were Calculated Using the Following Formulas

$$\%\text{Grafting} = \frac{W_2 - W}{W} \times 100 \quad (1)$$

$$\%\text{Grafting efficiency} = \left(\frac{W_2 - W}{W_1}\right) \times 100 \quad (2)$$

where W is the weight of AA, W_1 is the weight of AcA or McA, and W_2 is the weight of the grafted polymer.

4.2.3. Transform Infrared Spectroscopy (FTIR)

AA, AcA, McA, AA-g-AcA, and AA-g-McA, as well as 5-FU, ngAA-g-AcA, ngAA-g-McA, ngAA-g-McA/5-FU, and ngAA-g-McA/5-FU were characterized using FTIR-ATR spectroscopy with a Thermo-Nicolet FTIR instrument (Thermo Fischer Scientific, Waltham, MA, USA) in the range of 4000–400 cm^{-1} and with a resolution of 4 cm^{-1}.

4.2.4. Nuclear Magnetic Resonance (NMR)

A Bruker AV600 spectrometer (Bruker AV600, Berlin, Germany) was used to acquire 1H-NMR at 250 MHz. The 1H–NMR spectra were measured with solutions of approximately 0.03 M. DMSO-d6 was used as a solvent. As an internal standard, chemical shifts were expressed as δ values in parts per million (ppm) against tetramethylsilane (TMS).

4.2.5. Differential Scanning Calorimetry (DSC)

DSC curves of pure AA and graft copolymers AA-g-AcA and AA-g-McA were recorded using a differential scanning calorimeter PerkinElmer DSC-8500 (Waltham, MA, USA), equipped with an Intracooler 3 cooler. The samples were loaded into standard aluminum pans and were then scanned. The temperature range was from −50 °C to 180°C with a heat rate of 10 °C/min and from 180 °C to 20 °C with a cooling rate of 20 °C/min. The control of the device, data collection, and processing were performed with the help of specialized Pyris software v.10.1.0.0412.

4.2.6. Thermogravimetric Analysis (TGA)

Thermogravimetric measurements were performed by LABSYSEvo and SETARAM (Caluire, France) in an argon flow within a temperature range of 20–800 °C and a heating rate of 10 °C/min.

4.2.7. X-ray Diffraction Analysis (XRD)

Powder X-ray diffraction was performed on a Bruker D8 Advance diffractometer (Bruker AXS GmbH, Karlsruhe, Germany) with Cu K_α tube (λ = 0.15418 nm) equipped with LynxEye detector, with steps of 0.02°2θ in the region 5–80°2θ.

4.2.8. Viscosity Measurements

The specific viscosities of AA, AA-g-AcA, and AA-g-McA were determined with an Ubbelodhe viscometer (Laborxing, Shenzhen, China) at 35 °C according to the procedure suggested by Rochas and Lachaye [33] using the formula:

$$\eta_{sp} = \frac{t - t_0}{t_0} \quad (3)$$

where the specific viscosity is η_{sp}, t, and t_0 are the flow time in seconds of the polymer solution and pure solvent through the Ubbelohde viscometer.

Aqueous solutions of agar, AA-g-AcA, and AA-g-McA (0.005–0.05% *w/v*) were prepared using 0.75 M NaSCN to inhibit Agar aggregation.

Using $\eta_{sp,}$ the reduced viscosity was calculated using the following formula:

$$\eta_{red} = \frac{\eta_{sp}}{C} \quad (4)$$

where C is concentration in g/mL.

Extrapolating the reduced viscosity versus C (η_{sp}/C) at 0 concentration, the intrinsic viscosity was obtained.

The viscosity average molecular weight ($\overline{M\eta}$) was determined using the relation with intrinsic viscosity given by the Mark–Houwink–Sakurada equation:

$$\text{Intrinsic viscosity } [\eta] = K(\overline{M\eta}) \quad (5)$$

Based on literature data [33], the values of parameters α and K, depending on the polymer-solvent system, were taken assuming that they did not change with the grafting (K = 0.07 mL/g, α = 0.72).

4.2.9. Determination of Swelling Indices Q at Different pH

The grafted polymers' swelling behavior (pH-dependent) was studied at varying pH (1–9) and at 20 min time intervals from 20 to 180 min in deionized water, and swelling indices (Q%) were calculated.

Dried samples of the grafted polymer (10 mg) were placed in a previously weighed (M) tube, and 20 mL water or a series of buffer solutions with pH values ranging from 1 to 9 were added. At equal time intervals, the liquid was removed through suction, and the sample tube was then weighed to determine the quantity of water absorbed per mg of grafted polymer. The experiment continued until a constant mass was achieved (M_s), and the polymer swelled. The following formula is used to calculate the Q% value:

$$Q\% = \frac{M_s - M}{M} \times 100 \quad (6)$$

4.2.10. Determination of Swelling Behavior of the Nanoparticles

The equilibrium swelling behavior of AA-g-AcA and AA-g-McA nanogels were studied in buffer solutions with pH 5.0 and pH 7.4 at 37 °C by gravimetric analysis following a reported procedure with some modifications [54]. A sample of the nanogels was placed in a previously measured cellulose membrane bag. It was placed in 25 mL of the corresponding buffer solution until complete equilibration. Afterward, the excess liquid was removed, and the swollen nanoparticles were weighed. The percent Equilibrium swelling degree (ESD) was calculated as follows.

$$\text{ESD}\,\% = \frac{M_s - M_d}{M_d} \times 100 \tag{7}$$

where M_s is the mass of the nanogels in the swollen state, and M_d is the mass of the dry nanogels.

4.2.11. Transmission Electron Micrograph (TEM) Characterization

The size and structure of the samples were characterized using transmission electron microscopy (JEOL JEM 2100 h STEM (200 kV; point-resolution 0.23 nm) JEOL (Freising, Germany) Samples were prepared by placing a water suspension of the nanoparticles on a polymer microgrid supported on a Cu grid.

4.2.12. Dynamic Light Scattering (DLS)

The nanogel size, polydispersity index, and zeta potential were determined using a Zetasizer (Zetasizer Nano ZS, Malvern Instruments, Worcestershire, UK). Immediately after purification and before vacuum drying, the aqueous dispersion samples were measured at a scattering angle of 90° and 25 °C.

4.2.13. Drug Loading Efficiency and Release

The 5-FU content of nanogels was estimated in deionized water using the extraction method [61]. The desired amount of 5-FU-loaded dry nanoparticles was extracted in 100 mL of deionized water at room temperature under stirring until all the drug was removed from the nanogels. After that, the polymer suspension was removed by passing it through a filter paper, and the filtrate was collected. The 5-FU content in the filtrate was measured with a UV spectrophotometer (Thermo Scientific Evolution 300, Madison, WI, USA) at a wavelength of 270 nm using the calibration curve of the series of 5-FU solutions with standard concentrations. The percent of entrapment efficiency was calculated according to the following equation.

$$\text{Encapsulation Efficiency}\,\%\,(\text{EE}\%) = \frac{5-\text{FU total} - 5-\text{FU filtrate}}{5\quad\text{FU total}} \times 100 \tag{8}$$

The in vitro drug release from the nanoparticles was studied in a phosphate buffer pH 7.4 and 5.0, the physiological pH of blood serum, and the intercellular pH of neoplastic cells, respectively. The freshly prepared nanogel dispersion was introduced into a dialysis membrane bag (MW = 6000–8000 kDa), placed into 40 mL of release medium, and incubated in a shaking water bath at 37 °C with a speed of 100 rpm. At appropriate time intervals, an aliquot of the samples was withdrawn, and the amount of 5-FU released from the nanogels was evaluated using a UV spectrophotometer at a λmax of 270 nm. Then, an equal volume of fresh dissolution medium was added back to maintain a constant volume. All the experiments were performed in triplicate, and the presented results show the mean value.

4.2.14. Hemolysis Assay

The test substances were assessed for their hemolytic potential using the methodology outlined in Evans et al.'s 2013 protocol [62]. Blood specimens from healthy volunteers were acquired from a certified clinical laboratory (Bodimed, Sofia, Bulgaria). All experimental procedures were performed in accordance with the rules of the Institutional Ethics Committee (KENIMUS) at the Medical University—Sofia, Sofia, Bulgaria [63]. Erythrocytes were isolated from blood through successive centrifugation in 0.9% NaCl buffer. The blood cells underwent resuspension in phosphate buffer (pH 7.4). In 96-well plates, test substances (empty AA-g-McA (0.03–10 mg/mL), empty AA-g-AcA (0.03–10 mg/mL); 5-FU (0.1–30 μM) and 5-FU in AA-g-McA, and 5-FU in AA-g-AcA, loaded in corresponding equimolar concentrations), along with 20% Triton X-100 (utilized as a positive control), and phosphate buffer (utilized as a negative control) were dispensed. Then, the erythrocyte suspension in phosphate buffer was introduced to these plates. Incubation at 37 °C was performed for 1 h, followed by centrifugation at 500× *g* for 5 min. The resulting supernatant was transferred to new 96-well plates, and the hemoglobin absorbance was measured at 430 nm using a Synergy 2 plate reader (BioTek Instruments, Inc, Highland Park, Winooski, VT, USA). The outcomes were expressed as the percentage of hemolysis concerning the hemoglobin absorbance values in the positive controls, with the hemoglobin absorbance of negative controls regarded as zero hemolysis. Substances causing hemolysis below 5% (in accordance with the acceptable hemolytic threshold defined by ISO 10993-4) were considered biocompatible [57]

4.2.15. Statistical Analysis

One-way ANOVA followed by a Dunnett post-hoc test was employed for statistical analysis. On GraphPad Prism software (version 8, Informer Technologies, Inc., CA, USA). The significance was determined at $p < 0.001$. These data are presented as mean ± SD (n = 3) derived from three independent experiments.

Author Contributions: Conceptualization, I.I. and C.V.; methodology, C.V., D.S. and V.T.; software, I.S. and D.K.; formal analysis, M.S. and B.T.; investigation, T.P. and D.T.; writing—original draft preparation, I.I., M.S., I.S., D.K. and D.S.; writing—review and editing, M.S.; visualization, C.V.; supervision, V.T., B.T. and C.V.; project administration, C.V.; funding acquisition, M.S. All authors have read and agreed to the published version of the manuscript.

Funding: This research was funded by the European Union—Next Generation EU, through the National Recovery and Resilience Plan of the Republic of Bulgaria, Project No. BG-RRP-2.004-0004-C01.

Institutional Review Board Statement: Ethical review and approval were waived for this study due to the hemolysis test being a standard in vitro procedure without clinical or invasive procedures on humans and all samples are subsequently destroyed.

Informed Consent Statement: Not applicable.

Data Availability Statement: The original contributions presented in the study are included in the article, further inquiries can be directed to the corresponding authors.

Conflicts of Interest: The authors declare no conflicts of interest. The funders had no role in the design of the study, in the collection, analyses, or interpretation of data, in the writing of the manuscript, or in the decision to publish the results.

References

1. Jeevanandam, J.; Dable-Tupas, G.; De Guzman, M.B. Chapter 6—Applications of Nanotechnology in Pharmaceutical Products. In *Applications of Nanotechnology in Drug Discovery and Delivery*; Egbuna, C., Găman, M.-A., Jeevanandam, J., Eds.; Drug Discovery Update; Elsevier: Amsterdam, The Netherlands, 2022; pp. 119–156. ISBN 978-0-12-824408-1.
2. Jung, T.; Kamm, W.; Breitenbach, A.; Kaiserling, E.; Xiao, J.; Kissel, T. Biodegradable Nanoparticles for Oral Delivery of Peptides: Is There a Role for Polymers to Affect Mucosal Uptake? *Eur. J. Pharm. Biopharm. Off. J. Arbeitsgemeinschaft Für Pharm. Verfahrenstechnik EV* **2000**, *50*, 147–160. [CrossRef]
3. Garg, A.; Shah, K.; Chauhan, C.S.; Agrawal, R. Ingenious Nanoscale Medication Delivery System: Nanogel. *J. Drug Deliv. Sci. Technol.* **2024**, *92*, 105289. [CrossRef]
4. Soni, K.S.; Desale, S.S.; Bronich, T.K. Nanogels: An Overview of Properties, Biomedical Applications and Obstacles to Clinical Translation. *J. Control. Release* **2016**, *240*, 109–126. [CrossRef] [PubMed]
5. Yin, Y.; Hu, B.; Yuan, X.; Cai, L.; Gao, H.; Yang, Q. Nanogel: A Versatile Nano-Delivery System for Biomedical Applications. *Pharmaceutics* **2020**, *12*, 290. [CrossRef] [PubMed]
6. Attama, A.A.; Nnamani, P.O.; Onokala, O.B.; Ugwu, A.A.; Onugwu, A.L. Nanogels as Target Drug Delivery Systems in Cancer Therapy: A Review of the Last Decade. *Front. Pharmacol.* **2022**, *13*, 874510. [CrossRef]

7. Oh, J.K.; Drumright, R.; Siegwart, D.J.; Matyjaszewski, K. The Development of Microgels/Nanogels for Drug Delivery Applications. *Prog. Polym. Sci.* **2008**, *33*, 448–477. [CrossRef]
8. Siraj, S.; Sudhakar, P.; Rao, U.S.; Sekharnath, K.V.; Rao, K.C.; Subha, M.C.S. Interpenetrating polymer network microspheres of poly (vinyl alcohol)/methyl cellulose for controlled release studies of 6-thioguanine. *Int. J. Pharm. Pharm. Sci.* **2014**, *6*, 101–106.
9. Rejinold, N.S.; Chennazhi, K.P.; Nair, S.V.; Tamura, H.; Jayakumar, R. Biodegradable and Thermo-Sensitive Chitosan-g-Poly(N-Vinylcaprolactam) Nanoparticles as a 5-Fluorouracil Carrier. *Carbohydr. Polym.* **2011**, *83*, 776–786. [CrossRef]
10. Avramović, N.; Mandić, B.; Savić-Radojević, A.; Simić, T. Polymeric Nanocarriers of Drug Delivery Systems in Cancer Therapy. *Pharmaceutics* **2020**, *12*, 298. [CrossRef]
11. Bawa, P.; Pillay, V.; Choonara, Y.E.; du Toit, L.C. Stimuli-Responsive Polymers and Their Applications in Drug Delivery. *Biomed. Mater. Bristol Engl.* **2009**, *4*, 022001. [CrossRef] [PubMed]
12. Zha, L.; Banik, B.; Alexis, F. Stimulus Responsive Nanogels for Drug Delivery. *Soft Matter* **2011**, *7*, 5908–5916. [CrossRef]
13. Maghsoudi, M.; Abbasian, M.; Farhadi, K.; Mahmoodzadeh, F.; Ghorbani, M.; Hoseinzadeh, M. Mesoporous Si-MCM-41/Polymer as a pH-Responsive Drug Delivery System for Cancer Therapy. *ChemistrySelect* **2020**, *5*, 11901–11909. [CrossRef]
14. Shymborska, Y.; Budkowski, A.; Raczkowska, J.; Donchak, V.; Melnyk, Y.; Vasiichuk, V.; Stetsyshyn, Y. Switching It Up: The Promise of Stimuli-Responsive Polymer Systems in Biomedical Science. *Chem. Rec.* **2023**, e202300217. [CrossRef] [PubMed]
15. Liu, Z.; Jiao, Y.; Wang, Y.; Zhou, C.; Zhang, Z. Polysaccharides-Based Nanoparticles as Drug Delivery Systems. *Adv. Drug Deliv. Rev.* **2008**, *60*, 1650–1662. [CrossRef] [PubMed]
16. Durkut, S.; Elçin, Y.M. Synthesis and Characterization of Thermosensitive Poly(N-Vinylcaprolactam)-g-Collagen. *Artif. Cells Nanomed. Biotechnol.* **2017**, *45*, 1665–1674. [CrossRef]
17. Bhattacharya, A.; Rawlins, J.W.; Ray, P. *Polymer Grafting and Crosslinking*; Wiley: Hoboken, NJ, USA, 2008; ISBN 978-0-470-41480-4.
18. Huh, K.M.; Kang, H.C.; Lee, Y.J.; Bae, Y.H. pH-Sensitive Polymers for Drug Delivery. *Macromol. Res.* **2012**, *20*, 224–233. [CrossRef]
19. Wuttisela, K.; Panijpan, B.; Triampo, W.; Triampo, D. Optimization of the Water Absorption by Crosslinked Agar-g-Poly (Acrylic Acid). *Polym. Korea* **2008**, *32*, 537–543.
20. Betraoui, A.; Seddiki, N.; Souag, R.; Guerfi, N.; Semlali, A.; Aouak, T.; Aliouche, D. Synthesis of New Hydrogels Involving Acrylic Acid and Acrylamide Grafted Agar-Agar and Their Application in the Removal of Cationic Dyes from Wastewater. *Gels* **2023**, *9*, 499. [CrossRef]
21. Pino-Ramos, V.H.; Duarte-Peña, L.; Bucio, E. Highly Crosslinked Agar/Acrylic Acid Hydrogels with Antimicrobial Properties. *Gels* **2021**, *7*, 183. [CrossRef]
22. Rani, G.U.; Konreddy, A.K.; Mishra, S.; Sen, G. Synthesis and Applications of Polyacrylamide Grafted Agar as a Matrix for Controlled Drug Release of 5-ASA. *Int. J. Biol. Macromol.* **2014**, *65*, 375–382. [CrossRef]
23. Yin, Z.-C.; Wang, Y.-L.; Wang, K. A pH-Responsive Composite Hydrogel Beads Based on Agar and Alginate for Oral Drug Delivery. *J. Drug Deliv. Sci. Technol.* **2018**, *43*, 12–18. [CrossRef]
24. Zhang, N.; Yin, Y.; Xu, S.-J.; Chen, W.-S. 5-Fluorouracil: Mechanisms of Resistance and Reversal Strategies. *Mol. Basel Switz.* **2008**, *13*, 1551–1569. [CrossRef] [PubMed]
25. Meyerhardt, J.A.; Mayer, R.J. Systemic Therapy for Colorectal Cancer. *N. Engl. J. Med.* **2005**, *352*, 476–487. [CrossRef] [PubMed]
26. Longley, D.B.; Harkin, D.P.; Johnston, P.G. 5-Fluorouracil: Mechanisms of Action and Clinical Strategies. *Nat. Rev. Cancer* **2003**, *3*, 330–338. [CrossRef] [PubMed]
27. Sun, X.; Liu, C.; Omer, A.M.; Lu, W.; Zhang, S.; Jiang, X.; Wu, H.; Yu, D.; Ouyang, X.-K. pH-Sensitive ZnO/Carboxymethyl Cellulose/Chitosan Bio-Nanocomposite Beads for Colon-Specific Release of 5-Fluorouracil. *Int. J. Biol. Macromol.* **2019**, *128*, 468–479. [CrossRef] [PubMed]
28. Farjadian, F.; Ghasemi, A.; Gohari, O.; Roointan, A.; Karimi, M.; Hamblin, M.R. Nanopharmaceuticals and Nanomedicines Currently on the Market: Challenges and Opportunities. *Nanomedicine* **2019**, *14*, 93–126. [CrossRef]
29. Krishnaiah, Y.S.R.; Satyanarayana, V.; Dinesh Kumar, B.; Karthikeyan, R.S.; Bhaskar, P. In Vivo Pharmacokinetics in Human Volunteers: Oral Administered Guar Gum-Based Colon-Targeted 5-Fluorouracil Tablets. *Eur. J. Pharm. Sci. Off. J. Eur. Fed. Pharm. Sci.* **2003**, *19*, 355–362. [CrossRef]
30. Sirdah, M.; Reading, N.S.; Perkins, S.L.; Shubair, M.; Aboud, L.; Prchal, J.T. Hemolysis and Mediterranean G6PD Mutation (c.563 C>T) and c.1311 C>T Polymorphism among Palestinians at Gaza Strip. *Blood Cells. Mol. Dis.* **2012**, *48*, 203–208. [CrossRef]
31. Jeswani, G.; Alexander, A.; Saraf, S.; Saraf, S.; Qureshi, A.; Ajazuddin. Recent Approaches for Reducing Hemolytic Activity of Chemotherapeutic Agents. *J. Control. Release Off. J. Control. Release Soc.* **2015**, *211*, 10–21. [CrossRef]
32. SRani, G.U.; Mishra, S.; Sen, G.; Jha, U. Polyacrylamide Grafted Agar: Synthesis and Applications of Conventional and Microwave Assisted Technique. *Carbohydr. Polym.* **2012**, *90*, 784–791. [CrossRef]
33. Rochas, C.; Lahaye, M. Average Molecular Weight and Molecular Weight Distribution of Agarose and Agarose-Type Polysaccharides. *Carbohydr. Polym.* **1989**, *10*, 289–298. [CrossRef]
34. Sadeghi, M.; Soleimani, F. Synthesis of pH-Sensitive Hydrogel Based on Starch-Polyacrylate Superabsorbent. *J. Biomater. Nanobiotechnology* **2012**, *2012*, 310–314. [CrossRef]
35. Tripathy, J.; Mishra, D.K.; Yadav, M.; Sand, A.; Behari, K. Modification of κ-Carrageenan by Graft Copolymerization of Methacrylic Acid: Synthesis and Applications. *J. Appl. Polym. Sci.* **2009**, *114*, 3896–3905. [CrossRef]

36. Shabir, F.; Erum, A.; Tulain, U.R.; Hussain, M.A.; Ahmad, M.; Akhter, F. Preparation and Characterization of pH Sensitive Crosslinked Linseed Polysaccharides-Co-Acrylic Acid/Methacrylic Acid Hydrogels for Controlled Delivery of Ketoprofen. *Des. Monomers Polym.* **2017**, *20*, 485–495. [CrossRef] [PubMed]
37. Baranovskii, V.Y.; Ganev, V.G.; Petkova, V.B.; Voicheva, K.C.; Dimitrov, M.V. Hydrogels Based on Polycarboxylic Acid-Agar-Agar Complexes. *Colloid J.* **2012**, *74*, 645–648. [CrossRef]
38. Sharma, K.; Kumar, V.; Swart-Pistor, C.; Chaudhary, B.; Swart, H.C. Synthesis, Characterization, and Anti-Microbial Activity of Superabsorbents Based on Agar–Poly(Methacrylic Acid-Glycine). *J. Bioact. Compat. Polym.* **2017**, *32*, 74–91. [CrossRef]
39. Gottlieb, H.E.; Kotlyar, V.; Nudelman, A. nmR Chemical Shifts of Common Laboratory Solvents as Trace Impurities. *J. Org. Chem.* **1997**, *62*, 7512–7515. [CrossRef] [PubMed]
40. C, A.L.; Flores, M.; Arroyo, R.; Caldiño, U. Optical Spectroscopy of Er3+ Ions in Poly(Acrylic Acid). *Opt. Mater.* **2006**, *28*, 1171.
41. Fernando, I.P.S.; Sanjeewa, K.K.A.; Samarakoon, K.W.; Lee, W.W.; Kim, H.-S.; Kim, E.-A.; Gunasekara, U.K.D.S.S.; Abeytunga, D.T.U.; Nanayakkara, C.; de Silva, E.D.; et al. FTIR Characterization and Antioxidant Activity of Water Soluble Crude Polysaccharides of Sri Lankan Marine Algae. *Algae* **2017**, *32*, 75–86. [CrossRef]
42. Ferraz, C.C.; Varca, G.H.C.; Ruiz, J.-C.; Lopes, P.S.; Mathor, M.B.; Lugão, A.B.; Bucio, E. Radiation-Grafting of Thermo- and pH-Responsive Poly(N-Vinylcaprolactam-Co-Acrylic Acid) onto Silicone Rubber and Polypropylene Films for Biomedical Purposes. *Radiat. Phys. Chem.* **2014**, *97*, 298–303. [CrossRef]
43. Prachayawarakorn, J.; Limsiriwong, N.; Kongjindamunee, R.; Surakit, S. Effect of Agar and Cotton Fiber on Properties of Thermoplastic Waxy Rice Starch Composites. *J. Polym. Environ.* **2012**, *1*, 88–95. [CrossRef]
44. Ouyang, Q.-Q.; Hu, Z.; Li, S.-D.; Quan, W.-Y.; Wen, L.-L.; Yang, Z.-M.; Li, P.-W. Thermal Degradation of Agar: Mechanism and Toxicity of Products. *Food Chem.* **2018**, *264*, 277–283. [CrossRef]
45. Grant, D.H.; Grassie, N. The Thermal Decomposition of Polymethacrylic Acid. *Polymer* **1960**, *1*, 125–134. [CrossRef]
46. Schild, H. Thermal Degradation of Poly(Methacrylic Acid): Further Studies Applying TGA/FTIR. *J. Polym. Sci. Polym. Chem.* **1993**, *31*, 2403–2405. [CrossRef]
47. Fyfe, C.A.; McKinnon, M.S. Investigation of the Thermal Degradation of Poly(Acrylic Acid) and Poly(Methacrylic Acid) by High-Resolution Carbon-13 CP/MAS nmR Spectroscopy. *Macromolecules* **1986**, *19*, 1909–1912. [CrossRef]
48. Singh, S.; Nwabor, O.F.; Ontong, J.C.; Kaewnopparat, N.; Voravuthikunchai, S.P. Characterization of a Novel, Co-Processed Bio-Based Polymer, and Its Effect on Mucoadhesive Strength. *Int. J. Biol. Macromol.* **2020**, *145*, 865–875. [CrossRef]
49. Freile-Pelegrín, Y.; Madera-Santana, T.; Robledo, D.; Veleva, L.; Quintana, P.; Azamar, J.A. Degradation of Agar Films in a Humid Tropical Climate: Thermal, Mechanical, Morphological and Structural Changes. *Polym. Degrad. Stab.* **2007**, *92*, 244–252. [CrossRef]
50. Swamy, B.Y.; Chang, J.H.; Ahn, H.; Lee, W.-K.; Chung, I. Thermoresponsive N-Vinyl Caprolactam Grafted Sodium Alginate Hydrogel Beads for the Controlled Release of an Anticancer Drug. *Cellulose* **2013**, *20*, 1261–1273. [CrossRef]
51. Pérez-Álvarez, L.; Laza, J.M.; Álvarez-Bautista, A. Covalently and Ionically Crosslinked Chitosan Nanogels for Drug Delivery. *Curr. Pharm. Des.* **2016**, *22*, 3380–3398. [CrossRef] [PubMed]
52. Yusefi, M.; Shameli, K.; Jahangirian, H.; Teow, S.-Y.; Umakoshi, H.; Saleh, B.; Rafiee-Moghaddam, R.; Webster, T.J. The Potential Anticancer Activity of 5-Fluorouracil Loaded in Cellulose Fibers Isolated from Rice Straw. *Int. J. Nanomed.* **2020**, *15*, 5417–5432. [CrossRef] [PubMed]
53. Reddy, Y.D.; Dhachinamoorthi, D.; Ch, K.B.; Sekhar, R. Formulation and in Vitro Evaluation of Antineoplastic Drug Loaded Nanoparticles as Drug Delivery System. *Afr. J. Pharm. Pharmacol.* **2013**, *7*, 1592–1604. [CrossRef]
54. Geyik, G.; Güncüm, E.; Işıklan, N. Design and Development of pH-Responsive Alginate-Based Nanogel Carriers for Etoposide Delivery. *Int. J. Biol. Macromol.* **2023**, *250*, 126242. [CrossRef]
55. Khan, N.R.; Wong, T.W. Microwave-Aided Skin Drug Penetration and Retention of 5-Fluorouracil-Loaded Ethosomes. *Expert Opin. Drug Deliv.* **2016**, *13*, 1209–1219. [CrossRef]
56. Su, S.; Wang, H.; Liu, X.; Wu, Y.; Nie, G. iRGD-Coupled Responsive Fluorescent Nanogel for Targeted Drug Delivery. *Biomaterials* **2013**, *34*, 3523–3533. [CrossRef] [PubMed]
57. *ISO, P. 10993-4: 2017*; Biological Evaluation of Medical Devices. Part 4: Selection of Tests for Interaction with Blood, Technical Committee ISO/TC 194, ICS: 11.100. 20. ISO/TC 194: Geneva, Switzerland, 2017.
58. Belman-Flores, C.E.; Herrera-Kao, W.; Vargas-Coronado, R.F.; May-Pat, A.; Oliva, A.I.; Rodríguez-Fuentes, N.; Vázquez-Torres, H.; Cauich-Rodríguez, J.V.; Cervantes-Uc, J.M. Synthesis and Characterization of pH Sensitive Hydrogel Nanoparticles Based on Poly(N-Isopropyl Acrylamide-Co-Methacrylic Acid). *J. Mater. Sci. Mater. Med.* **2020**, *31*, 61. [CrossRef] [PubMed]
59. Zhang, Y.; Cai, J.; Li, C.; Wei, J.; Liu, Z.; Xue, W. Effects of Thermosensitive Poly(N-Isopropylacrylamide) on Blood Coagulation. *J. Mater. Chem. B* **2016**, *4*, 3733–3749. [CrossRef] [PubMed]
60. Hudiță, A.; Radu, I.C.; Zaharia, C.; Ion, A.C.; Ginghină, O.; Gălățeanu, B.; Măruțescu, L.; Grama, F.; Tsatsakis, A.; Gurevich, L.; et al. Bio- and Hemo-Compatible Silk Fibroin PEGylated Nanocarriers for 5-Fluorouracil Chemotherapy in Colorectal Cancer: In Vitro Studies. *Pharmaceutics* **2021**, *13*, 755. [CrossRef] [PubMed]
61. Suhail, M.; Fang, C.-W.; Chiu, I.-H.; Khan, A.; Wu, Y.-C.; Lin, I.-L.; Tsai, M.-J.; Wu, P.-C. Synthesis and Evaluation of Alginate-Based Nanogels as Sustained Drug Carriers for Caffeine. *ACS Omega* **2023**, *8*, 23991–24002. [CrossRef]

62. Evans, B.C.; Nelson, C.E.; Yu, S.S.; Beavers, K.R.; Kim, A.J.; Li, H.; Nelson, H.M.; Giorgio, T.D.; Duvall, C.L. Ex Vivo Red Blood Cell Hemolysis Assay for the Evaluation of pH-Responsive Endosomolytic Agents for Cytosolic Delivery of Biomacromolecular Drugs. *J. Vis. Exp.* **2013**, *9*, e50166. [CrossRef]
63. Medical University, Sofia, Bulgaria. Available online: https://mu-sofia.bg/nauka/nauka/etika-nauchni-izsledvania/ (accessed on 20 December 2023).

gels

MDPI

Article

Nanocomposite Gels Loaded with Flurbiprofen: Characterization and Skin Permeability Assessment in Different Skin Species

Sheimah El Bejjaji [1,†], Gladys Ramos-Yacasi [2,†], Joaquim Suñer-Carbó [1,3,*], Mireia Mallandrich [1,3,*], Lara Goršek [1], Chandler Quilchez [4] and Ana Cristina Calpena [1,3]

1 Department of Pharmacy, Pharmaceutical Technology and Physical Chemistry, Faculty of Pharmacy and Food Sciences, University of Barcelona, 08028 Barcelona, Spain; sheimah.el@gmail.com (S.E.B.); gorseklara@gmail.com (L.G.); anacalpena@ub.edu (A.C.C.)
2 Facultad de Ciencias Farmacéuticas, Bioquímicas y Biotecnológicas, Universidad Católica de Santa María (UCSM), Arequipa 04001, Peru; glramos011@hotmail.es
3 Institute of Nanoscience and Nanotechnology (IN2UB), University of Barcelona, 08028 Barcelona, Spain
4 Department of Biology, University of Texas at Arlington, Arlington, TX 76019, USA; chandler.quilchez@gmail.com
* Correspondence: jsuner@ub.edu (J.S.-C.); mireia.mallandrich@ub.edu (M.M.)
† These authors contributed equally to this work.

Abstract: Nanocomposite gels consist of nanoparticles dispersed in a gel matrix. The main aim of this work was to develop nanocomposite gels for topical delivery of Flurbiprofen (FB) for humans and farm animals. Nanocomposite gels were prepared stemming from nanoparticles (NPs) freeze-dried with two different cryoprotectants, D-(+)-trehalose (NPs-TRE) and polyethylene glycol 3350 (NPs-PEG), sterilized by gamma (γ) irradiation, and gelled with Sepigel® 305. Nanocomposite gels with FB-NPs-TRE and FB-NPs-PEG were physiochemically characterized in terms of appearance, pH, morphological studies, porosity, swelling, degradation, extensibility, and rheological behavior. The drug release profile and kinetics were assessed, as well as, the ex vivo permeation of FB was assessed in human, porcine and bovine skin. In vivo studies in healthy human volunteers were tested without FB to assess the tolerance of the gels with nanoparticles. Physicochemical studies demonstrated the suitability of the gel formulations. The ex vivo skin permeation capacity of FB-NPs nanocomposite gels with different cryoprotectants allowed us to conclude that these formulations are suitable topical delivery systems for human and veterinary medicine. However, there were statistically significant differences in the permeation of each formulation depending on the skin. Results suggested that FB-NPs-PEG nanocomposite gel was most suitable for human and porcine skin, and the FB-NPs-TRE nanocomposite gel was most suitable for bovine skin.

Keywords: flurbiprofen; nanocomposite gel; nanoparticles; drug permeation; polyethylene glycol 3350; D-(+)-trehalose; human skin; bovine skin; porcine skin

Citation: El Bejjaji, S.; Ramos-Yacasi, G.; Suñer-Carbó, J.; Mallandrich, M.; Goršek, L.; Quilchez, C.; Calpena, A.C. Nanocomposite Gels Loaded with Flurbiprofen: Characterization and Skin Permeability Assessment in Different Skin Species. *Gels* **2024**, *10*, 362. https://doi.org/10.3390/gels10060362

Academic Editors: Daihua Fu and Jieyu Zhang

Received: 17 April 2024
Revised: 17 May 2024
Accepted: 21 May 2024
Published: 24 May 2024

1. Introduction

NSAIDs (non-steroidal anti-inflammatory drugs) are commonly used to manage pain, inflammation, and febrile processes in both human and veterinary patients. The ease of their accessibility is directly proportional to the increase in their consumption, and consequently, an increment in episodes of adverse gastrointestinal, cardiovascular, and renal reactions, among others. Given these circumstances, studies have assessed new routes of administration for NSAIDs that reduce the prevalence of these adverse effects [1–3].

Flurbiprofen (FB), 2-(2-fluoro-4-biphenylyl) propionic acid, inhibits the cyclooxygenase enzymes (COXs), decreasing the production of prostaglandins as all NSAIDs do, and shows a greater inhibition of the COX 1 enzyme than COX 2. This affinity for COX 1 implies that this active ingredient is more likely to be gastro and kidney injurious because it is the isoform responsible for promoting the production of protective prostaglandins in the gastric mucosal cells, kidneys, and platelets. Hence, FB is currently used to treat gout, relieve sore throat pain in the short term, suppress the onset of edema, and decrease ocular postoperative intense inflammation in both animals and humans. Also, it has been shown to be effective in preventing pain associated with chronic diseases such as osteoarthritis and rheumatoid arthritis. Moreover, Carprofen, ketoprofen, and FB are the most widely reported from 2-arylpropionic acids in veterinary studies. They are commonly prescribed for an antipyretic effect, perioperative pain, osteoarthritis, and orthopedic procedures in veterinary medicine. There is a relative lack of information about FB use in animals. Likewise, it must be taken into account that animals are more susceptible than humans to the adverse effects, thus accurate dosing is absolutely necessary [4–7]. However, the widespread use of FB oral therapy requires frequent dosing given that it has a short half-life (4 h) and therefore, is not practical [8–11]. Under these circumstances, it is interesting to assess new routes of administration for FB that reduce the prevalence of adverse effects.

In this study we raise the possibility of considering the dermal route as an alternative since it allows a rapid action of a drug with localized absorption, avoids the first pass effect, improves the drug bioavailability, and provides fewer fluctuation in plasma drug levels, and subsequently side effects are reduced [10]. Only a few substances with a specific set of criteria have this ability, such as small molecular weight, adequate ionization, water solubility, and being lipophilic [11–13].

FB is classified as class II by BCS (Biopharmaceutical Classification System). It has low solubility and high permeability; thus, its dissolution as well as its absorption into skin are a challenge. Achieving sufficient bioavailability for a water-insoluble drug in a new dosage form is generally a challenge from a medical, industrial, and scientific point of view. To overcome these hurdles, FB can be produced in polymeric nanoparticles (NPs) as a nanosuspension (NS). These carriers are known as permeation enablers and exhibit prolonged retention with minimal systemic toxicity. Moreover, to improve its conservation and prevent contamination by microorganisms, FB nanoparticles can be freeze-dried and γ-irradiated [12–18].

Pharmaceutical hydrogels are semisolid dosage forms whose topical delivery popularity is increasing due to their simple application and resistance to physiological stress, adopting the shape of the applied area by not disrupting the skin's flexion [19]. Among other properties, it also provides modulation of drug solubility and release [9,16,20]. Emphasis is often placed on nanocomposite gels, which are gel materials with nanoparticles dispersed into their structure. This combination of gel formulations and nanomaterials leads to enhanced formulations with unique physiochemical characteristics [21,22] and synergic effects [23]. That makes them suitable for diverse biomedical applications such as drug delivery systems, tissue engineering, and wound healing due to their biocompatibility and controlled release capabilities [20,24–26]. Among the diverse biomedical applications, Mostafa et al. prepared a nanocomposite formulation composed of levofloxacin carried into a chitosan and zeolite-A system [27]; Li et al. developed a nanocomposite gel dexamethasone and imidazole for treating periodontitis [28]; Moghaddam evaluated the tolerance and anti-inflammatory activity of ibuprofen loaded in liposomes embedded in a Carbopol gel [29]; and Pramanik et al. elaborated a nanocomposite hydrogel with dexamethasone for ocular delivery [30].

Sepigel 305® is a gelling agent. It is composed of polymer and surfactant that facilitates the inclusion of non-water-soluble substances. Moreover, it is pre-neutralized and is effective over a wide pH range. Sepigel 305® does not need to be pre-moistened and its formulations turn out to be lightweight with light shades, therefore, it gives a better appearance than other polymers used in gelation (Carbopol®, PemulenTM) [31,32].

In this study, we started with the synthesis of flurbiprofen nanoparticles (FB-NPs) which had already been characterized for ocular administration. FB was encapsulated in poly-ε-caprolactone, one of the most often investigated synthetic biomedical polymers due to its biodegradability, biocompatibility, and good encapsulation capacity and release, particularly of hydrophobic drugs [33,34]. These nanoparticles were synthesized with D-(+)-trehalose (TRE) and polyethylene glycol 3350 (PEG) as protectant agents [15,17].

After confirming their ability to permeate bovine, porcine, and human skin, we incorporated them into a gel for topical application. Sepigel® 305 was selected for being an excellent stabilizer and texturizing agent. These gel formulations were characterized, assessed for permeation on the three skin types, and had the role of cryoprotectant agents being investigated. A schematic illustration of the approach of this work can be seen in Figure 1.

Overall, our work focused on developing nanocomposite gel formulations for topical administration for human and farm animals, specifically bovine and porcine. As a secondary aim, we evaluated whether the cryoprotectant used during the nanoparticle elaboration impacted the nanocomposite gel features. This work is also a contribution to counteract the relative lack of studies of NSAIDs in the veterinary field.

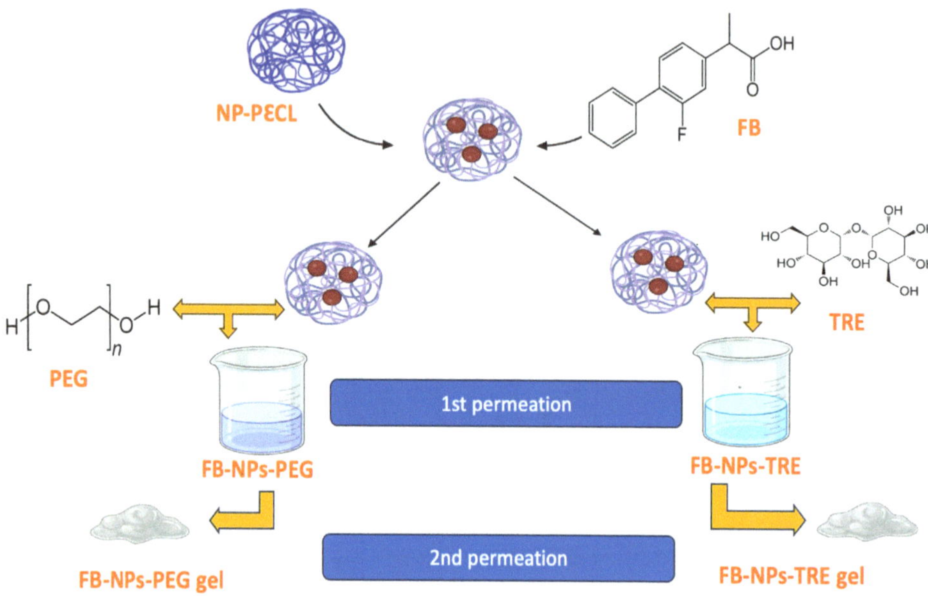

Figure 1. Schematic approach of the preparation of flurbiprofen-loaded nanocomposite gels. PεCL: poly(ε-caprolactone).

2. Results and Discussion

At the beginning of the research work, the permeation capacity of flurbiprofen-loaded polymeric nanoparticles in the skin was investigated. To this end, nanoparticles optimized by Ramos et al. [14,17] were synthesized by the solvent displacement technique, then lyophilized with either TRE or PEG, and lastly, sterilized by γ-irradiation (Section 2.1).

When the permeation capacity of the nanoparticles through human, porcine, and bovine skin was confirmed, the nanoparticles were incorporated with Sepigel® 305 to obtain a nanocomposite gel suitable for dermal delivery. The results related to the nanocomposite gels are presented from Section 2.3. onwards.

2.1. Morphological Analysis of Flurbiprofen Nanoparticles (FB-NPs)

The surface morphology and size study of the FB-NPs was carried out using Transmission Electron Microscopy (TEM). As can be seen in Figure 2, freeze-dried and sterilized nanoparticles showed slightly oval regular shapes with uniform distribution. Some grainy surface was observed in nanoparticles containing TRE, specifically after irradiation sterilization. TEM images showed sizes smaller than 200 nm and no particle aggregation phenomena were observed.

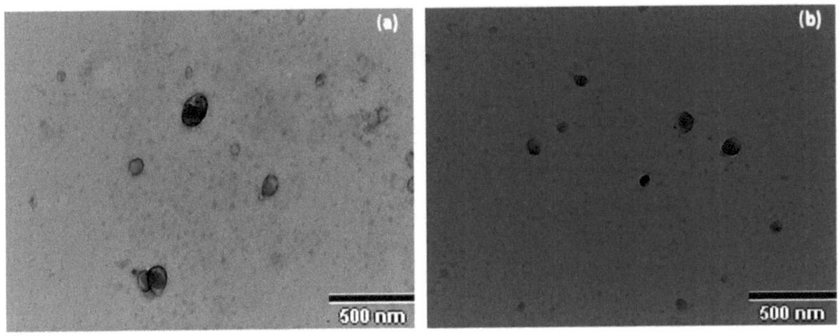

Figure 2. Morphological study of the nanoparticles: (**a**) Transmission Electron Microscopy image of NPs-PEG, (**b**) Transmission Electron Microscopy image of NPs-TRE.

2.2. Ex Vivo Permeation Studies of the Nanoparticle in Suspension

Initially, permeation studies were conducted using nanoparticle suspensions. The first stage involved assessing the permeation potential of FB in porcine, human, and bovine skin. Permeation parameters are summarized in Tables 1 and 2. Table 1 shows results from the intrinsic permeation capacity of FB in PBS solution on the three types of skin and the statistical differences. The saturated solution of FB stood out for its high permeation parameters values on bovine skin. Also, bovine skin as a barrier strength was found to be weaker than human and porcine skin for FB. Similar results were obtained by Parra et al. [35] for the carprofen permeation on ex vivo bovine skin.

Skin permeation of FB from the two nanoparticle suspensions and free drug solution with TRE and PEG added were compared, and the parameter values are summarized in Table 2.

The Kp value obtained from porcine and human formulations was higher for FB-NPs-TRE and Free drug + TRE than the formulation containing PEG, with significant statistical differences. Regarding bovine skin, NPs-TRE showed a notoriously high Kp parameter. The J values of bovine permeation obtained from all formulations were the most prominent compared to porcine and human skin data.

It is probable that TRE's features could facilitate FB permeation in human and porcine stratum corneum due to their similar surface lipids, barrier thickness, and morphological aspects [10]. In general, the higher FB permeability of bovine skin could be associated with higher follicular transport. Bovine udder skin has a greater number of hair follicles (207–338 follicles/cm^2), than human skin (~6/cm^2) and porcine skin (30–36 follicles/cm^2) [12]. Qr values indicated Free drug + TRE has greater retention of the drug. Both nanoparticle formulations showed acceptable permeation [36,37].

Table 1. FB skin permeation parameters of the intrinsic permeation, which was evaluated with a saturated solution of FB in PBS.

	FB Saturated Solution		
	Bovine	**Porcine**	**Human**
Kp × 10^2 (cm/h)	2.646 ± 0.218	0.152 ± 0.009 [a]	0.512 ± 0.041 [a,b]
J (μg/cm^2/h)	31.440 ± 2.820	1.810 ± 0.132 [a]	6.070 ± 0.525 [a,b]
TL (h)	10.0 ± 0.9	4.7 ± 0.7 [a]	5.9 ± 0.5 [a]
Qr (μg/cm^2/g)	10.346 ± 0.987	4.721 ± 0.387 [a]	3.492 ± 0.378 [a,b]

[a] difference statistically significant to Bovine; [b] difference statistically significant to porcine; Kp: Permeability coefficient; J: Flux; TL: Lag time; Qr: Retained FB amount.

Table 2. FB skin permeation parameters from irradiated freeze-dried formulations and free drug solution dissolved with TRE and PEG.

	Parameter	Nanoparticles Suspension		Solutions	
		FB-NPs-TRE	FB-NPs-PEG	Free Drug + TRE	Free Drug + PEG
Bovine	Kp × 10^2 (cm/h)	8.846 ± 0.070	1.032 ± 0.097 [a]	0.822 ± 0.043 [a,b]	1.050 ± 0.091 [a,c]
	J (µg/cm^2/h)	8.470 ± 0.420	10.320 ± 0.980 [a]	8.230 ± 0.038 [b]	10.500 ± 0.961 [a,c]
	TL (h)	4.1 ± 0.2	5.6 ± 0.3 [a]	2.5 ± 0.1 [a,b]	6.8 ± 0.3 [a,b,c]
	Qr (µg/cm^2/g)	5.346 ± 0.253	1.763 ± 0.095 [a]	2.046 ± 0.091 [a]	3.446 ± 0.161 [a,b,c]
Porcine	Kp × 10^2 (cm/h)	0.236 ± 0.016	0.039 ± 0.008 [a]	0.093 ± 0.004 [a,b]	0.047 ± 0.003 [a,c]
	J (µg/cm^2/h)	2.362 ± 0.192	0.387 ± 0.032 [a]	0.926 ± 0.042 [a,b]	0.473 ± 0.034 [a,c]
	TL (h)	1.2 ± 0.1	1.1 ± 0.1	2.7 ± 0.2 [a,b]	12.2 ± 1.1 [a,b,c]
	Qr (µg/cm^2/g)	2.208 ± 0.218	1.980 ± 0.128	10.098 ± 1.026 [a,b]	1.968 ± 0.121 [c]
Human	Kp × 10^2 (cm/h)	0.965 ± 0.068	0.516 ± 0.231 [a]	0.888 ± 0.412	0.169 ± 0.017 [a,c]
	J (µg/cm^2/h)	9.650 ± 0.821	5.156 ± 0.421 [a]	8.876 ± 0.711 [b]	1.694 ± 0.982 [a,b,c]
	TL (h)	5.5 ± 0.4	2.1 ± 0.2 [a]	4.9 ± 0.4 [b]	5.9 ± 0.4 [b,c]
	Qr (µg/cm^2/g)	1.605 ± 0.097	0.432 ± 0.038 [a]	2.439 ± 0.199 [a,b]	1.054 ± 0.096 [a,b,c]

[a]: difference statistically significant to FB-NPs-TRE; [b]: difference statistically significant to FB-NPs-PEG; [c]: difference statistically significant to Free drug + TRE solution. Kp: Permeability coefficient; J: Flux; TL: Lag time; Qr: Retained FB amount.

As nanoparticles exhibited capacity to promote the penetration of FB into the different skin species, the nanoparticles were further dispersed in a Sepigel® matrix and the resulting nanocomposite gels were evaluated, the results of which are presented in the following sections.

2.3. *Nanocomposite Gels' Physicochemical Characterization*

2.3.1. Appearance and pH Evaluation

FB-NPs-TRE and FB-NPs-PEG nanocomposite gels showed a translucent to opaque fluid appearance, white or slightly yellow (Figure 3). FB-NPs-TRE registered a pH of 4.0, while FB-NPs-PEG nanocomposite gel had a pH of 4.6.

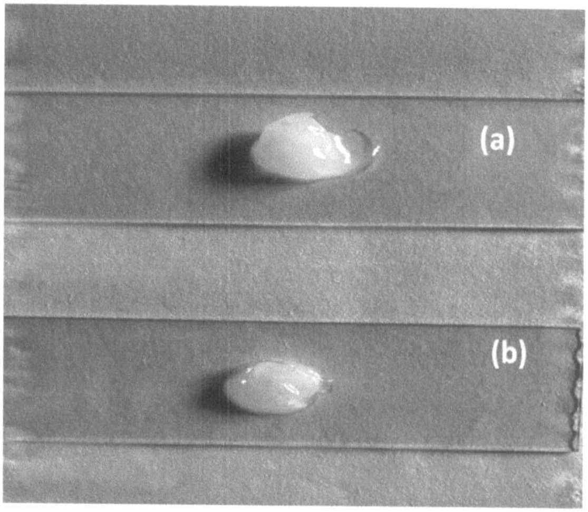

Figure 3. Physical appearance of the nanocomposite gels at room temperature: (**a**) FB-NPs-TRE gel; (**b**) FB-NPs-PEG gel.

These values are within the pH range tolerated by human skin 4.1–5.8 [36]; hence, we would not expect irritation due to the pH of the formulations. On the other hand, the pH value of porcine skin is between 5.3–7.2 and the pH value of bovine udder skin is between 6.3–7.1 [38,39]. These skins tolerate a wider range of pH well, and the formulations could be also used in the veterinary field.

2.3.2. Fourier Transform Infrared (FT-IR)

Fourier Transform Infrared (FT-IR) analysis was conducted to explore potential interactions between the drug and the gel matrix. Figure S1, in the Supplementary Material, shows the characteristic carbonyl group. FB undergoes a slight shift when interacting with the Sepigel® 305 matrix and PEG; it has a more pronounced shift when interacting with the TRE formulation. The range of the –C=O– peak is observed between approximately 1850 and 1650 cm^{-1}, as can be verified in Figure S1 [38,40,41].

2.3.3. Morphological Analysis of Nanocomposite Gels Loading Flurbiprofen

The morphological study of the nanocomposite gels was carried out using Scanning Electron Microscopy (SEM). SEM images of FB-NPs-TRE and FB-NPs-PEG nanocomposite gels are shown in Figure 4. Both nanocomposite gels showed a dense structure with no pores; however, slight differences in the gels' structure were observed when using different cryoprotectants. While FB-NPs-TRE gel depicted a smooth surface arrangement, FB-NPs-PEG appeared with a foliage-like pattern, a well-ordered structure attributable to PEG crystallization might have crystallized during solvent evaporation of the sample for SEM observation. PEG is a semicrystalline polymer and [42] is widely used in pharmaceutical and biomedical sciences because it is a biocompatible and hydrophilic compound [43]. For instance, Burdick et al. developed PEG-diacrylate hydrogels as a scaffold for bone tissue engineering [44]. Wu et al. characterized the morphology of a chitosan–PEG hydrogel for nasal delivery; the SEM images revealed a non-porous and smooth surface [45].

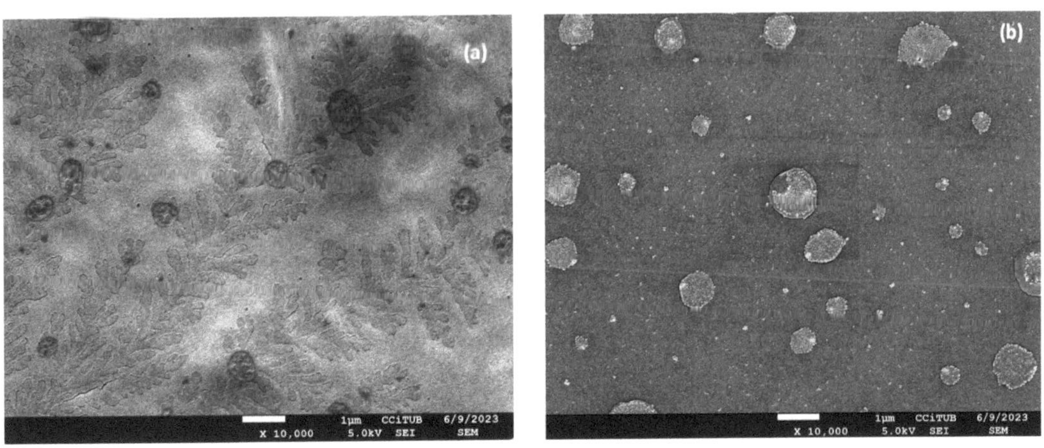

Figure 4. Morphological study of the nanocomposite gels: (**a**) Scanning Electron Microscopy image of PEG gelled with Sepigel® 305 (FB-NPs-PEG), and (**b**) Scanning Electron Microscopy image of TRE gelled with Sepigel® 305 (FB-NPs-TRE). Both images are taken at 10,000× magnification.

Moreover, the crystallization of PEG has been extensively studied. Golitsyn et al. prepared different PEG networks and investigated the formation of crystals by deep characterization of the networks. They used different techniques such as differential scanning calorimetry, NMR spectroscopy, and X-ray scattering, among others [46]. According to Bilal et al., PEG-based polymer networks show promising properties for biomedical applications. Their study revealed that molar mass had an impact on cross-links which, in turn, affected strength and stoichiometry [47]. In the same vein, Van Duong et al. investigated the microstructure of semicrystalline solid dispersions of PEG with different molecular weights.

The authors concluded that the conformation of the polymer significantly influences the microstructure of semicrystalline dispersions, impacting their stability, dissolution behavior, and pharmaceutical performance [48].

2.3.4. Porosity and Swelling Studies of the Nanocomposite Gels

Swelling and porosity studies for gels are essential for understanding and characterizing the physical and functional properties of gel materials. In pharmaceuticals, gels are often used as drug delivery systems. By studying their swelling and porosity, researchers can optimize the release kinetics of drugs from the gel matrix. Controlling the gel's ability to swell and release drugs at a specific rate is crucial for effective drug delivery [49]. The porosity of the nanocomposite gels prepared with two different cryoprotectants was similar: 39.3% for FB-NPs-PEG and 38.4% for FB-NPs-TRE. Hence, the cryoprotectants do not seem to have an impact on the porosity. Additionally, the low porosity of both formulations is in line with the dense structure observed by SEM. However, Sepigel® 305 also provides formulations with highly porous structures; Ahmadi et al. developed a Sepigel® 305 containing pranoprofen (an NSAID) encapsulated in nanostructured lipid carriers, and the authors observed a porosity of about 84% [50].

The swelling capacity was evaluated for both nanocomposite gels at three different pH values. The results show that the PEG formulation swelled the most at pH 7.4 and the least at pH 5.5 (Figure 5a). The opposite occurs with the TRE nanocomposite gel, which swelled the most at a pH of 5.5, increasing its volume almost eight times (Figure 5b). It is notable how differently the nanocomposite gels responded to various pH levels depending on which cryoprotectant (PEG or TRE) was added to the formulation.

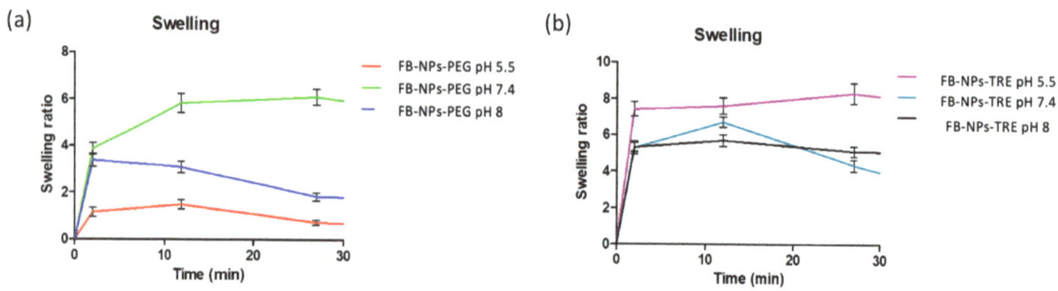

Figure 5. Swelling ratio of dried nanocomposite gels upon immersion in PBS at different pH levels (n = 3): (**a**) FB-NPs-PEG, (**b**) FB-NPs-TRE.

The results of swelling the PEG nanocomposite gel are in line with those obtained by Berenguer et al. who assessed the swelling ratio (SR) of a Sepigel® 305 loading meglumine antimoniate at a pH of 5.5 observing an SR of about 1 [32]. Similar results were obtained by Ahmadi et al. for the swelling capacity of a Sepigel® 305 loading pranoprofen nanostructured lipid carriers in a medium at pH 5.5 [50]. Altogether, this suggests that TRE increases the capacity of the formulation by up-taking solvent.

2.3.5. Degradation Studies of the Nanocomposite Gels Loading Nanoparticles

Both nanocomposite gels degraded the fastest at a basic pH and slowest at pH 5.5. Formulations containing PEG tended to exhibit higher degradation (80.34% at pH 5.5, 94.23% at pH 7.4, and 97.26% at pH 8 in 22 min.) than formulations containing TRE (70.49% at pH 5.5, 70.30% at pH 7.4, and 90.09% at pH 8 in 22 min) as demonstrated in Figure 6. Consequently, this suggests that the FB-NPs-PEG formulation is more prone to degradation under these conditions. Figure 6 presents the degradation of both nanocomposite gels. The degradation process appears to be pH-dependent.

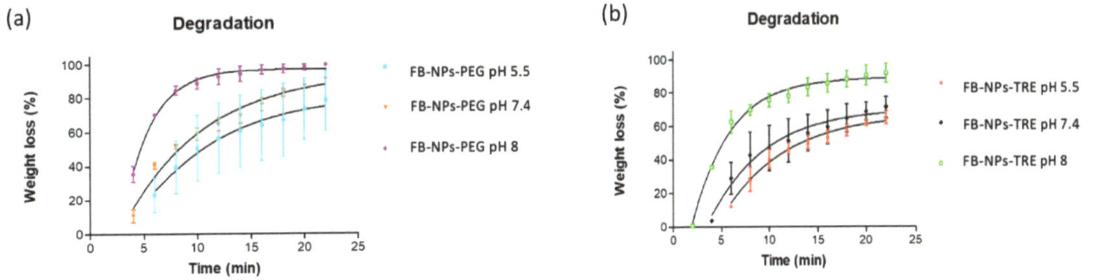

Figure 6. Degradation of the nanocomposite gels in PBS tested at varied pH levels: (**a**) FB-NPs-PEG, and (**b**) FB-NPs-TRE.

Our results are in line with those observed by Berenguer et al. of the degradation of a Sepigel® 305 formulation with meglumine antimoniate. The formulation was evaluated for degradation at pH 5.5, and the authors observed that about 89% of the gel degraded in 20 min [32].

2.3.6. Extensibility Studies of the Nanocomposite Gels

Extensibility provides information about how the formulation spreads after weight is applied. The nanocomposite gels hyperbolic model (Figure 7) results demonstrate an elevated extensibility capacity for the PEG formulation compared to the TRE formulation, as the former covered an area of 100 cm^2 while the latter covered less than 20 cm^2. This could be due to the higher viscosity exhibited by the PEG formulation (Section 2.3.7). It is important to assess the extensibility of topical formulations since the ease of spreading helps to apply the formulation uniformly to the skin [51,52] using gentle movements and preventing the need to add pressure to the inflamed skin.

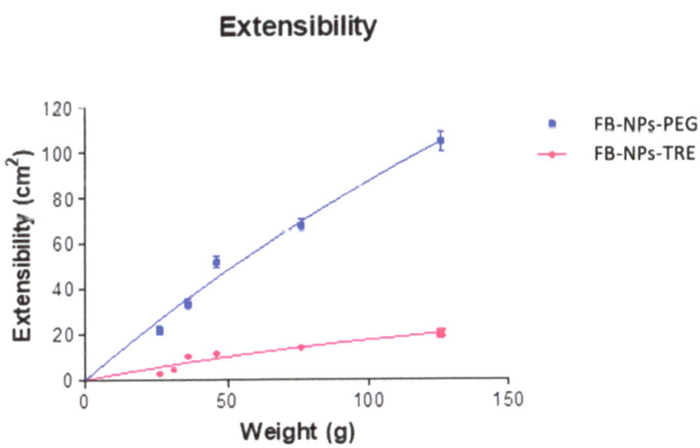

Figure 7. Extensibility FB-NPs-TRE and FB-NPs-PEG formulations.

2.3.7. Rheological Study of the Nanocomposite Gels

Rheological nanocomposite gel measurements showed that formulations displayed a non-Newtonian behavior. The formulations exhibited pseudoplastic flow and shear thinning behaviors since the viscosity decreased with an increase in the shear rate from 0.1 to 100 s^{-1} (Figure 8). The mathematical model that best fit the experimental data was the Cross equation which describes a general model for pseudoplastic materials (Equation (1)):

$$\tau = \dot{\gamma} \cdot (\eta_\infty + (\eta_0 - \eta_\infty))/(1 + (\dot{\gamma}/\dot{\gamma}_0)^n) \quad (1)$$

where τ is the shear stress (Pa), $\dot{\gamma}$ is the shear rate (1/s), $\dot{\gamma}_0$ is the zero shear rate (1/s), η_0 is the zero shear rate viscosity (Pa·s), η_∞ is the infinity shear rate viscosity (Pa·s), n is the flow index.

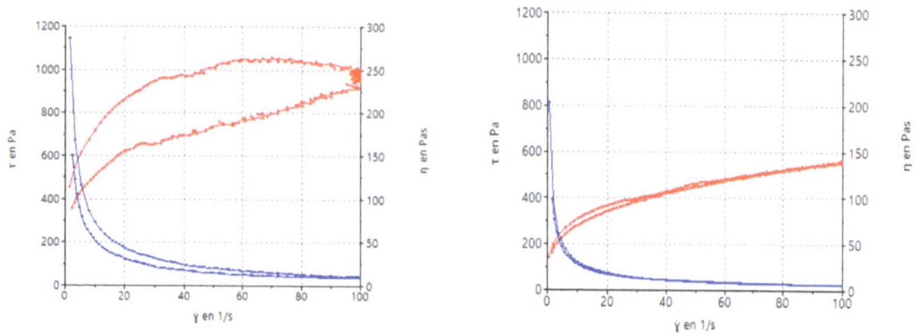

(**a**) (**b**)

Figure 8. Rheograms of FB-NPs nanocomposite gels: (**a**) FB-NPs-PEG; (**b**) FB-NPs-TRE. The red line corresponds to the flow curve and the blue line represents the viscosity curve.

Concerning viscosity measurements, at 100 s^{-1}, the addition of TRE resulted in a viscosity value of 9.82 ± 0.21 Pa·s and the formulation with PEG showed a lower viscosity 5.60 ± 0.03 Pa·s than the TRE one.

Pseudoplastic behavior is important due to its ability to facilitate a smooth and effortless application without requiring excessive pressure, making the process painless for inflamed skin. The pseudoplastic behavior from the galenic point of view is noteworthy because the formulation has to retain its consistency during the storage of the product [53].

Additionally, the formulation with PEG (Figure 8a) shows thixotropy, presenting a hysteresis loop that indicates a dependence on viscosity over time.

2.4. Drug-Release Kinetics of the Nanocomposite Gels

In vitro release studies of FB and the nanoparticles in suspension were published by Ramos et al. [14,17]. The release studies of the nanocomposite gels were executed by Franz diffusion cells to measure the drug release using a dialysis membrane at a cutaneous temperature (32 °C), this gave the cumulative amount of FB released as a function of time [17,20,35,36], depicted in Figure 9.

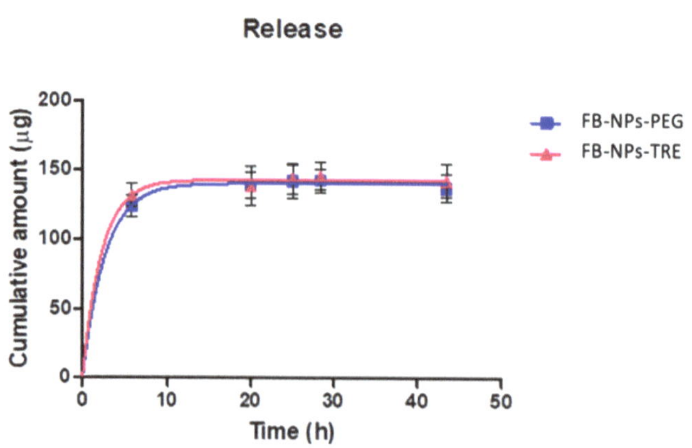

Figure 9. Cumulative amount of FB released from the nanocomposite gel with either PEG or TRE through a dialysis membrane (n = 5 each). The in vitro release test was conducted at a pH of 7.4 using PBS as the receptor medium.

Flurbiprofen was rapidly released from the matrix for both formulations according to a one-phase exponential association model. This model was selected among Higuchi, Korsmeyer-Peppas, and Weibull based on the Akaike Information Criterion (AIC), which considers the number of parameters of the models, penalizing complex models to avoid overfitting. Lower AIC values indicate a better fit. Table S1 reports the results of the kinetic modelling. One-phase exponential association model describes the in vitro drug release from the topical formulation as a process where the drug release rate is dependent on the concentration of the drug remaining in the system, resulting in a fast drug release at early times and decreasing the release rate as the drug depletes. The Korsmeyer-Peppas model is often used to analyze the mechanism of drug release kinetics from systems. The n exponential (Table S1) provides information about the release kinetics, values below 0.45 indicate that the mechanism of the drug release is predominantly controlled by Fickian diffusion, where the drug molecules diffuse through the matrix proportionally to the concentration gradient [54], which is consistent with the one-phase exponential association model.

Table 3 shows the results of fitting one-phase exponential association obtained for FB-NPs-PEG and FB-NPs-PEG nanocomposite gels.

Table 3. Parameter values of the in vitro drug release data fitted one-phase exponential model and the goodness of fit. Amax: maximum cumulative amount released estimated by the model; K: release rate.

Parameter	FB-NPs-PEG	FB-NPs-TRE	*p*-Value
Best-fit values			
Amax (μg)	142.3	146.9	0.1494
K (h^{-1})	0.3615	0.3961	0.6889
Half-life (h)	1.917	1.750	-
Standard error			
Amax	1.6	2.4	-
K	0.0410	0.0726	-
R^2	0.9947	0.9889	-

Both nanocomposite gels presented similar release profiles and similar values for the kinetic parameters, suggesting that the use of cryoprotectant either PEG or TRE does not impact the release of FB. However, when the release of FB from the nanoparticles was investigated, results revealed that the nanoparticles with TRE as cryoprotectant exhibited a higher release of FB than the nanoparticles with PEG (Figure S2), thus refuting the assumption that the cryoprotectant did not impact the drug release, but rather the opposite. Our results are in line with previous observations. Ramos et al. assessed the release of FB from the lyophilized and irradiated nanoparticles and observed a slower release rate for the formulation containing PEG as the cryoprotectant. The authors concluded that on one hand, an increase in viscosity in the medium, caused by PEG, could slow down the drug release process and, on the other hand, an apparent increase in the nanoparticles' porosity caused by TRE might cause the differences in drug release [17]. When the nanoparticles were incorporated into the Sepigel® 305, the differences in the release rate vanished, possibly due to the increase in viscosity for both formulations leading to similar drug release profiles. Table S2 reports the kinetics of nanoparticles. According to the determination coefficient, the one-phase exponential association model and Weibull model fitted the release data well; actually, the Weibull model had slightly higher R^2 values, yet showed higher AIC values; therefore, the one-phase exponential model was selected as the simplest model that best described the release of FB from the nanoparticles.

Other researchers have also investigated the incorporation of nanoparticles in gel formulations; for instance, Abrantes et al. encapsulated mosquito-repellent ingredients in PCL nanoparticles which were further dispersed in a poloxamer-based hydrogel and the formulations were tested for drug release. The authors observed remarkably higher drug release for IR3535 from the nanoparticles than from the nanocomposite gel, whereas slight differences were observed when assessing geraniol from the nanoparticles and the nanocomposite gels. However, similar profiles were observed in the permeation study [55]. Bini et al. prepared nanocomposite formulations in which nanoparticles were embedded in gelatin gels. They assessed the release of curcumin from the nanoparticles and the nanoparticles dispersed in the gelatin gel; a lower release was observed for the nanocomposite gel. The researchers also evaluated the release of sodium naproxen from the nanocomposite gel, which was slower compared to the free drug [56]. Momekova et al. assessed the drug release of cannabidiol loaded in polymeric micelles which had been vehiculized in a Hydroxyethyl Cellulose gel. The authors observed a sustained release of cannabidiol from the nanocomposite cryogel with respect to conventional gel [57].

2.5. *Ex vivo Permeation of Nanocomposite Gels in Bovine, Porcine, and Human Skin*

Similar to the in vitro drug release study, the amount of FB that was capable of permeating through the porcine, human, and bovine skin was evaluated by Franz diffusion cells. Table 4 shows the results obtained for the permeability coefficient, flux, lag-time, and amount of FB retained in the skin after 24 h of exposure to the nanocomposite gels. There are significant differences between gels with different cryoprotectants for each species. The formulation with TRE shows better permeability in bovine skin, while the formulation with PEG shows better permeation in porcine and human skin. While the retained amount of FB (Qr) in human skin is similar in both formulations, the retained amounts of FB in bovine and porcine skin show the same trend as the permeability studies.

Table 4. Permeation parameters for FB from the nanocomposite gels, with either TRE or PEG as cryoprotectant for the lyophilization of the nanoparticles.

	Parameter	FB-NPs-TRE Gel	FB-NPs-PEG Gel
Bovine	$Kp \times 10^2$ (cm/h)	0.400 ± 0.040	0.100 ± 0.012 [a]
	J (μg/cm^2/h)	4.070 ± 0.440	1.100 ± 0.120 [a]
	TL (h)	Not applicable	2.3 ± 0.2 [a]
	Qr (μg/cm^2/g)	1.864 ± 0.182	0.893 ± 0.092 [a]
Porcine	$Kp \times 10^2$ (cm/h)	0.200 ± 0.020	0.300 ± 0.031 [a]
	J (μg/cm^2/h)	1.790 ± 0.200	2.840 ± 3.100 [a]
	TL (h)	13.3 ± 1.3	8.3 ± 0.8 [a]
	Qr (μg/cm^2/g)	1.269 ± 0.118	3.866 ± 0.038 [a]
Human	$Kp \times 10^2$ (cm/h)	0.045 ± 0.004	0.220 ± 0.022 [a]
	J (μg/cm^2/h)	0.454 ± 0.040	2.030 ± 0.220 [a]
	TL (h)	12.4 ± 1.3	5.6 ± 0.6 [a]
	Qr (μg/cm^2/g)	0.636 ± 0.060	0.653 ± 0.068

[a]: statistical significance $p < 0.0001$; Kp: Permeability coefficient; J: Flux; TL: Lag time; Qr: Retained FB amount.

When comparing the permeation of FB in nanocomposite gels to that of FB in the nanoparticles (Table 2), it is noticeable that the Sepigel® 305 matrix modulates the permeation in both directions. It either promotes the permeation, as observed in porcine skin for FB-NPs-PEG gel, or constrains it, as seen in human skin for FB-NPs-TRE gel. It is known that drug diffusion across a biological membrane depends not only on the physicochemical properties of the drug but also on how the formulation interacts with the skin. This is characterized by the partition coefficient [52]. Considering that skin from different species may possess distinct attributes, these differences may influence the partition coefficient and consequently contribute to variations in permeation between skin types.

Ex vivo models provide close insights into drug permeation behavior before testing in humans. This allows researchers to evaluate how dosage forms interact with skin and biological membranes, and to select and optimize formulations that save time and costs in further clinical studies. Ternullo et al. developed hydrogels loading curcumin in deformable liposomes with different surface charges, which were tested on ex vivo human skin. Results showed a slightly higher permeation of curcumin from the hydrogel with neutral-charged liposomes compared to the curcumin hydrogel. Higher amounts of the drug retained in the skin were also observed from the hydrogel containing liposomes with respect to the curcumin hydrogel [58]. Shebata et al. evaluated the performance of gels containing insulin niosomes through rat skin. The in vitro drug release study revealed that the gels released curcumin following a Higuchi model and the permeation of insulin was much superior from the noisome gels than from the conventional gels. The results from the ex vivo permeation test were consistent with those from the in vivo in rats [59]. Khan et al. compared the permeation of ketoconazole loaded in nanoparticles and vehiculized in a Carbopol-based gel to ketoconazole nanoparticles and a dispersion of the drug. Nanoparticles were the formulation that most permeated through rat skin, followed by the nanostructured hydrogel and finally the dispersion. The authors conclude that the nanostructured hydrogel was a potential candidate for topical delivery since the formulation exhibited significant activity in the in vitro antifungal study [60].

After the permeation study, the amount of FB that remained in the skin discs was extracted. The highest amount of FB was recovered from the porcine skin when exposed to the nanocomposite gel composed of FB-NPs-PEG. The cryoprotectant used in the preparation of the nanoparticles affected the amount of drug retained in the skin. The same donor skin saw a threefold decrease in the amount of drug retained when exposed to the nanocomposite gel composed of nanoparticles with TRE as cryoprotectant. Opposite results were observed in bovine skin samples. The human skin showed the lowest amount of drug retained within the tissue with no significant differences regarding the cryoprotectant used.

2.6. Evaluation of the Biomechanical Properties of Skin

Skin hydration is essential for maintaining overall skin health. Adequately hydrated skin is softer, smoother, and more flexible [61], while dehydrated skin is more prone to irritation and inflammation, which can be exacerbated by the application of topical products [62]. Additionally, the skin's hydration level may have an impact on the percutaneous absorption of topical products [63].

Stratum Corneum Hydration (SCH) and Trans-Epidermal Water Loss (TEWL) were measured to characterize the biomechanical properties of both nanocomposite gels without active ingredient (NPs-PEG and. NPs-TRE) in healthy humans (Figure 10). Both formulations presented comparable results.

At the time 0 (prior to the formulation's application) the skins show a basal hydration between 30–40 AU; this is considered dry skin [64]. The Corneometer initially measured a decrease in the SCH value for both formulations (Figure 10a,b). This initial decrease could be due to the nanocomposite gel absorbing water from the skin because of its swelling capacity; this phenomenon was more remarkable in the nanocomposite gel formulated with TRE, as evidenced in Figure 6b at pH 5.5. NPs-PEG formulation showed statistically significant differences from the baseline values at all times (Figure 10a). Within the first hour, the hydration values decreased, but then, gradually increased, trending towards the baseline value. Despite this, even at 4 h, a statistically significant difference was observed ($p < 0.01$). In contrast, with the NPs-TRE formulation (Figure 10b), the decrease was observed only in the first 15 min showing a recovery afterwards. Compared to the baseline, statistically significant differences were observed during the first 30 min, but were no longer present beyond one hour; suggesting that baseline hydration had been restored. Other researchers have prepared Sepigel®-based formulations and evaluated the impact of the formulation on skin hydration. Berenguer et al. developed two gel formulations for leishmaniasis treatment from a topical approach. One gel contained Amphotericin B [65] and the other gel was loaded with meglumine antimoniate (MA) [32]. Interestingly, the MA Sepigel showed a similar behavior on skin hydration to FB-NPs-TRE, a decrease in the hydration levels was observed at 15 min post-application followed by a gradual recovery of the SCH values [32], while the Sepigel loaded with Amphotericin B showed the opposite effect; it increased the SCH values at 15 min post-application, and afterwards the hydration returned to the basal levels [65]. In another work, Ahmadi et al. prepared a Sepigel-based nanocomposite gel loading nanostructured lipid carrier of pranoprofen. The authors monitored changes in the hydration of ear skin in mice. The results showed an increase in SCH levels after the application of the nanocomposite gel [50]. In all cases, the Sepigel-based formulations were well tolerated and the authors did not observe any signs of irritation during the tolerance study [32,50,65]. TEWL relates to skin integrity; values below 15 $g/h/m^2$ indicate that the barrier function of the skin is in good condition [66,67]. When the nanocomposite gels were applied, TEWL values did not increase (Figure 10c,d), signifying that the nanocomposite gels did not disrupt the stratum corneum, and therefore were well tolerated by the skin. Besides the skin integrity that remained intact during the experiments, no visible skin irritation or alterations were observed. No significant statistical differences were observed between the two gels when comparing all time points to time 0. Our results are in line with those obtained by Berenguer et al. [32]; the authors evaluated an MA Sepigel and no statistically significant differences were found after applying the formulation with respect to the basal values. Evaluating TEWL is a useful tool to detect irritant products for the skin because an increase in TEWL values when using a topical product may indicate skin damage since the increase of TEWL is proportional to the skin barrier impairment [68]. The assessment of the biomechanical properties of the skin led to the conclusion that the nanocomposite gels were well tolerated and would not disrupt the skin barrier function.

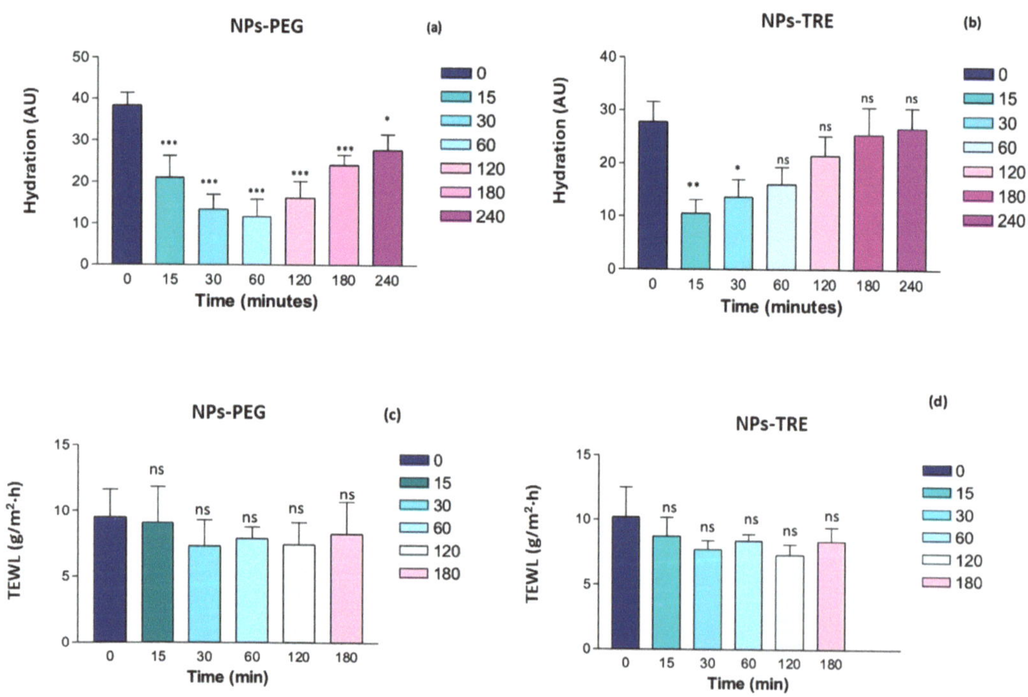

Figure 10. Monitoring of the skin hydration (**a**,**b**) and barrier function integrity (**c**,**d**) after the application of NPs-PEG and NPs-TRE (gels with and drug-free nanoparticles lyophilized with the cryoprotectants, and γ irradiated) on healthy human skin. *: $p < 0.01$; **: $p < 0.001$; ***: $p < 0.0001$; ns: non-significant statistical differences.

3. Conclusions

Two nanocomposite gels have been developed and characterized. The formulations were prepared stemming from polymeric nanoparticles loading flurbiprofen. These nanoparticles were lyophilized using two different cryoprotectants, and sterilized by γ-irradiation in previous work by our research group. First, the permeation capacity of the nanoparticles was assessed through three skin species: human, porcine, and bovine. Once it was clear that nanoparticles were a suitable carrier for FB topical delivery, they were incorporated in Sepigel® resulting in nanocomposite gels, which were physicochemically characterized and biopharmaceutically evaluated. Both nanocomposite gels exhibited similar porosity and pH-dependent degradation patterns, showing a higher degradation in alkaline medium, whereas the swelling behavior was different between the two nanocomposite gels, the highest swelling capacity was observed at pH 8 for FB-NPs-TRE, while FB-NPs-PEG swelled most at the physiological pH. When analyzing the extensibility of the nanocomposite gels, FB-NPs-PEG showed five-fold higher spreadability than FB-NPs-TRE. FB-NPs-PEG nanocomposite gel also showed higher viscosity values and thixotropy, suggesting that these parameters were affected by the different cryoprotectants used in the lyophilization of the nanoparticles. The release study revealed that the incorporation of the nanoparticles in Sepigel® modulated the release of FB, minimizing the differences observed in the release of FB from the nanoparticles. Yet, when the nanocomposite gels were tested on skin from three different species, it was observed that FB-NPs-PEG nanocomposite gel was most suitable for human and porcine skin, and FB-NPs-TRE nanocomposite gel formulation was most suitable for bovine skin. Both nanocomposite gels had an initial drying effect on the skin. However, the formulation with TRE as cryoprotectant tended to revert the dehydration at earlier times than the formulation with PEG. Despite this effect, the nanocomposite gels were well tolerated since no signs of irritation were observed and no statistically significant changes were observed on TEWL values with respect to the basal values. In conclusion, the formulations may contribute to increasing the human and veterinary medicinal products available for the management the inflammation in skin disorders.

4. Materials and Methods

4.1. Chemicals and Reagents

Flurbiprofen, Poly(ε-caprolactone) with a molecular weight (Mw) ~14,000 g/mol and a number-average molecular weight (Mn) of ~10,000 g/mol and dispersity of 1.4, PEG-3350, D-(+)-trehalose and Acetone were purchased from Sigma-Aldrich Co. (St. Louis, MO, USA). The Poloxamer 188 (P188; Lutrol® F68) used was purchased from BASF (Barcelona, Spain). Sepigel® 305 (Polyacrylamide, C13-14 Isoparaffin Laureth-7) was purchased from Acofarma (Barcelona, Spain). Phosphate-buffered saline tablets (PBS) were acquired from Sigma-Aldrich Chemie (Steinheim, Germany) and processed according to the manufacturer's specifications and then refrigerated for later use, ensuring optimal storage conditions. The double distilled water was filtered using the Millipore® system (EMD Millipore, Billerica, MA, USA). The chemicals and reagents used for high-performance (HPLC) were purchased from Fisher Scientific (Leicestershire, UK).

4.2. NPs Preparation

Flurbiprofen nanoparticles were developed by Ramos et al. in previous works [10,11]. Those previous studies on flurbiprofen polymeric nanoparticles, which were prepared by solvent displacement technique, provided specific materials and processes employed to produce two highly reliable formulations [12,38]. Ramos et al. used 1 mg/mL of drug, 3.3 mg/mL of PεCL, 16.6 mg/mL of P188 and 100 mg/mL of TRE; and 1 mg/mL of drug, 3.3 mg/mL of PεCL, 35.0 mg/mL of P188 and 160 mg/mL of PEG to produce two nanoparticle suspensions. TRE and PEG were utilized as stabilizers to support the morphometrical property conservation of the nanoparticles during freeze-drying. They were selected from factorial designs according to small particle size (Zav), a low polydispersity index (PI), a great entrapment efficiency (EE) and an appropriate zeta potential (ZP) after re-dispersion and the right conditions in the freezing and drying stages [12].

The resultant nanoparticles of the study of Ramos et al. [14] were sterilized for future studies using a γ ray dose of 25 KGy to get NPs-TRE and NPs-PEG, the irradiated formulation prepared with TRE and PEG, respectively. Likewise, they described the physicochemical characterization of sterilized flurbiprofen nanoparticles within the following values: NPs-TRE showed 187.5 $\pm$ 1.2 nm, 0.131 $\pm$ 0.015, 86.0 $\pm$ 0.2% and $-$13.20 $\pm$ 0.17 mV of Zav, PI, EE and ZP, respectively and NPs-PEG showed 192.5 $\pm$ 2.0 nm, 0.091 $\pm$ 0.028, 85.1 $\pm$ 1.0% and $-$15.30 $\pm$ 0.37 mV of Zav, PI, EE and ZP, respectively [14,17].

4.3. *Morphological Analysis of the Nanoparticles in Suspension*

A transmission electron microscope (TEM) JEOL 1010 (JEOL Inc., Peabody, MA, USA) was used to further assess the morphology and particle size of the optimized FB-NPs using 40,000 to 60,000$\times$ magnification. A sample drop (without previous dilution) was placed onto a copper TEM grid coated with carbon film and negative stained with uranyl acetate solution (1%, *w*/*v*). The grids were left to dry at room temperature.

4.4. *Biological Tissue for Ex Vivo Permeation Study*

Human, bovine, and porcine skin were used for permeation study. The study protocol was approved by the Bioethics Committee of the Barcelona SCIAS Hospital in Spain with reference number 0012016. Fresh samples of udder skin from healthy Holstein Frisian bovines that had been legally butchered were collected at a nearby slaughterhouse in Barcelona, Spain. The Yorkshire-Landrace pigs' flank skin was collected from the animal facility at the Bellvitge Campus of University of Barcelona (Barcelona, Spain) right after the animals had been sacrificed for various reasons. The studies were carried out in accordance with protocol that was approved by the Committee of Animal Experimentation of the Regional Autonomous Government of Catalonia (Spain) and the Animal Experimentation Ethics Committee of the University of Barcelona (Barcelona, Spain) with number 7428. The skin samples were sliced using a dermatome GA 630 (Aesculap, Tuttlingen, Germany) at varied thicknesses depending on the type of skin—400, 700, and 1000 µm for human, porcine, and bovine udder skin after being frozen to a temperature of $-$20 °C [35].

4.5. *Ex Vivo Permeation Profile Analysis of the Nanoparticles*

The experiments were conducted as described in Section 2.2., in independent vertical Franz diffusion cells with a diffusional surface area of 0.64 cm^2. Biological tissues were positioned between the two compartments of a Franz cell with the dermal side in contact with the receptor medium and the epidermis side in contact with the donor chamber, covered with a laboratory film (Parafilm®, Chicago, IL, USA) to prevent evaporation during the study. Phosphate-buffered saline (PBS) solution at pH 7.4 was used as the receptor medium. The permeation study was conducted for 24 h at 32 $\pm$ 0.5 °C under continuous stirring, keeping sink conditions throughout the tests to avoid the medium being saturated. For the donor compartment, 500 µL of the test formulations were applied once the temperature of the skin surface had equilibrated to 32 $\pm$ 0.5 °C [35]. A saturated solution of FB in PBS was also assayed. At each sampling interval up to 24 h, a volume of 300 µL was withdrawn and an equal volume of fresh PBS solution was added. Samples were analyzed in triplicate by RP-HPLC for the cumulative amount of drug permeated. Figure 11 summarizes the permeation studies on different skin types.

Permeation parameters such as permeability coefficient (Kp, cm/h), flux (J, µg/cm^2/h) and lag time (TL, h) were calculated by linear regression analysis using the GraphPad Prism® software v. 5.0 (GraphPad Software Inc., San Diego, CA, USA) and Laplace software (Scientist 2.01, Micromath. Inc., Salt Lake City, UT, USA) [39]. FB retained was extracted with acetonitrile/water (50:50, *v*:*v*) under sonication for 15 min using an ultrasound bath. Non-exposed skin around the diffusion area was removed prior to assay. The resulting FB solutions were determined by RP-HPLC described in Section 4.9, yielding to the amount of FB extracted from the skin Qr (μg/cm^2/g). The results are reported as the median and range of six replicates (n = 6).

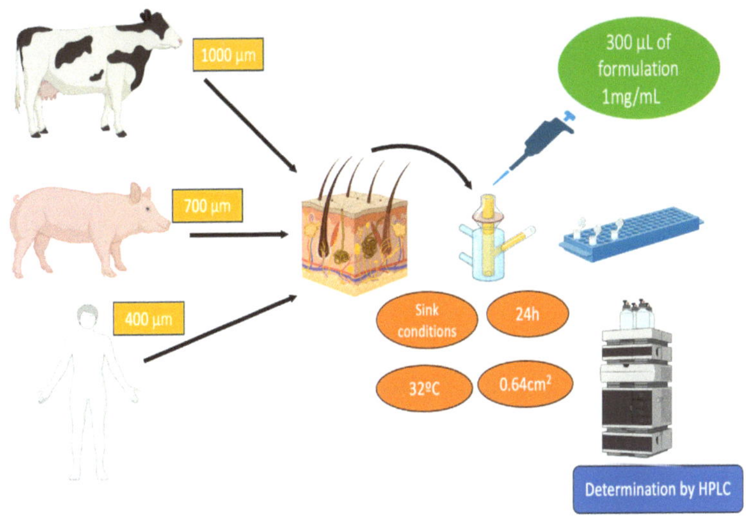

Figure 11. Scheme of permeation studies carried out on the three skin samples.

The permeability coefficient (*Kp*, cm/h) was determined as the ratio between the flux and the formulation's concentration (Equation (2)):

$$K_p - \frac{J}{C_0}, \quad (2)$$

where J (μg/cm^2/h) is the flux across the skin sample and C_0 (μg/cm^3) is the drug concentration in the formulation applied to the donor compartment.

Starting from t amount of drug extracted from the tissues, the amount of drug retained in the skin discs (Q_R (μg/cm^2/g) according to the following formula:

$$Q_R = \frac{\left(\frac{Ex}{Px}\right) \times R}{A \times 100}, \quad (3)$$

where Ex (μg) is the quantity of drug extracted, Px (g) the weight of the skin discs that have been permeated, A (cm^2) is the effective surface area accessible for diffusion, and R the recovery percentage of the drug as outlined previously [50,69].

4.6. Preparation and Characterization of the Gels Loading Flurbiprofen Nanoparticles

NPs-TRE and NPs-PEG lyophilized and irradiated were aqueously reconstituted to prepare two nanocomposite gels. To complete the gelation process, 0.55 g of Sepigel® 305 was added to 5 mL of each nanoparticle in suspension and agitated to ensure a homogeneous mixture. This led to the formation of thin yet independent semisolid nanocomposite gel structures.

4.6.1. pH Measurements

The pH determination was measured using a pH meter Micro-pH 2000 (Crison Instruments S.A., Alella, Spain). Measurements were conducted in triplicate at room temperature with the recently prepared nanocomposite gels, and repeated two months later.

4.6.2. Fourier Transform Infrared (FT-IR)

The FB-NPs-TRE and FB-NPs-PEG nanocomposite gels, and FLB samples were examined with Fourier Transform Infrared Spectroscopy (FT-IR). Before the measurements, the samples were dried in an oven at 55 °C. A Nicolet iZ10 spectrometer (Thermo Scientific, Waltham, MA, USA) was used to obtain the FT-IR spectra. With a DTGS detector and a spectral resolution of 4 cm^{-1}, the measurements were carried out in the 4000–525 cm^{-1} range, yielding 32 scans per spectrum. Attenuated total reflectance (ATR) was used to record the spectra using a diamond crystal [39,70].

4.6.3. Morphological Analysis of the Nanocomposite Gels

The nanocomposite gels' microstructure was evaluated by Scanning Electron Microscopy (SEM). When using nonconductive samples, the sample must be dried and covered in carbon or metal. For this reason, the nanocomposite gels were dried in an oven at 55 °C, coated with a thin layer of gold, and examined using a JSM-7001F (JEOL, Inc., Peabody, MA, USA). A small amount of material was deposited on a glass coverslip, quickly immersed in absolute ethanol, and dried using the critical point technique (replacing ethanol with CO_2). Then the coverslips were mounted on the microscope slides and coated with a thin layer of gold to improve their electrical conductivity.

4.6.4. Porosity and Swelling Studies

FB-NPs-TRE and FB-NPs-PEG nanocomposite gels were added to vials for porosity experimentation and placed in an oven at 55 °C to dry out until a constant weight was achieved.

To evaluate the porosity of the nanocomposite gels, ethanol was added to a weighted amount of dried nanocomposite gel. The sample was shaken by hand and added to a bath at a cutaneous temperature (32 °C) for 2 min. After the bath, the vials were centrifuged at 3000 revolutions per minute (rpm) for 3 min. Excess liquid was removed from the sample and the remaining nanocomposite gel was weighed and recorded. The experiment concluded when there was no longer a weight increase; either the weight remained consistent or decreased. At this point, the porosity was determined using Equation (4):

$$\mathrm{P} = \frac{Ws - Wd}{\rho - Vs} \tag{4}$$

W_S is the weight of a swollen nanocomposite gel, Wd is the weight of the dried nanocomposite gel, ρ is the density of ethanol, and Vs is the volume of the swollen nanocomposite gel determined by a pycnometer (Vidra Foc, Barcelona, Spain) [50].

FB-NPs-TRE and FB-NPs-PEG nanocomposite gels were added to vials for swelling experimentation and placed in an oven at 55 °C to dry out until a constant weight. The experiment was conducted by adding solutions of 0.5 mL of PBS with pHs of 5.5, 7.4, and 8.0 to dried PEG and TRE nanocomposite gels. The same as in the porosity experiment, samples were shaken by hand before being added to a 32 °C bath for 2 min, then centrifuged at 3000 rpm for 3 min. The supernatant PBS was removed, and the weight was noted to quantify the amount of solvent absorbed by the nanocomposite gel. This process was repeated until the amount of liquid absorbed by the nanocomposite gels over set intervals of time was constant. Equation (5) was used to obtain the Swelling (SR):

$$\mathrm{SR} = \frac{Ws - Wd}{Wd} \tag{5}$$

Ws is the weight of a swollen nanocomposite gel and Wd is the weight of a dried nanocomposite gel [50,69].

4.6.5. Degradation Studies

The experiment was conducted by adding 0.5 mL of PBS with pH's of 5.5, 7.4, and 8 to fresh PEG and TRE nanocomposite gels, and the weight loss as a time function was recorded. The process was the same as in porosity and swelling experimentation. Excess PBS was removed every time after centrifugation and the nanocomposite gels weight was observed over a few hours until it was constant or completely degraded [24]. Degradation was calculated according to Equation (6):

$$WL\ (\%) = \frac{Wi - Wd}{Wi} \times 100\% \quad (6)$$

where *WL* is weight loss, *Wi* is the initial mass (weight), *Wd* is the weight of the gel at each interval [50,71].

4.6.6. Extensibility Studies

A volume of 30 µL of FB-NPs-TRE and FB-NPs-PEG nanocomposite gel samples were placed between two glass plates. The formulations' surface areas were then measured when various weights (5 g, 10 g, 20 g, 50 g, and 100 g) were placed to the upper plate with a weight of 26 g. The weights were removed after 60 s and the diameter of the spread was recorded. At room temperature, each sample was measured by triplicate for each weight [38,72]. The increased spreading areas were plotted as a function of the increasing weights applied. The extensibility was obtained with Equation (7):

$$Extensibility = \frac{\pi \times d^2}{4} \quad (7)$$

d represents the mean diameter assessed across various orientations.

Different mathematical functions were evaluated using GraphPad Prism® software v. 5.0 (GraphPad Software Inc., San Diego, CA, USA), following with the major R^2 value.

4.6.7. Rheological Behavior

Rheological rotational measurements were performed using a Haake Rheostress® 1 rheometer (Thermo Fisher Scientific, Karlsruhe, Germany) connected to a thermostatic circulator Thermo® Haake Phoenix II and a computer PC provided with Haake Rheowin® Job Manager v. 4.91 software. Haake Rheowin® Data Manager v. 4.91 software (Thermo Electron Corporation, Karlsruhe, Germany) was used to perform the analyses of the obtained data. Steady-state measurements were performed using a cone (Haake C60/2° Ti, 60 mm diameter, 2° angle)-and-plate geometry (0.105 mm gap). To obtain viscosity curves ($\eta = f(\gamma)$) and flow curves ($\tau = f(\dot{\gamma})$) of nanocomposite gels, continuous rotational testing were recorded in triplicate at 25 ± 0.2 °C 24 h after preparation in the shear rate range of 0.1–100 s^{-1}. The shear rate ramp program included a 180 s ramp-up period from 0 to 100 s^{-1} (ascendant curve), 60 s constant shear rate period at 100 s^{-1}, and finally a 180 s ramp-down period from 100 to 0 s^{-1} (descendent curve). Both curves collected 100 data points. In order to describe the flow curves, five most commonly models were applied. Mathematical models were used for fitting data from the flow curves; the models included in the fitting were Bingham, Ostwald de Waele, Herschel-Bulkley, Casson, and Cross. The one which best described the rheological profile was selected on the basis of the correlation coefficient value (r) and chi-square value. The hysteresis loop area (S_R), known as apparent thixotropy (Pa/s), was determined to assess the microstructure disturbance during the test. Apparent viscosity (η, Pa s) was obtained interpolated at the share rate section at 100 s^{-1}.

4.7. *In Vitro Drug Release Study from the Nanocomposite Gels*

Franz-type diffusion cells, a diffusion area of 0.64 cm^2 and a receptor chamber of 4.9 mL), together with a dialysis membrane (from Sigma-Aldrich, Madrid, Spain), and a molecular weight of 14,000 Da, were used to study the FB drug release. The membrane was hydrated in methanol and water (1:1) for 24 h, rinsed, and then assembled into Franz diffusion cells. PBS solution with a pH of 7.4 was used as the receptor medium and stirred at 600 rpm to maintain the sink conditions. The experiment was carried out at 32 °C in a thermostatic water bath. A total of 0.25 g of PEG and TRE nanocomposite gels was added to the donor compartment. Throughout the experiment, 300 µL samples were extracted at predetermined time intervals, and PBS solution was added to the cells after each sample collection to keep the volume consistent. A validated HPLC-fluorescence method as described in Section 4.9 was used to analyze the collected samples. FB release profiles were described using various kinetic models that examined the cumulative amounts of FB released from each formulation over time [73].

4.8. *Ex vivo Permeation Profile of the Nanocomposite Gels*

The experiments were conducted as described in Section 4.5 for the nanoparticles by applying 500 mL of the nanocomposite gels to the skin. The sample collection and their analysis were in accordance with the methodology described in Section 4.5. The calculation of the permeation parameters also followed the same methodology.

4.9. *Flurbiprofen Determination by HPLC*

High-Performance Liquid Chromatography (HPLC) with a UV detector was used to determine the amount of FB in each sample.

The mobile phase was comprised of water and acetonitrile (35:60, *v:v*); water was previously acidified to a pH of 2.5 with orthophosphoric acid. The chromatographic column was a reverse phase C18 column (4.6 × 75 mm, 3.5 µm) and the detection of FB was set at the wavelength of 247 nm in the UV detector. The flux was 1 mL/min, and the injection volume was 10 µL. The retention time of FB was at approximately 4 min. [14].

4.10. *In Vivo Tolerance Study by Evaluating Biomechanical Properties of Human Skin*

An in vivo tolerance study was carried out on 10 female volunteers aged between 20–40 years old with healthy skin. The experimental procedure, in accordance with The Code of Ethics of the Declaration of Helsinki's requirements for experiments involving humans, was approved by the Local Ethics Committee of the University of Barcelona (iRB00003099).

Baseline values were recorded before application of the blank nanocomposite gels (only excipients, with unloaded nanoparticles) on the right and left forearm, and at determined time intervals after formulation application. Trans-Epidermal Water Loss (TEWL, $g/m^2/h$) and Stratum Corneum Hydration (SCH, arbitrary units, AU) were registered by using a Tewameter® TM and Corneometer® to evaluate the tolerability of the selected excipients formulations on the skin. Results were reported as the mean ± SD (n = 10).

4.10.1. Stratum Corneum Hydration (SCH)

NPs-PEG and NPs-TRE gels were applied on the skin of human volunteers. The measurements of the Stratum Corneum Hydration (SCH) over a period of 4 h were performed with Corneometer® 825 (Courage and Khazaka, Electronic GmbH, Koln, Germany). This was carried out using the capacitance technique, which utilizes water's relatively high dielectric constant in comparison to the dielectric constants of other skin-related substances [74].

4.10.2. Trans-Epidermal Water Loss (TEWL)

Nanocomposite gels without FB, containing NPs-PEG and NPs-TRE, were applied to the skin of human volunteers. TEWL values were measured by placing a Tewameter® TM 300 (Courage and Khazaka, Electronic GmbH, Koln, Germany) on the skin. The probe was placed on the skin for 2 min at each time interval, to allow equilibration before readings. The rate of water evaporation and diffusion from the epidermal layer to the surrounding atmosphere was recorded over 3 h [75].

4.11. Statistical Analysis

Results are reported as mean ± standard deviation. Statistical differences were determined conducting a one-way analysis of variance (ANOVA) in GraphPad Prism® software v. 5.0 (GraphPad Software Inc., San Diego, CA, USA). The Tukey post-hoc test was used in the permeation studies to determine significance differences between the mean of all groups to the mean of each group. A p-value < 0.05 was considered as statistically significant.

Supplementary Materials: The following supporting information can be downloaded at: https://www.mdpi.com/article/10.3390/gels10060362/s1, Figure S1: FITR spectra of the gels: (a) FB-NPs gel (red), (b) PEG with Sepigel ® (red), (c) TRE with Sepigel ® (red), (d) FB-NPs-PEG gel (red) and PEG with Sepigel ® (blue), (e) FB-NPs-TRE gel (red) and TRE with Sepigel ® (blue); Figure S2: In vitro drug release profiles of FB for the nanoparticles FB-NPs-PEG and FB-NPs-TRE in comparison to the nanocomposite gels FB-NPs-PEG and FB-NPs-TRE; Equation (S1): First-order kinetic model; Equation (S2): Korsmeyer-Peppas model; Equation (S3): Higuchi model; Equation (S4): Weibull model; Table S1: Estimated parameter values obtained by fitting of the data from the in vitro drug release for the nanocomposite gels FB-NPs-PEG and FB-NPs-TRE; and Table S2: Estimated parameter values obtained by fitting of the data from the in vitro drug release for the nanoparticles FB-NPs-PEG and FB-NPs-TRE.

Author Contributions: Conceptualization, G.R.-Y., M.M. and A.C.C.; methodology, A.C.C. and M.M.; validation A.C.C., M.M. and J.S.-C.; formal analysis S.E.B., G.R.-Y., L.G. and M.M.; investigation S.E.B., L.G., C.Q. and J.S.-C.; resources A.C.C., M.M. and J.S.-C.; data curation S.E.B., L.G. and M.M.; writing—original draft preparation S.E.B. and L.G.; writing—review and editing G.R.-Y., M.M., J.S.-C., A.C.C. and C.Q.; supervision G.R.-Y., J.S.-C. and A.C.C.; project administration J.S.-C. and A.C.C. All authors have read and agreed to the published version of the manuscript.

Funding: The authors wish to acknowledge the support of the Generalitat of Catalonia (2021 SGR 01085).

Institutional Review Board Statement: The study was conducted in accordance with the Declaration of Helsinki and approved by the Institutional Review Board (or Ethics Committee) of the University of Barcelona (IRB00003099, dated on 30 January 2019).

Informed Consent Statement: Informed consent was obtained from all subjects involved in the study.

Data Availability Statement: The data presented in this study are available on request from the corresponding author. The data are not publicly available due to their being used as a part of a doctoral thesis, and they will be available once the thesis has been published.

Acknowledgments: The authors would like to thank to AG Sanding from animal facility 758 (University of Barcelona) for his valuable help in animal studies. The authors also would like to acknowledge Antoni Boix for his support during the in vivo studies. The authors would like to acknowledge the Electronic Microscopy Unit of the Scientific and Technological Center of the University of Barcelona for their technical support.

Conflicts of Interest: The authors declare no conflicts of interest.

References

1. Oscanoa-Espinoza, T.; Lizaraso-Soto, F. Antiinflamatorios No Esteroides: Seguridad Gastrointestinal, Cardiovascular y Renal Nonsteroidal Antiinflammatory Drugs: Gastrointestinal and Cardiovascular and Renal Safety. *Rev. Gastroenterol. Perú* **2015**, *35*, 63–71. [PubMed]
2. Modi, C.M.; Mody, S.K.; Patel, H.B.; Dudhatra, G.B.; Kumar, A.; Avale, M. Toxicopathological Overview of Analgesic and Anti-Inflammatory Drugs in Animals. *J. Appl. Pharm. Sci.* **2012**, *02*, 149–157.
3. Clària, J. Los Nuevos Antiinflamatorios. *Med. Integral.* **2001**, *38*, 175–183.
4. Mathews, K.A. Non-Steroidal Anti-Inflammatory Analgesics: A Review of Current Practice. *J. Vet. Emerg. Crit. Care* **2002**, *12*, 89–97. [CrossRef]
5. Flood, J.; Stewart, A.J. Non-Steroidal Anti-Inflammatory Drugs and Associated Toxicities in Horses. *Animals* **2022**, *12*, 2939. [CrossRef] [PubMed]
6. Lascelles, B.D.X.; Court, M.H.; Hardie, E.M.; Robertson, S.A. Nonsteroidal Anti-Inflammatory Drugs in Cats: A Review. *Vet. Anaesth. Analg.* **2007**, *34*, 228–250. [CrossRef] [PubMed]

7. Dorbandt, D.M.; Labelle, A.L.; Mitchell, M.A.; Hamor, R.E. The Effects of Topical Diclofenac, Topical Flurbiprofen, and Humidity on Corneal Sensitivity in Normal Dogs. *Vet. Ophthalmol.* **2017**, *20*, 160–170. [CrossRef] [PubMed]
8. Fang, J.-Y.; Hwang, T.-L.; Leu, Y.-L. Effect of Enhancers and Retarders on Percutaneous Absorption of Flurbiprofen from Hydrogels. *Int. J. Pharm.* **2003**, *250*, 313–325. [CrossRef] [PubMed]
9. Gonzalez-Mira, E.; Egea, M.A.; Garcia, M.L.; Souto, E.B. Design and Ocular Tolerance of Flurbiprofen Loaded Ultrasound-Engineered NLC. *Colloids Surf. B Biointerfaces* **2010**, *81*, 412–421. [CrossRef]
10. Oktay, A.N.; Ilbasmis-Tamer, S.; Han, S.; Uludag, O.; Celebi, N. Preparation and in Vitro/in Vivo Evaluation of Flurbiprofen Nanosuspension-Based Gel for Dermal Application. *Eur. J. Pharm. Sci.* **2020**, *155*, 105548. [CrossRef]
11. Aljubailah, A.; Alqahtani, S.M.S.; Al-Garni, T.S.; Saeed, W.S.; Semlali, A.; Aouak, T. Naproxen-Loaded Poly(2-Hydroxyalkyl Methacrylates): Preparation and Drug Release Dynamics. *Polymers* **2022**, *14*, 450. [CrossRef] [PubMed]
12. Krishnaiah, Y.S.R. Pharmaceutical Technologies for Enhancing Oral Bioavailability of Poorly Soluble Drugs. *J. Bioequivalence Bioavailab.* **2010**, *2*, 28–36. [CrossRef]
13. Zeb, A.; Arif, S.T.; Malik, M.; Shah, F.A.; Din, F.U.; Qureshi, O.S.; Lee, E.-S.; Lee, G.-Y.; Kim, J.-K. Potential of Nanoparticulate Carriers for Improved Drug Delivery via Skin. *J. Pharm. Investig.* **2018**, *49*, 485–517. [CrossRef]
14. Ramos-Yacasi, G.R.; Calpena-Campmany, A.C.; Egea-Gras, M.A.; Espina-García, M.; García-López, M.L. Freeze Drying Optimization of Polymeric Nanoparticles for Ocular Flurbiprofen Delivery: Effect of Protectant Agents and Critical Process Parameters on Long-Term Stability. *Drug Dev. Ind. Pharm.* **2017**, *43*, 637–651. [CrossRef] [PubMed]
15. Silva, A.; Martins-Gomes, C.; Coutinho, T.; Fangueiro, J.; Sanchez-Lopez, E.; Pashirova, T.; Andreani, T.; Souto, E. Soft Cationic Nanoparticles for Drug Delivery: Production and Cytotoxicity of Solid Lipid Nanoparticles (SLNs). *Appl. Sci.* **2019**, *9*, 4438. [CrossRef]
16. Buwalda, S.J.; Boere, K.W.M.; Dijkstra, P.J.; Feijen, J.; Vermonden, T.; Hennink, W.E. Hydrogels in a Historical Perspective: From Simple Networks to Smart Materials. *J. Control. Release* **2014**, *190*, 254–273. [CrossRef] [PubMed]
17. Ramos Yacasi, G.R.; García Lopéz, M.L.; Espina García, M.; Parra Coca, A.; Calpena Campmany, A.C. The Influence of Freeze Drying and ϒ-Irradiation in Pre-Clinical Studies of Flurbiprofen Polymeric Nanoparticles for Ocular Delivery Using D-(+)-Trehalose and Polyethylene Glycol. *Int. J. Nanomed.* **2016**, *11*, 4093–4106. [CrossRef] [PubMed]
18. Chandel, A.K.S.; Nutan, B.; Raval, I.H.; Jewrajka, S.K. Self-Assembly of Partially Alkylated Dextran-Graft-Poly[(2-Dimethylamino)Ethyl Methacrylate] Copolymer Facilitating Hydrophobic/Hydrophilic Drug Delivery and Improving Conetwork Hydrogel Properties. *Biomacromolecules* **2018**, *19*, 1142–1153. [CrossRef] [PubMed]
19. Barreira, J.C.M.; Rodrigues, S.; Carvalho, A.M.; Ferreira, I.C.F.R. Development of Hydrosoluble Gels with Crataegus Monogyna Extracts for Topical Application: Evaluation of Antioxidant Activity of the Final Formulations. *Ind. Crops Prod.* **2013**, *42*, 175–180. [CrossRef]
20. Barrett-Catton, E.; Ross, M.L.; Asuri, P. Multifunctional Hydrogel Nanocomposites for Biomedical Applications. *Polymers* **2021**, *13*, 856. [CrossRef]
21. Topuz, F.; Bartneck, M.; Pan, Y.; Tacke, F. One-Step Fabrication of Biocompatible Multifaceted Nanocomposite Gels and Nanolayers. *Biomacromolecules* **2017**, *18*, 386–397. [CrossRef] [PubMed]
22. Gandla, K.; Kumar, K.P.; Rajasulochana, P.; Charde, M.S.; Rana, R.; Singh, L.P.; Haque, M.A.; Bakshi, V.; Siddiqui, F.A.; Khan, S.L.; et al. Fluorescent-Nanoparticle-Impregnated Nanocomposite Polymeric Gels for Biosensing and Drug Delivery Applications. *Gels* **2023**, *9*, 669. [CrossRef] [PubMed]
23. Hajareh Haghighi, F.; Binaymotlagh, R.; Fratoddi, I.; Chronopoulou, L.; Palocci, C. Peptide-Hydrogel Nanocomposites for Anti-Cancer Drug Delivery. *Gels* **2023**, *9*, 953. [CrossRef] [PubMed]
24. Hasnain, M.S.; Nayak, A.K. 7—Nanocomposites for Improved Orthopedic and Bone Tissue Engineering Applications. In *Applications of Nanocomposite Materials in Orthopedics*; Inamuddin, A.M., Mohammad-Asiri, A., Eds.; Woodhead Publishing Series in Biomaterials; Woodhead Publishing: Sawston, UK, 2019; pp. 145–177. ISBN 978-0-12-813740-6.
25. Karchoubi, F.; Afshar Ghotli, R.; Pahlevani, H.; Baghban Salehi, M. New Insights into Nanocomposite Hydrogels; a Review on Recent Advances in Characteristics and Applications. *Adv. Ind. Eng. Polym. Res.* **2024**, *7*, 54–78. [CrossRef]
26. Ding, W.; Ge, Y.; Zhang, T.; Zhang, C.; Yin, X. Advanced Construction Strategies to Obtain Nanocomposite Hydrogels for Bone Repair and Regeneration. *NPG Asia Mater.* **2024**, *16*, 14. [CrossRef]
27. Mostafa, M.; El-Meligy, M.A.; Sharaf, M.; Soliman, A.T.; AbuKhadra, M.R. Insight into Chitosan/Zeolite-A Nanocomposite as an Advanced Carrier for Levofloxacin and Its Anti-Inflammatory Properties; Loading, Release, and Anti-Inflammatory Studies. *Int. J. Biol. Macromol.* **2021**, *179*, 206–216. [CrossRef] [PubMed]
28. Li, N.; Xie, L.; Wu, Y.; Wu, Y.; Liu, Y.; Gao, Y.; Yang, J.; Zhang, X.; Jiang, L. Dexamethasone-Loaded Zeolitic Imidazolate Frameworks Nanocomposite Hydrogel with Antibacterial and Anti-Inflammatory Effects for Periodontitis Treatment. *Mater. Today Bio* **2022**, *16*, 100360. [CrossRef]
29. Moghaddam, A.A.; Ahad, A.; Aqil, M.; Ahmad, F.J.; Sultana, Y.; Ali, A. Ibuprofen Loaded Nano-Ethanolic Liposomes Carbopol Gel System: In Vitro Characterization and Anti-Inflammatory Efficacy Assessment in Wistar Rats. *J. Polym. Eng.* **2018**, *38*, 291–298. [CrossRef]
30. Pramanik, A.; Sahoo, R.N.; Nanda, A.; Mohapatra, R.; Singh, R.; Mallick, S. Ocular Permeation and Sustained Anti-Inflammatory Activity of Dexamethasone from Kaolin Nanodispersion Hydrogel System. *Curr. Eye Res.* **2018**, *43*, 828–838. [CrossRef]

31. Zamora, C.M.P.; Michaluk, A.G.; Torres, C.A.; Mouriño, V.; Chiappetta, D.A.; Nuñez, M.B. Influence of Herbal Extracts in Physicochemical Properties and Stability of Antibacterial Gels. *J. Adv. Pharm. Educ. Res.* **2023**, *13*, 16–24. [CrossRef]
32. Berenguer, D.; Sosa, L.; Alcover, M.; Sessa, M.; Halbaut, L.; Guillén, C.; Fisa, R.; Calpena-Campmany, A.C.; Riera, C. Development and Characterization of a Semi-Solid Dosage Form of Meglumine Antimoniate for Topical Treatment of Cutaneous Leishmaniasis. *Pharmaceutics* **2019**, *11*, 613. [CrossRef] [PubMed]
33. Dash, T.K.; Konkimalla, V.B. Poly-ϵ-Caprolactone Based Formulations for Drug Delivery and Tissue Engineering: A Review. *J. Control. Release* **2012**, *158*, 15–33. [CrossRef] [PubMed]
34. Shreiber-Livne, I.; Sulimani, L.; Shapira, A.; Procaccia, S.; Meiri, D.; Sosnik, A. Poly(Ethylene Glycol)-b-Poly(Epsilon-Caprolactone) Nanoparticles as a Platform for the Improved Oral Delivery of Cannabidiol. *Drug Deliv. Transl. Res.* **2023**, *13*, 3192–3203. [CrossRef] [PubMed]
35. Parra, A.; Clares, B.; Rosselló, A.; Garduño-Ramírez, M.L.; Abrego, G.; García, M.L.; Calpena, A.C. Ex Vivo Permeation of Carprofen from Nanoparticles: A Comprehensive Study through Human, Porcine and Bovine Skin as Anti-Inflammatory Agent. *Int. J. Pharm.* **2016**, *501*, 10–17. [CrossRef] [PubMed]
36. Netzlaff, F.; Schaefer, U.F.; Lehr, C.-M.; Meiers, P.; Stahl, J.; Kietzmann, M.; Niedorf, F. Comparison of Bovine Udder Skin with Human and Porcine Skin in Percutaneous Permeation Experiments. *Altern. Lab. Anim.* **2006**, *34*, 499–513. [PubMed]
37. Godin, B.; Touitou, E. Transdermal Skin Delivery: Predictions for Humans from in Vivo, Ex Vivo and Animal Models. *Adv. Drug Deliv. Rev.* **2007**, *59*, 1152–1161. [CrossRef] [PubMed]
38. Cecone, C.; Hoti, G.; Caldera, F.; Ginepro, M.; Matencio, A.; Trotta, F. Evaluation of the Swelling Properties and Sorption Capacity of Maltodextrin-Based Cross-Linked Polymers. *Gels* **2024**, *10*, 232. [CrossRef] [PubMed]
39. Hamann, J.; Hansen, S.; Lacy-Hulbert, J.; Wollford, M.W. Measurement of Bovine Teat Skin PH. *Milchwissenschaft* **2002**, *57*, 490–493.
40. Stoilova, S.; Georgieva, D.; Mihaylova, R.; Petrov, P.D.; Kostova, B. Nanogels Based on N,N-Dimethylacrylamide and β-Cyclodextrin Triacrylate for Enhanced Solubility and Therapeutic Efficacy of Aripiprazole. *Gels* **2024**, *10*, 217. [CrossRef]
41. Zhang, J.; Froelich, A.; Michniak-Kohn, B. Topical Delivery of Meloxicam Using Liposome and Microemulsion Formulation Approaches. *Pharmaceutics* **2020**, *12*, 282. [CrossRef]
42. Xu, W.; Zheng, Y.; Pan, P. Crystallization-Driven Self-Assembly of Semicrystalline Block Copolymers and End-Functionalized Polymers: A Minireview. *J. Polym. Sci.* **2022**, *60*, 2136–2152. [CrossRef]
43. Rahmani, S.; Maroufkhani, M.; Mohammadzadeh-Komuleh, S.; Khoubi-Arani, Z. Chapter 7—Polymer Nanocomposites for Biomedical Applications. In *Fundamentals of Bionanomaterials*; Barhoum, A., Jeevanandam, J., Danquah, M.K., Eds.; Elsevier: Amsterdam, The Netherlands, 2022; pp. 175–215. ISBN 978-0-12-824147-9.
44. Burdick, J.A.; Anseth, K.S. Photoencapsulation of Osteoblasts in Injectable RGD-Modified PEG Hydrogels for Bone Tissue Engineering. *Biomaterials* **2002**, *23*, 4315–4323. [CrossRef]
45. Wu, J.; Wei, W.; Wang, L.-Y.; Su, Z.-G.; Ma, G.-H. A Thermosensitive Hydrogel Based on Quaternized Chitosan and Poly(Ethylene Glycol) for Nasal Drug Delivery System. *Biomaterials* **2007**, *28*, 2220–2232. [CrossRef]
46. Golitsyn, Y.; Pulst, M.; Samiullah, M.H.; Busse, K.; Kressler, J.; Reichert, D. Crystallization in PEG Networks: The Importance of Network Topology and Chain Tilt in Crystals. *Polymer* **2019**, *165*, 72–82. [CrossRef]
47. Bilal, M.H.; Mahmood, N.; Samiullah, M.H.; Busse, K.; Kressler, J. Physico-Mechanical Properties of Poly(Ethylene Glycol) Based Polymer Networks. *J. Appl. Polym. Sci.* **2023**, *140*, e54726. [CrossRef]
48. Van Duong, T.; Goderis, B.; Van Humbeeck, J.; Van Den Mooter, G. Microstructure of Pharmaceutical Semicrystalline Dispersions: The Significance of Polymer Conformation. *Mol. Pharm.* **2018**, *15*, 629–641. [CrossRef]
49. Suflet, D.M.; Constantin, M.; Pelin, I.M.; Popescu, I.; Rimbu, C.M.; Horhogea, C.E.; Fundueanu, G. Chitosan–Oxidized Pullulan Hydrogels Loaded with Essential Clove Oil: Synthesis, Characterization, Antioxidant and Antimicrobial Properties. *Gels* **2024**, *10*, 227. [CrossRef]
50. Ahmadi, N.; Rincón, M.; Silva-Abreu, M.; Sosa, L.; Pesantez-Narvaez, J.; Calpena, A.C.; Rodríguez-Lagunas, M.J.; Mallandrich, M. Semi-Solid Dosage Forms Containing Pranoprofen-Loaded NLC as Topical Therapy for Local Inflammation: In Vitro, Ex Vivo and In Vivo Evaluation. *Gels* **2023**, *9*, 448. [CrossRef]
51. Dantas, M.G.B.; Reis, S.A.G.B.; Damasceno, C.M.D.; Rolim, L.A.; Rolim-Neto, P.J.; Carvalho, F.O.; Quintans-Junior, L.J.; Almeida, J.R.G. da S. Development and Evaluation of Stability of a Gel Formulation Containing the Monoterpene Borneol. *Sci. World J.* **2016**, *2016*, 7394685. [CrossRef]
52. Sandri, G.; Ruggeri, M.; Rossi, S.; Bonferoni, M.C.; Vigani, B.; Ferrari, F. (Trans)Buccal Drug Delivery. In *Nanotechnology for Oral Drug Delivery*; Elsevier: Amsterdam, The Netherlands, 2020; pp. 225–250.
53. Moussaoui, S.E.; Fernández-Campos, F.; Alonso, C.; Limón, D.; Mallandrich, M. Topical Mucoadhesive Alginate-Based Hydrogel Loading Ketorolac for Pain Management after Pharmacotherapy, Ablation, or Surgical Removal in Condyloma Acuminata. *Gels* **2021**, *7*, 8. [CrossRef]
54. De Grau-Bassal, G.; Mallandrich, M.; Sosa, L.; Espinoza, L.; Calpena, A.C.; Bozal-de Febrer, N.; Rodríguez-Lagunas, M.J.; Garduño-Ramírez, M.L.; Rincón, M. A Novel Approach for Dermal Application of Pranoprofen-Loaded Lipid Nanoparticles for the Treatment of Post-Tattoo Inflammatory Reactions. *Pharmaceutics* **2024**, *16*, 643. [CrossRef]

55. Abrantes, D.C.; Rogerio, C.B.; Campos, E.V.R.; Germano-Costa, T.; Vigato, A.A.; Machado, I.P.; Sepulveda, A.F.; Lima, R.; de Araujo, D.R.; Fraceto, L.F. Repellent Active Ingredients Encapsulated in Polymeric Nanoparticles: Potential Alternative Formulations to Control Arboviruses. *J. Nanobiotechnol.* **2022**, *20*, 520. [CrossRef]
56. Bini, R.A.; Silva, M.F.; Varanda, L.C.; da Silva, M.A.; Dreiss, C.A. Soft Nanocomposites of Gelatin and Poly(3-Hydroxybutyrate) Nanoparticles for Dual Drug Release. *Colloids Surf. B Biointerfaces* **2017**, *157*, 191–198. [CrossRef] [PubMed]
57. Momekova, D.; Ivanov, E.; Konstantinov, S.; Ublekov, F.; Petrov, P.D. Nanocomposite Cryogel Carriers from 2-Hydroxyethyl Cellulose Network and Cannabidiol-Loaded Polymeric Micelles for Sustained Topical Delivery. *Polymers* **2020**, *12*, 1172. [CrossRef] [PubMed]
58. Ternullo, S.; Werning, L.V.S.; Holsæter, A.M.; Škalko-Basnet, N. Curcumin-in-Deformable Liposomes-in-Chitosan-Hydrogel as a Novel Wound Dressing. *Pharmaceutics* **2020**, *12*, 8. [CrossRef]
59. Shehata, T.M.; Nair, A.B.; Al-Dhubiab, B.E.; Shah, J.; Jacob, S.; Alhaider, I.A.; Attimarad, M.; Elsewedy, H.S.; Ibrahim, M.M. Vesicular Emulgel Based System for Transdermal Delivery of Insulin: Factorial Design and in Vivo Evaluation. *Appl. Sci.* **2020**, *10*, 5341. [CrossRef]
60. Khan, M.U.; ur Rehman, A.; Khan, M.F.A.; Ahmed, N.; Rashid, S.A.; Munawar, K.S. Novel Hybrid Nanostructure Hydrogel for Treating Fungal Infections: Design and Evaluation. *Bionanoscience* **2024**. [CrossRef]
61. Załęcki, P.; Rogowska, K.; Wąs, P.; Łuczak, K.; Wysocka, M.; Nowicka, D. Impact of Lifestyle on Differences in Skin Hydration of Selected Body Areas in Young Women. *Cosmetics* **2024**, *11*, 13. [CrossRef]
62. Kang, S.Y.; Um, J.Y.; Chung, B.Y.; Lee, S.Y.; Park, J.S.; Kim, J.C.; Park, C.W.; Kim, H.O. Moisturizer in Patients with Inflammatory Skin Diseases. *Medicina* **2022**, *58*, 888. [CrossRef]
63. Yu, Y.Q.; Yang, X.; Wu, X.F.; Fan, Y. Bin Enhancing Permeation of Drug Molecules Across the Skin via Delivery in Nanocarriers: Novel Strategies for Effective Transdermal Applications. *Front. Bioeng. Biotechnol.* **2021**, *9*, 646554. [CrossRef]
64. Constantin, M.M.; Poenaru, E.; Poenaru, C.; Constantin, T. Skin Hydration Assessment through Modern Non-Invasive Bioengineering Technologies. *Maedica* **2014**, *9*, 33–38.
65. Berenguer, D.; Alcover, M.M.; Sessa, M.; Halbaut, L.; Guillen, C.; Boix-Montanes, A.; Fisa, R.; Calpena-Campmany, A.C.; Riera, C.; Sosa, L. Topical Amphotericin B Semisolid Dosage Form for Cutaneous Leishmaniasis: Physicochemical Characterization, Ex Vivo Skin Permeation and Biological Activity. *Pharmaceutics* **2020**, *12*, 149. [CrossRef] [PubMed]
66. Nitiyarom, R.; Siriwat, N.; Wisuthsarewong, W. Transepidermal Water Loss after Water Immersion. *Siriraj Med. J.* **2021**, *73*, 386–390. [CrossRef]
67. Dey, S.; Rothe, H.; Page, L.; O'Connor, R.; Farahmand, S.; Toner, F.; Marsh, R.; Wehmeyer, K.; Zhou, S. An In Vitro Skin Penetration Model for Compromised Skin: Estimating Penetration of Polyethylene Glycol [(1)(4)C]-PEG-7 Phosphate. *Skin. Pharmacol. Physiol.* **2015**, *28*, 12–21. [CrossRef] [PubMed]
68. Roelandt Truus and Hachem, J.-P. Practical Use and Significance of Transepidermal Water Loss Measurements. In *Practical Aspects of Cosmetic Testing: How to Set Up a Scientific Study in Skin Physiology*; Fluhr, J.W., Ed.; Springer International Publishing: Cham, Switzerland, 2020; pp. 297–304. ISBN 978-3-030-44967-4.
69. Zia, S.; Khan, S.M.; Butt, M.T.Z.; Gull, N. Insight into CMC-PVA-FHNTs Nanocomposite Hydrogel as an Advance Carrier for Cefadroxil Monohydrate: Fabrication and Characterization/Angiogenic Potential Analysis. *Gels* **2024**, *10*, 235. [CrossRef] [PubMed]
70. Matousek, J.L.; Campbell, K.L. A Comparative Review of Cutaneous PH. *Vet. Dermatol.* **2002**, *13*, 293–300. [CrossRef] [PubMed]
71. Khorshidian, A.; Sharifi, N.; Choupani Kheirabadi, F.; Rezaei, F.; Sheikholeslami, S.A.; Ariyannejad, A.; Esmaeili, J.; Basati, H.; Barati, A. In Vitro Release of Glycyrrhiza Glabra Extract by a Gel-Based Microneedle Patch for Psoriasis Treatment. *Gels* **2024**, *10*, 87. [CrossRef] [PubMed]
72. Pérez-González, N.; Bozal-de Febrer, N.; Calpena-Campmany, A.C.; Nardi-Ricart, A.; Rodríguez-Lagunas, M.J.; Morales-Molina, J.A.; Soriano-Ruiz, J.L.; Fernández-Campos, F.; Clares-Naveros, B. New Formulations Caspofungin for Topical of Vulvovaginal Candidiasis. *Gels* **2021**, *7*, 259. [CrossRef] [PubMed]
73. Rincón, M.; Calpena, A.C.; Clares, B.; Espina, M.; Garduño-Ramírez, M.L.; Rodríguez-Lagunas, M.J.; García, M.L.; Abrego, G. Skin-Controlled Release Lipid Nanosystems of Pranoprofen for the Treatment of Local Inflammation and Pain. *Nanomedicine* **2018**, *13*, 2397–2413. [CrossRef]
74. Anthonissen, M.; Daly, D.; Peeters, R.; Van Brussel, M.; Fieuws, S.; Moortgat, P.; Flour, M.; den Kerckhove, E. Reliability of Repeated Measurements on Post-Burn Scars with Corneometer CM825®. *Skin Res. Technol.* **2014**, *21*, 302–312. [CrossRef]
75. Green, M.; Kashetsky, N.; Feschuk, A.; Maibach, H.I. Transepidermal Water Loss (TEWL): Environment and Pollution—A Systematic Review. *Skin Health Dis.* **2022**, *2*, e104. [CrossRef]

gels

MDPI

Article

Molecular Recognition between Carbon Dioxide and Biodegradable Hydrogel Models: A Density Functional Theory (DFT) Investigation

Domingo Cesar Carrascal-Hernandez [1], Maximiliano Mendez-Lopez [1], Daniel Insuasty [1], Samira García-Freites [2], Marco Sanjuan [2] and Edgar Márquez [1,*]

1 Departamento de Química y Biología, Facultad de Ciencias Básicas, Universidad del Norte, Barranquilla 080020, Colombia; domingoh@uninorte.com (D.C.C.-H.); maximilianom@uninorte.edu.co (M.M.-L.); insuastyd@uninorte.edu.co (D.I.)

2 Centro de Investigación e Innovación en Energía y Gas—CIIEG, Promigas S.A. E.S.P., Barranquilla 11001, Colombia; samira.garcia@promigas.com (S.G.-F.); marco.sanjuan@promigas.com (M.S.)

* Correspondence: ebrazon@uninorte.edu.co

Abstract: In this research, we explore the potential of employing density functional theory (DFT) for the design of biodegradable hydrogels aimed at capturing carbon dioxide (CO_2) and mitigating greenhouse gas emissions. We employed biodegradable hydrogel models, including polyethylene glycol, polyvinylpyrrolidone, chitosan, and poly-2-hydroxymethacrylate. The complexation process between the hydrogel and CO_2 was thoroughly investigated at the ωB97X-D/6-311G(2d,p) theoretical level. Our findings reveal a strong affinity between the hydrogel models and CO_2, with binding energies ranging from −4.5 to −6.5 kcal/mol, indicative of physisorption processes. The absorption order observed was as follows: chitosan > PVP > HEAC > PEG. Additionally, thermodynamic parameters substantiated this sequence and even suggested that these complexes remain stable up to 160 °C. Consequently, these polymers present a promising avenue for crafting novel materials for CO_2 capture applications. Nonetheless, further research is warranted to optimize the design of these materials and assess their performance across various environmental conditions.

Keywords: carbon dioxide; CO_2-capture; frontier molecular orbitals; green-hydrogen; DFT

Citation: Carrascal-Hernandez, D.C.; Mendez-Lopez, M.; Insuasty, D.; García-Freites, S.; Sanjuan, M.; Márquez, E. Molecular Recognition between Carbon Dioxide and Biodegradable Hydrogel Models: A Density Functional Theory (DFT) Investigation. *Gels* **2024**, *10*, 386. https://doi.org/10.3390/gels10060386

Academic Editors: Daihua Fu and Jieyu Zhang

Received: 5 May 2024
Revised: 25 May 2024
Accepted: 28 May 2024
Published: 5 June 2024

1. Introduction

With the rapid development of industrialization and urbanization, there has been a significant increase in the emission of greenhouse gases into the atmosphere [1]. This increase in emissions is largely due to human activities such as burning fossil fuels, deforestation, and changing agricultural and industrial practices. As a result, the buildup of greenhouse gases has caused global warming and climate change, which have had severe consequences such as rising sea levels, extreme weather events, and loss of biodiversity. The impact of climate change is not limited to the environment but also affects human welfare, causing public health issues and economic losses. It is essential to take coordinated action at the global, regional, national, and local levels to mitigate the impact of climate change and make cities an integral part of the solution [2].

To address the challenge of climate change, the technology known as carbon capture, utilization, and storage (CCUS) has been proposed. This technique allows for the capture of CO_2 from emission sources, including power plants, industrial facilities, and the atmosphere. Once captured, CO_2 can be utilized as a raw material for chemical synthesis or stored deep underground in a safe and permanent manner [3]. This innovative approach offers the potential to achieve net-zero emissions on a large scale. CCUS, which includes biomass power with carbon capture and storage (BECCUS) when biomass is used, can be implemented in existing coal and gas power plants. By capturing CO_2 from these emission sources, as well as directly from the atmosphere, it becomes possible to utilize the captured CO_2 for chemical synthesis or store it securely underground for long-term storage [4]. In addition to contributing to the electricity supply sector, CCUS is possibly the only scalable and cost-effective option to achieve a deep decarbonization of certain industries such as steel, cement, glass, and ceramics, as well as the manufacture of chemical products that generate CO_2 during production processes [5]. The analyses carried out by the Intergovernmental Panel on Climate Change (IPCC) and the International Energy Agency (IEA) have shown that CCUS will be key to achieving "Net Zero" in 2050, which contributes to a sixth of the reduction of global CO_2 emissions in order to limit the increase in global temperature to less than 1.5 °C, as established in the Paris Agreement [6].

Carbon capture technologies have been developed in recent decades, and can be divided into three pathways: post-combustion capture, oxy-combustion, and pre-combustion capture [7]. Various physical and chemical processes have been reported to be used in carbon capture technology, such as solvent-based adsorption, solid solvent adsorption/absorption, membranes, cryogenics, and chemical loops for CO_2 separation [8–10]. Among these methods, chemical absorption is currently one of the most widely used and commercially available techniques [11,12]. However, there are several limitations, such as the high energy consumption required for solvent regeneration, as well as the corrosiveness, high toxicity, volatility and high cost of solvents, which are the main barriers to the deployment of carbon capture technology carbon [13].

In this sense, the use of biodegradable hydrogels for CO_2 absorption is an interesting approach that has attracted attention in recent research. The advantages of using biodegradable hydrogels in CO_2 absorption include their biodegradability, which makes them more environmentally friendly than traditional materials [14]. Furthermore, hydrogels can be synthesized to exhibit CO_2 responsive properties, and swell or deflate in the presence of CO_2, further enhancing their potential for CO_2 capture applications [13]; however, the selection of the most suitable absorbent hydrogels remains a challenge.

In light of the aforementioned challenges, our study introduces a novel approach to CO_2 absorption using Density Functional Theory (DFT) to predict the most effective absorbent hydrogels. This innovative method stands out as it allows for a more precise and efficient selection of hydrogels, thereby overcoming the existing challenge in the field [15,16]. By leveraging DFT, we are able to model and predict the interaction between CO_2 and potential absorbent materials at the atomic and molecular levels. This provides a more detailed understanding of the absorption process, enabling the design of more efficient and effective hydrogels. Furthermore, our approach also considers the environmental impact and cost-effectiveness of the absorbent materials, ensuring a sustainable and economically viable solution to CO_2 capture [17,18]. This work, therefore, not only contributes to the advancement of CCUS technology but also paves the way for the development of next-generation absorbents for greenhouse gas capture.

Molecular recognition is a pivotal factor influencing the selectivity of CO_2-absorbent materials. The paramount significance of frontier molecular orbitals (FMOs) in molecular recognition cannot be overstated. These FMOs precisely delineate the regions where chemical bonds exhibit heightened reactivity owing to their associated orbital energies. Furthermore, molecular orbital theory appropriately elucidates the selectivity of chemical reactions and the formation of discrete product species [19,20].

Given the significant role of molecular frontier orbitals in molecular recognition, their utility in elucidating the selectivity of CO_2-absorbent materials is substantial. This understanding serves as the foundation for our research, which endeavors to engineer biodegradable hydrogels for CO_2 absorption using Density Functional Theory (DFT).

Our study employs concepts such as binding energy, frontier molecular orbitals, and chemical bonds, and utilizes poly(2-hydroxyethylmethacrylate), poly(ethylene glycol), chitosan, and polyvinylpyrrolidone as models of CO_2 bio-sorbent hydrogels. These hydrogels have demonstrated a commendable capacity for CO_2 absorption [21–23].

This methodological approach will facilitate the validation of DFT as a predictive tool and stimulate the synthesis of hydrogel matrices with enhanced CO_2 absorption capacity.

DFT, a widely accepted method for understanding and predicting the adsorption of a specific molecule on a polymeric structure, has been shown to yield results that correlate satisfactorily with experimental outcomes [24,25]

Therefore, the study of the relative absorption capacity of biodegradable hydrogels for CO_2 absorption was carried out using the concepts of interaction energy and Gibbs energy. The nature of molecular interactions was studied using frontier molecular theory, binding energy, and the Gibbs energy equation. This work underscores the critical role of frontier molecular orbitals in designing effective CO_2-absorbent materials, thereby emphasizing the importance of this study.

2. Results and Discussion

2.1. Minimum-Energy Structures

The minimum-energy structures for the dimer were selected from the results of the systematic rotor search; the global minimum-energy structures were optimized at ωB97X-D/6-311G(2d,p) and verified using frequency calculations. The global minimum structures for dimer along with CO_2 are shown in Figure 1 embedded in their electrostatic potential (ESP) surface. The electrostatic surface potential is crucial for CO_2 molecular recognition by hydrogels. It enables the design of hydrogel materials that can selectively and efficiently capture CO_2, offering promising solutions for carbon capture and reduction of greenhouse gas emissions [26,27].

As shown by the results, except for carbon dioxide, all the compounds studied herein are polar in nature. Even though CO_2 are non-polar, the molecule is polarizable due to the presence of O in the extreme of the molecule; therefore, in the presence of polar compounds, CO_2 could polarize its electron-cloud and participate in dipole-induced dipole interaction or hydrogen bond acceptor [28,29].

Chitosan, polyvinylpyrrolidone, and 2-hydroxyethylacrylamide showed electron density distributions characterized by pronounced asymmetry. This property significantly enhances their propensity for dipole–dipole interactions. Conversely, polyethene glycol exhibits a more symmetrical electron density distribution, marked by the presence of hydrogen atoms bearing partial positive charges (depicted in blue). These hydrogen atoms are pivotal for hydrogen bonding interactions. Collectively, these findings suggest a heightened likelihood of induced dipole–dipole interactions with carbon dioxide (CO_2), thereby promoting its absorption.

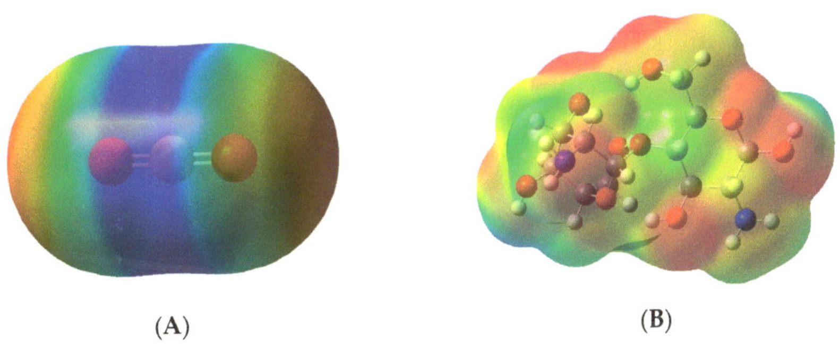

Figure 1. *Cont.*

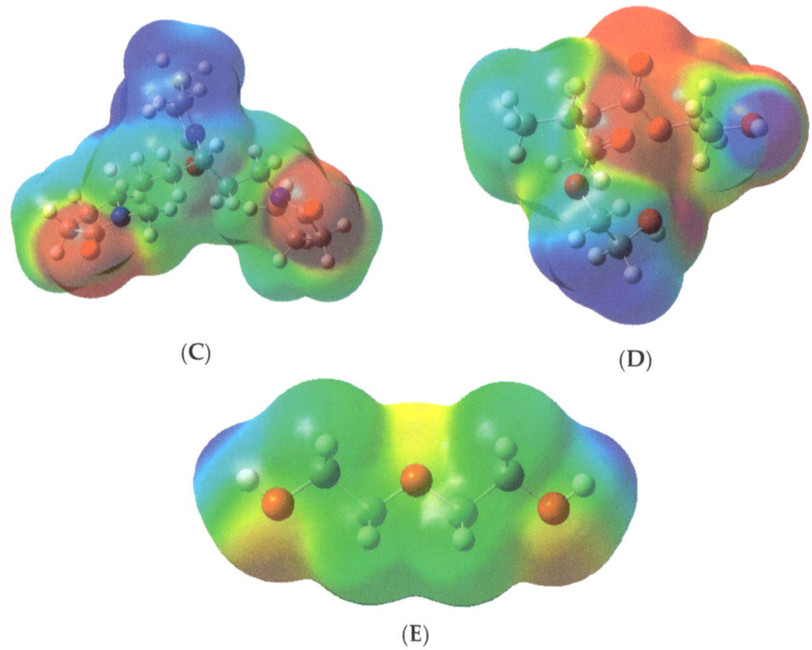

Figure 1. Electrostatic surface potential (ESP). (A): carbon dioxide; (B): chitosan; (C): polyvinylpyrrolidone; (D): 2-hydroxyethyl acrylamide and (E): polyethyleneglycol.

2.2. Minimum-Energy Structures for Dimer and CO_2-Hydrogel Complexes: Molecular Recognition

Molecular recognition is a fundamental concept in the chemical industry that involves specific interactions between molecules based on complementary shapes, sizes, charges, and functional groups [30,31]. It plays a crucial role in various aspects of the chemical industry, primarily in the field of drug discovery, material and engineering design, catalysis, separation technology, sensor development, and supramolecular chemistry, among others [31,32]. The relationship between frontier molecular orbitals and molecular recognition lies in how the electronic structure of molecules, described by these orbitals, influences their ability to interact and recognize each other specifically [33,34]. Frontier molecular orbitals, consisting of the highest occupied molecular orbital (HOMO) and the lowest unoccupied molecular orbital (LUMO), play a crucial role in molecular recognition [35,36].

The energy gap between the HOMO and LUMO orbitals determines the ease of electron transfer between molecules. In molecular recognition, this energy gap influences the strength of the interactions between molecules. When two molecules approach each other, their HOMO and LUMO orbitals can overlap, allowing charge transfer interactions. A smaller HOMO–LUMO gap improves electron transfer, affecting the strength of recognition interactions [36].

In this context, the investigation of frontier orbitals at the interface between dimers and CO_2 emerges as a critical analytical approach. This exploration provides invaluable insights into the feasibility of CO_2 absorption within a given polymeric matrix, constituting a fundamental step in the selection of the optimal absorbent material.

Figure 2 illustrates the frontier orbitals of the candidate absorptive compounds, including vinylpyrrolidone, ethylene glycol, hydroxyethyl methacrylate, and chitosan, alongside the target absorbate, CO_2. Notably, the frontier orbitals of CO_2 exhibit a remarkable degree of uniformity, with electronic density uniformly distributed throughout its molecular structure. This characteristic renders CO_2 highly versatile, and capable of functioning as both an electron acceptor and donor—a trait that augments its propensity for interaction with a wide array of chosen biodegradable matrices.

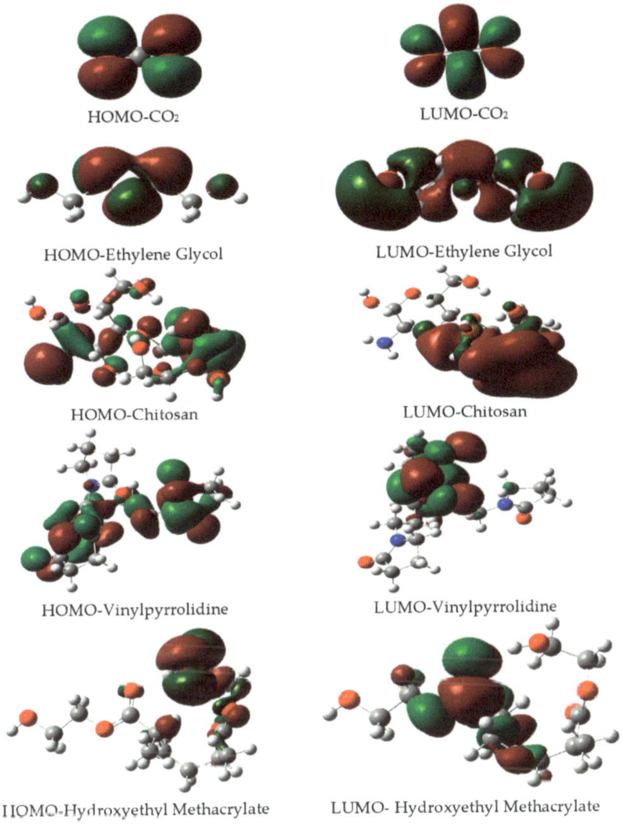

Figure 2. Frontier molecular orbitals (HOMO and LUMO) for all compounds studied herein, obtained at ωB97X D/6-311G(2d,p).

Notably, the frontier orbitals of chitosan, polyvinylpyrrolidone, and ethylene glycol share striking similarities with those of CO_2. This observation underscores the potential for efficient absorption of CO_2 gas onto the polymeric matrices formed by these compounds. Such molecular-level compatibility hints at the possibility of establishing strong intermolecular interactions between the polymeric matrices and CO_2 molecules, suggesting these materials as promising candidates for CO_2 capture and storage applications. This intricate understanding of frontier orbital interactions not only facilitates the selection of the most suitable absorbent but also contributes significantly to the advancement of environmentally sustainable solutions for managing greenhouse gas emissions.

2.3. Hydrogels Model CO_2 Complexation

Once it was established that symmetry existed between the frontier orbitals, the search for the minimum energy configuration in the formation of the complex between CO_2 (the absorbate) and the hydrogel model (the absorbent) was studied. To attain this minimum energy configuration, a systematic rotor search was conducted in tandem with a genetic algorithm to identify the global minimum. Subsequently, these complexes were saved in .gjf file format and optimized using Gaussian 16. Their optimization was verified through a frequency calculation at the ωB97X-D/6-311G(2d,p) theory level. The structures with the lowest energy levels are depicted in Figure 3. In all instances, the CO_2 molecule is oriented such that the carbon atom is proximate to an electronegative atom, such as oxygen or nitrogen, at interatomic distances of 2.8–3.6 Å. This suggests short-range interactions, such as dipole–dipole or hydrogen bonding; therefore, a Van der Waals interaction seemed imminent. It is noteworthy that in the case of chitosan and PVP, the interatomic distances are smaller, which aligns with the Highest Occupied Molecular Orbitals (HOMO) shape in both cases. This implies a superior molecular interaction with the CO_2 molecule, facilitating its absorption.

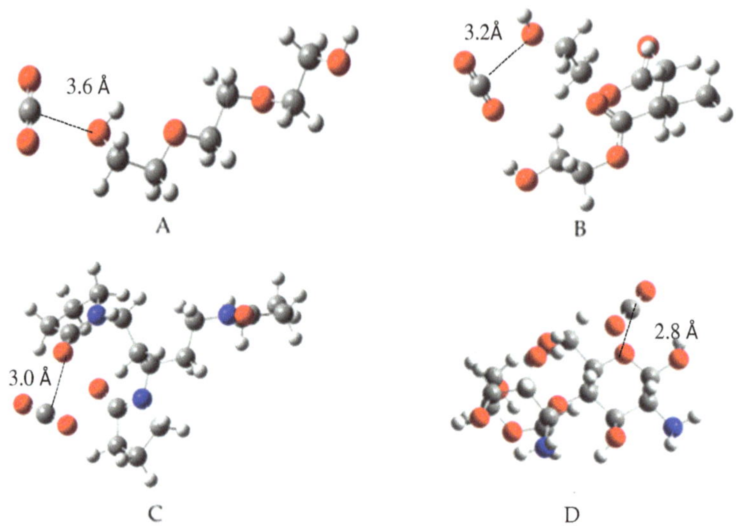

Figure 3. Minimum molecular structure for complexes of: (**A**): CO_2–polyethyleneglycol; (**B**): CO_2–2-hydroximethacrylate; (**C**): CO_2–polyvinylpyrrolidone; (**D**): CO_2–chitosan.

2.4. Thermodynamics of Molecular Complexation

Molecular recognition occurs through an orbital interaction, specifically, the Highest Occupied Molecular Orbital (HOMO) of one participant must interact with the Lowest Unoccupied Molecular Orbital (LUMO) of the other species. Once the structures of the frontier orbitals have been examined, suggesting a probable molecular recognition, it becomes necessary to determine the interaction energy. To do this, as explained in the methodology, we used binding energy. A negative value implies a spontaneous attraction between the species, signifying that molecular recognition takes place spontaneously. The feasibility of this interaction and the potential adsorption of the absorbate into the absorbent matrix depend on the magnitude of this parameter. Thus, values between -50 and -1 kJ/mol would indicate physisorption, meaning CO_2 is absorbed without the presence of a covalent bond [37,38]. These results are qualitatively in line with some experimental works [21–23].

Table 1 shows that all ΔEb (binding energy difference) values are negative, indicating an attractive interaction. The enthalpy (H) values suggest that the absorbent–absorbate complex formation is exothermic; the entropy change (S) suggests a diminishing of disorder during the complexes' formation; chitosan and PVP are the dimers with the major molar volume; in addition are the models with the more negative values of H, suggesting a greater molecular interaction between CO_2 therefore, As expected, these two complexes (chitosan–CO_2 and PVP–CO_2) showed the greater change in entropy values.

Table 1. Binding energy, free energy, enthalpy, and interaction entropy for studied complexes at ωB97X-D/6-311G(2d,p).

Complex	ΔΔG (kcal/mol)	ΔΔH (kcal/mol)	ΔΔS (Cal/mol)	ΔEb (kcal/mol)	Predicted Absorption Yes/No	Experimental Work
Chitosan–CO_2	−3.50	−11.21	−25.88	−5.41	Yes	[24]
PVP–CO_2	−3.12	−9.99	−23.07	−4.83	Yes	[25]
EG–CO_2	−1.98	−6.34	−14.64	−3.06	Yes	[21]
HEAC–CO_2	−2.78	−8.91	−20.56	−4.29	Yes	No

The Gibbs free energy (G) values confirm that the processes are spontaneous.

In other words, there is a high probability that the hydrogels constructed from these monomers will absorb CO_2 through physisorption, as corroborated by the magnitude of these values [39]. It is important to note that, based on these results, and knowing that since Gibbs energy is connected to the equilibrium constant through the formula $\Delta G^\circ = -RTlnKe$ [40], the order of CO_2 absorption would be chitosan > PVP > HEAC > PEG. Interestingly, all the first three compounds in the mentioned order contain nitrogen atoms in their molecular structure, demonstrating the significance of this atom in the absorption and transformation of CO_2.

The last two columns are related to the absorption predicted for the methods used herein and their experimental reports as CO_2-absorbent. Interesting, chitosan, PVP, and PEG are reported as CO_2-absorbent; therefore, the methodology used herein qualitatively reproduce the experimental results, suggesting a good starting point to predict new hydrogel matrices for CO_2 absorption. On the other hand, in the best of our knowledge, HEAC is not reported as a single hydrogel matrix for CO_2 absorption; however, the results presented herein suggest it could be a good CO_2 absorbent.

Computational calculations identify chitosan as the most promising absorbent. Chitosan offers the advantage of having more nitrogen groups in its structure, and possesses a more suitable reticular structure that facilitates the diffusion of CO_2 through the polymeric matrix [41]. Additionally, it is obtained from natural sources, making its natural degradation much more spontaneous than the other monomers listed in Table 1. However, the use of other monomers is not ruled out, and it could be interesting to explore their potential in CO_2 capture applications in future.

2.5. Temperature Dependence of Complexation Gibbs Energy

Given that CO_2 is a gas, it is anticipated that its absorption capacity will diminish as the temperature rises [37,42]. This phenomenon can be attributed to the thermodynamics of the exothermic CO_2 absorption system, which may induce reversible reactions at elevated temperatures. Consequently, it becomes pertinent to calculate the transition temperature, denoted as the temperature at which ΔG (Gibbs free energy change) equals zero.

The Gibbs free energy, which determines the equilibrium conditions of chemical reactions and materials stability, depends on temperature, and changing the temperature can change the sign of ΔG; in this sense, the variation of free energy with temperature for the complexation of CO_2 absorbance was calculated by the already known Helmholtz equation, and the results are shown in Table 2.

Table 2. Temperature dependence of Gibbs energy at 1 atm, calculated at ωB97X-D/6-311G(2d,p).

Complex/T (°C)	25 °C	100 °C	200 °C	Transition T °C
Chitosan–CO_2	−3.5	−1.55	1.04	160.00
PVP–CO_2	−3.12	−1.38	0.93	159.88
PEG–CO_2	−1.98	−0.88	0.59	159.91
HEAC–CO_2	−2.78	−1.24	0.82	160.22

Table 2 shows the values of Gibbs free energy in a temperature range of 25 to 200 °C. The Gibbs free energy decreases as the temperature increases and becomes zero (transition temperature), around 160 °C in each case. This result means that CO_2 is released above this temperature; that is, after this temperature, the complex becomes unstable. The value of T of transition (the temperature at which the process begins to become non-spontaneous), around 160 °C, indicates that from this temperature, any amount of CO_2 begins to separate from the polymeric matrix. This temperature seems suitable for the use of these polymeric matrices as absorbents in the post-combustion process of green hydrogen.

3. Conclusions

In conclusion, the application of density functional theory (DFT) in designing biodegradable hydrogels for carbon dioxide (CO_2) capture presents a significant advancement in the field of materials science. The results obtained in this work are qualitatively in line with experimental results. Through the exploration of molecular interactions, frontier molecular theory, and thermodynamic parameters, this study has shed light on the potential of polymeric matrices such as poly(2-hydroxyethyl methacrylate), poly(ethylene glycol), chitosan, and polyvinylpyrrolidone in effectively absorbing CO_2.

The findings of this research not only suggest the stability of the complexes formed between the hydrogel models and CO_2 but also suggest promising prospects for the development of novel materials for CO_2 capture applications. The calculated binding energies and Gibbs energy equations not only are qualitatively in line with experimental results but provide valuable insights into the affinity between the hydrogel models and CO_2, emphasizing the critical role of frontier molecular orbitals in designing efficient CO_2-absorbent materials. These methodological approaches offer a theoretical framework for future research and development in the design of eco-friendly materials for mitigating greenhouse gas emissions.

Overall, the outcomes of this study contribute to the growing body of literature on carbon capture technologies and highlight the potential of biodegradable hydrogels as sustainable solutions for addressing climate change. Further optimization and exploration of these materials across various environmental conditions are warranted to maximize their performance and applicability in real-world CO_2 capture scenarios.

By emphasizing the significance of this research in advancing the development of environmentally friendly materials and its implications for CO_2 mitigation strategies, this study paves the way for future investigations and innovations in the field of materials science and sustainable technology.

4. Materials and Methods

4.1. Molecular Modelling for Polymers with High CO_2 Absorption Capacity

The study aimed to design biodegradable hydrogels for CO_2 absorption using Density Functional Theory (DFT). The polymeric matrices, including poly(2-hydroxyethyl methacrylate), poly(ethylene glycol), chitosan, and polyvinylpyrrolidone, were modelled by searching for their 2D structures on the PubChem website (https://pubchem.ncbi.nlm.nih.gov/; accessed on 18 August 2023) as shown in Figure 4. To determine the lowest energy conformer in polymeric structures such as dimers and trimers, an advanced systematic rotor search was performed using Avogadro 1.2. software [43]. This study allowed us to explore the conformational space of polymeric structures and identify the arrangements of low-energy atoms. Once the global minimal structures were identified, they were saved in .gjf format and further optimized using density functional theory (DFT). Because Standard DFT methods often struggle to accurately describe dispersion interactions, which are important in systems like hydrogels, in this study the exchange-correlation functional ωB97X-D, which includes both large and short dispersion energy corrections [30], and the base ensemble 6-311G(2d,p) in Gaussian view 6 were used.

The use of DFT in this study provides a theoretical framework for designing hydrogels with optimal CO_2 absorption capacity, paving the way for further research and development in this field. It is important to note that the specific details and methodologies of the study may vary depending on the research context and objectives [28].

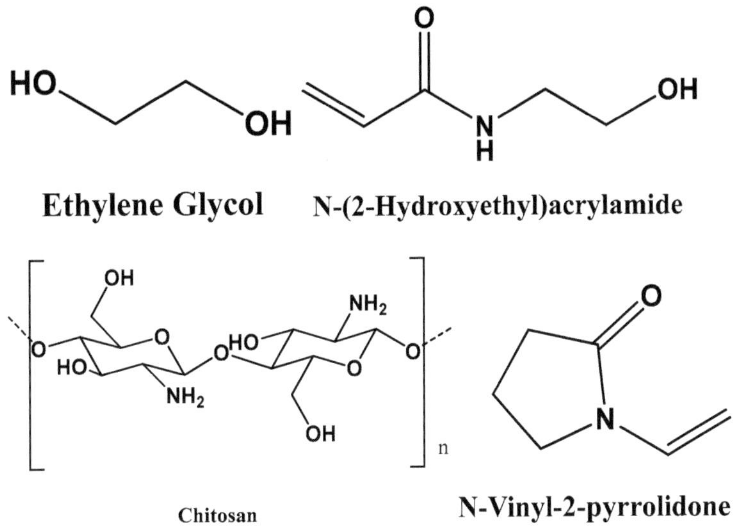

Figure 4. 2D molecular structures of monomers.

4.2. Computational Analysis of Polymers

The structural stability of the systems was ensured by conducting vibrational frequency calculations to confirm the absence of imaginary frequencies within all compounds. Following this validation, optimization of these structures was performed at the ωB97XD/6-311++G(2d,2p) level of theory to facilitate the formation of complexes with carbon dioxide (CO_2) at the same theoretical level. The determination of the minimum energy complex involved molecular dynamics calculations, employing the Autodock Vina tool [43]. To calculate the total binding energies of the optimized polymers, the following equation was used:

$$\Delta E_T = E^{(A,B)} - \left(E^A - E^B\right) + BSSE \quad (1)$$

where E^A and E^B represent the electronic energy of molecules A and B, respectively; $E^{(A,B)}$ represents the electronic energy of the A, B complex, and BSSE represents the overlap error of the base set. A negative value for the change in total energy (ΔET) indicates a favorable interaction between two molecules and an exothermic process [44,45]. Similarly, the thermodynamics of complexations was studied using the Gibbs equation as follows:

$$\Delta G_{binding} = \Delta H_{binding} - T\Delta S_{binding} \quad (2)$$

$$\Delta H_{binding} = H_{complex} - (H_{monom} + H_{dimer}) \quad (3)$$

$$\Delta S_{binding} = S_{complex} - (S_{monom} + S_{dimer}) \quad (4)$$

In this context, H and S represent the enthalpy and entropy values of the complex, drug, and dimer, respectively. These parameters are calculated at a standard temperature of 298.15 K and a pressure of 1 atm, with data extracted from the results of frequency calculations. A negative ΔE_T value signifies a favorable interaction between two molecules, indicative of an exothermic process. Likewise, H and S pertain to the enthalpy and entropy values of the complex, CO_2, and the hydrogel, all computed under the same standard conditions. A negative ΔG value denotes the spontaneous formation of the complex at the specified temperature, ensuring the spontaneous absorption of CO_2 within the hydrogel. A negative ΔH value indicates an exothermic process, while positive entropy values suggest an increase in degrees of freedom, signifying that the inclusion complex involves low-energy molecular interactions.

This comprehensive study encompasses a range of temperatures, enabling the construction of a free energy profile that will pinpoint the optimal temperature gradient for absorption and desorption processes. Additionally, the stability of these complexes in solution was assessed through simulation utilizing the CPCM (Conductor Polarizable Continuum Model) solvation method, ensuring a thorough examination of their behavior in a liquid environment [37].

Author Contributions: Conceptualization, E.M., D.I. and M.M.-L.; methodology, E.M., D.I., M.M.-L. and D.C.C.-H.; formal analysis, E.M., D.I., M.M.-L., D.C.C.-H., S.G.-F. and M.S.; investigation, E.M., D.I., M.M.-L., D.C.C.-H., S.G.-F. and M.S.; resources, S.G.-F. and M.S.; data curation, E.M., D.I., M.M.-L., D.C.C.-H., S.G.-F. and M.S.; E.M., S.G.-F. and M.S.; writing—review and editing, E.M., D.I., M.M.-L., D.C.C.-H., S.G.-F. and M.S.; supervision, E.M., D.I., M.M.-L., D.C.C.-H., S.G.-F. and M.S.; project administration, E.M., S.G.-F. and M.S.; funding acquisition, S.G.-F. and M.S. All authors have read and agreed to the published version of the manuscript.

Funding: This research was funded by green-Gas, with project number P.2.0.001-2022 by Centro de Investigación e Innovación en Energía y Gas—CIIEG, Promigas S.A. E.S.P., Barranquilla, 11001, Colombia.

Institutional Review Board Statement: Not applicable.

Informed Consent Statement: Not applicable.

Data Availability Statement: The data presented in this study are openly available in article.

Acknowledgments: All individual authors would like to extend their profound gratitude to the Promigas Building enterprise for their invaluable support towards this project.

Conflicts of Interest: The authors Domingo Cesar Carrascal-Hernandez, Maximiliano Mendez-Lopez, Daniel Insuasty, Samira García-Freites, Marco Sanjuan and Edgar Márquez, declare no conflicts of interest.

References

1. Luqman, M.; Rayner, P.J.; Gurney, K.R. On the Impact of Urbanisation on CO_2 Emissions. *Npj Urban Sustain.* **2023**, *3*, 6. [CrossRef]
2. Shivanna, K.R. Climate Change and Its Impact on Biodiversity and Human Welfare. *Proc. Indian Natl. Sci. Acad.* **2022**, *88*, 160–171. [CrossRef]
3. Ren, Q.; Wei, S.; Du, J.; Wu, P. Research Progress and Perspectives on Carbon Capture, Utilization, and Storage (CCUS) Technologies in China and the USA: A Bibliometric Analysis. *Environ. Sci. Pollut. Res.* **2023**, *30*, 76437–76454. [CrossRef] [PubMed]
4. Dziejarski, B.; Krzyżyńska, R.; Andersson, K. Current Status of Carbon Capture, Utilization, and Storage Technologies in the Global Economy: A Survey of Technical Assessment. *Fuel* **2023**, *342*, 127776. [CrossRef]
5. Pal, M.; Karaliūtė, V.; Malik, S. Exploring the Potential of Carbon Capture, Utilization, and Storage in Baltic Sea Region Countries: A Review of CCUS Patents from 2000 to 2022. *Processes* **2023**, *11*, 605. [CrossRef]
6. O'Neill, S. Global CO_2 Emissions Level Off in 2019, with a Drop Predicted in 2020. *Engineering* **2020**, *6*, 958–959. [CrossRef] [PubMed]
7. Szulczewski, M.L.; MacMinn, C.W.; Herzog, H.J.; Juanes, R. Lifetime of Carbon Capture and Storage as a Climate-Change Mitigation Technology. *Proc. Natl. Acad. Sci. USA* **2012**, *109*, 5185–5189. [CrossRef] [PubMed]
8. Boot-Handford, M.E.; Abanades, J.C.; Anthony, E.J.; Blunt, M.J.; Brandani, S.; Mac Dowell, N.; Fernández, J.R.; Ferrari, M.-C.; Gross, R.; Hallett, J.P.; et al. Carbon Capture and Storage Update. *Energy Environ. Sci.* **2014**, *7*, 130–189. [CrossRef]
9. Yu, W.; Wang, T.; Park, A.-H.A.; Fang, M. Review of Liquid Nano-Absorbents for Enhanced CO_2 Capture. *Nanoscale* **2019**, *11*, 17137–17156. [CrossRef] [PubMed]
10. Wang, T.; Liu, F.; Ge, K.; Fang, M. Reaction Kinetics of Carbon Dioxide Absorption in Aqueous Solutions of Piperazine, N-(2-Aminoethyl) Ethanolamine and Their Blends. *Chem. Eng. J.* **2017**, *314*, 123–131. [CrossRef]
11. Mantripragada, H.C.; Zhai, H.; Rubin, E.S. Boundary Dam or Petra Nova—Which Is a Better Model for CCS Energy Supply? *Int. J. Greenh. Gas Control* **2019**, *82*, 59–68. [CrossRef]
12. Stéphenne, K. Start-up of World's First Commercial Post-Combustion Coal Fired CCS Project: Contribution of Shell Cansolv to SaskPower Boundary Dam ICCS Project. *Energy Procedia* **2014**, *63*, 6106–6110. [CrossRef]
13. Yu, X.; Catanescu, C.O.; Bird, R.E.; Satagopan, S.; Baum, Z.J.; Lotti Diaz, L.M.; Zhou, Q.A. Trends in Research and Development for CO_2 Capture and Sequestration. *ACS Omega* **2023**, *8*, 11643–11664. [CrossRef] [PubMed]
14. Wu, X.; Wang, M.; Liao, P.; Shen, J.; Li, Y. Solvent-Based Post-Combustion CO_2 Capture for Power Plants: A Critical Review and Perspective on Dynamic Modelling, System Identification, Process Control and Flexible Operation. *Appl. Energy* **2020**, *257*, 113941. [CrossRef]

15. Ahmadi, M.; Ghaemi, A.; Qasemnazhand, M. Lithium Hydroxide as a High Capacity Adsorbent for CO_2 Capture: Experimental, Modeling and DFT Simulation. *Sci. Rep.* **2023**, *13*, 7150. [CrossRef] [PubMed]
16. Stankovic, B.; Barbarin, I.; Sanz, O.; Tomovska, R.; Ruipérez, F. Experimental and Theoretical Study of the Effect of Different Functionalities of Graphene Oxide/Polymer Composites on Selective CO_2 Capture. *Sci. Rep.* **2022**, *12*, 15992. [CrossRef]
17. Torkzadeh, M.; Moosavi, M. CO_2 Capture Using Dicationic Ionic Liquids (DILs): Molecular Dynamics and DFT-IR Studies on the Role of Cations. *J. Chem. Phys.* **2023**, *158*. [CrossRef]
18. Torkzadeh, M.; Moosavi, M. DFT and COSMO-RS Studies on Dicationic Ionic Liquids (DILs) as Potential Candidates for CO_2 Capture: The Effects of Alkyl Side Chain Length and Symmetry in Cations. *RSC Adv.* **2022**, *12*, 35418–35435. [CrossRef]
19. Scivetti, I.; Persson, M. Frontier Molecular Orbitals of a Single Molecule Adsorbed on Thin Insulating Films Supported by a Metal Substrate: Electron and Hole Attachment Energies. *J. Phys. Condens. Matter* **2017**, *29*, 355002. [CrossRef]
20. Zagni, C.; Coco, A.; Dattilo, S.; Patamia, V.; Floresta, G.; Fiorenza, R.; Curcuruto, G.; Mecca, T.; Rescifina, A. HEMA-Based Macro and Microporous Materials for CO_2 Capture. *Mater. Today Chem.* **2023**, *33*, 101715. [CrossRef]
21. Li, X.; Hou, M.; Zhang, Z.; Han, B.; Yang, G.; Wang, X.; Zou, L. Absorption of CO_2 by Ionic Liquid/Polyethylene Glycol Mixture and the Thermodynamic Parameters. *Green Chem.* **2008**, *10*, 879. [CrossRef]
22. Foong, S.Y.; Chan, Y.H.; Yiin, C.L.; Lock, S.S.M.; Loy, A.C.M.; Lim, J.Y.; Yek, P.N.Y.; Wan Mahari, W.A.; Liew, R.K.; Peng, W.; et al. Sustainable CO_2 Capture via Adsorption by Chitosan-Based Functional Biomaterial: A Review on Recent Advances, Challenges, and Future Directions. *Renew. Sustain. Energy Rev.* **2023**, *181*, 113342. [CrossRef]
23. Sharma, T.; Joshi, A.; Jain, A.; Chaturvedi, K.R. Enhanced Oil Recovery and CO_2 Sequestration Potential of Bi-Polymer Polyvinylpyrrolidone-Polyvinyl Alcohol. *J. Pet. Sci. Eng.* **2022**, *211*, 110167. [CrossRef]
24. Goeminne, R.; Vanduyfhuys, L.; Van Speybroeck, V.; Verstraelen, T. DFT-Quality Adsorption Simulations in Metal—Organic Frameworks Enabled by Machine Learning Potentials. *J. Chem. Theory Comput.* **2023**, *19*, 6313–6325. [CrossRef] [PubMed]
25. Makkar, P.; Ghosh, N.N. A Review on the Use of DFT for the Prediction of the Properties of Nanomaterials. *RSC Adv.* **2021**, *11*, 27897–27924. [CrossRef] [PubMed]
26. Vogel, P.; Möller, N.; Qaisrani, M.N.; Bista, P.; Weber, S.A.L.; Butt, H.-J.; Liebchen, B.; Sulpizi, M.; Palberg, T. Charging of Dielectric Surfaces in Contact with Aqueous Electrolytes—The Influence of CO_2. *J. Am. Chem. Soc.* **2022**, *144*, 21080–21087. [CrossRef] [PubMed]
27. Haskopoulos, A.; Maroulis, G. Carbon Dioxide Interacting with Rare Gases: Insights from High-Level Ab Initio Calculations of Polarizability and Hyperpolarizability Effects. *Chem. Phys.* **2016**, *475*, 90–103. [CrossRef]
28. Beil, R.L.; Hinde, R.J. Ab Initio Electrical Properties of CO_2: Polarizabilities, Hyperpolarizabilities, and Multipole Moments. *Theor. Chem. Acc.* **2021**, *140*, 120. [CrossRef]
29. Ariga, K.; Ito, H.; Hill, J.P.; Tsukube, H. Molecular Recognition: From Solution Science to Nano/Materials Technology. *Chem. Soc. Rev.* **2012**, *41*, 5800. [CrossRef]
30. Yang, W.; He, H.; Drueckhammer, D.G. Computer-Guided Design in Molecular Recognition: Design and Synthesis of a Glucopyranose Receptor. *Angew. Chem.* **2001**, *113*, 1764–1768. [CrossRef]
31. Fiedler, D.; Leung, D.H.; Bergman, R.G.; Raymond, K.N. Selective Molecular Recognition, C−H Bond Activation, and Catalysis in Nanoscale Reaction Vessels. *Acc. Chem. Res.* **2005**, *38*, 349–358. [CrossRef] [PubMed]
32. Yan, Z.; Wang, J. Specificity Quantification of Biomolecular Recognition and Its Implication for Drug Discovery. *Sci. Rep.* **2012**, *2*, 309. [CrossRef]
33. Shawon, J.; Khan, A.M.; Rahman, A.; Hoque, M.M.; Khan, M.A.K.; Sarwar, M.G.; Halim, M.A. Molecular Recognition of Azelaic Acid and Related Molecules with DNA Polymerase I Investigated by Molecular Modeling Calculations. *Interdiscip. Sci.* **2018**, *10*, 525–537. [CrossRef]
34. Furusho, Y.; Endo, T. Reversible Capture and Release of Carbon Dioxide by Binary System of Polyamidine and Polyethylene Glycol. *Polym. Bull.* **2017**, *74*, 1207–1219. [CrossRef]
35. Farmanzadeh, D.; Ghazanfary, S. DFT Studies of Functionalized Zigzag and Armchair Boron Nitride Nanotubes as Nanovectors for Drug Delivery of Collagen Amino Acids. *Struct. Chem.* **2014**, *25*, 293–300. [CrossRef]
36. Seeman, J.I. Kenichi Fukui, Frontier Molecular Orbital Theory, and the Woodward-Hoffmann Rules. Part II. A Sleeping Beauty in Chemistry [†**]. *Chem. Rec.* 2022; 22, e202100300. [CrossRef]
37. Akkus, E.; Tayfuroglu, O.; Yildiz, M.; Kocak, A. Accurate Binding Free Energy Method from End-State MD Simulations. *J. Chem. Inf. Model.* **2022**, *62*, 4095–4106. [CrossRef]
38. Li, J.; Sode, O.; Voth, G.A.; Hirata, S. A Solid–Solid Phase Transition in Carbon Dioxide at High Pressures and Intermediate Temperatures. *Nat. Commun.* **2013**, *4*, 2647. [CrossRef]
39. Hensley, A.J.R.; Ghale, K.; Rieg, C.; Dang, T.; Anderst, E.; Studt, F.; Campbell, C.T.; McEwen, J.-S.; Xu, Y. DFT-Based Method for More Accurate Adsorption Energies: An Adaptive Sum of Energies from RPBE and VdW Density Functionals. *J. Phys. Chem. C* **2017**, *121*, 4937–4945. [CrossRef]
40. Middelburg, J.J. The Gibbs Free Energy. In *Thermodynamics and Equilibria in Earth System Sciences: An Introduction*; Springer: Cham, Switzerland, 2024; pp. 35–46.
41. Aranaz, I.; Alcántara, A.R.; Civera, M.C.; Arias, C.; Elorza, B.; Heras Caballero, A.; Acosta, N. Chitosan: An Overview of Its Properties and Applications. *Polymers* **2021**, *13*, 3256. [CrossRef] [PubMed]

42. Rashid, H.; Hasan, N.; Mohamad Nor, M.I. Temperature Peak Analysis and Its Effect on Absorption Column for CO_2 Capture Process at Different Operating Conditions. *Chem. Prod. Process Model.* **2014**, *9*, 105–115. [CrossRef]
43. Eberhardt, J.; Santos-Martins, D.; Tillack, A.F.; Forli, S. AutoDock Vina 1.2.0: New Docking Methods, Expanded Force Field, and Python Bindings. *J. Chem. Inf. Model.* **2021**, *61*, 3891–3898. [CrossRef]
44. Ouyang, J.F.; Bettens, R.P.A. Many-Body Basis Set Superposition Effect. *J. Chem. Theory Comput.* **2015**, *11*, 5132–5143. [CrossRef]
45. Miliordos, E.; Aprà, E.; Xantheas, S.S. Benchmark Theoretical Study of the π–π Binding Energy in the Benzene Dimer. *J. Phys. Chem. A* **2014**, *118*, 7568–7578. [CrossRef]

gels

Article

Brown Adipose Stem Cell-Loaded Resilin Elastic Hydrogel Rebuilds Cardiac Function after Myocardial Infarction via Collagen I/III Reorganisation

Le Zhao [1,†], Huaying Liu [1,†], Rui Gao [1,2], Kaihui Zhang [1,3], Yuxuan Gong [1,4], Yaya Cui [1], Shen Ke [1], Jing Wang [5,*] and Haibin Wang [1,*]

1 College of Life Sciences and Bioengineering, School of Physical Science and Engineering, Beijing Jiaotong University, Beijing 100044, China; 23121758@bjtu.edu.cn (Y.C.)
2 Department of Wound Infection and Drug, Army Medical Center of PLA (Daping Hospital), Army Medical University, Chongqing 400042, China
3 School of Life Sciences, Inner Mongolia University, Hohhot 010000, China
4 Institute of Health Service and Transfusion Medicine, Academy of Military Medical Sciences, Beijing 100850, China
5 State Key Laboratory of Toxicology and Medical Countermeasures, Beijing Institute of Pharmacology and Toxicology, Beijing 100850, China
* Correspondence: jingw_biomed@163.com (J.W.); wanghb@bjtu.edu.cn (H.W.)
† These authors contributed equally to this work.

Abstract: Irreversible fibrosis following myocardial infarction (MI) stiffens the infarcted myocardium, which remains challenging to restore. This study aimed to investigate whether the injectable RLP12 hydrogel, derived from recombinant resilin protein, could serve as a vehicle for stem cells to enhance the function of the infarcted myocardium. The RLP12 hydrogel was prepared and injected into the myocardium of rats with MI, and brown adipose-derived mesenchymal stem cells (BADSCs) were loaded. The survival and differentiation of BADSCs in vivo were investigated using immunofluorescence one week and four weeks after treatment, respectively. The heart function, MI area, collagen deposition, and microvessel density were further assessed four weeks after treatment through echocardiography, histology, immunohistochemistry, and immunofluorescence. The RLP12 hydrogel was prepared with a shear modulus of 10–15 kPa. Four weeks after transplantation, the RLP12 hydrogel significantly improved cardiac function by increasing microvessel density and reducing infarct area size and collagen deposition in MI rats. Furthermore, the distribution ratio of collagen III to I increased in both the centre and edge areas of the MI, indicating the improved compliance of the infarct heart. Moreover, the RLP12 hydrogel also promoted the survival and differentiation of BADSCs into cardiac troponin T- and α-smooth muscle-positive cells. The RLP12 hydrogel can be utilised as an injectable vehicle of BADSCs for treating MI and regulating collagen I and III expression profiles to improve the mechanical microenvironment of the infarct site, thereby restoring heart function. The study provides novel insights into the mechanical interactions between the hydrogel and the infarct microenvironment.

Keywords: resilin; injectable hydrogel; myocardial tissue engineering; brown adipose-derived stem cells; myocardial infarction

Citation: Zhao, L.; Liu, H.; Gao, R.; Zhang, K.; Gong, Y.; Cui, Y.; Ke, S.; Wang, J.; Wang, H. Brown Adipose Stem Cell-Loaded Resilin Elastic Hydrogel Rebuilds Cardiac Function after Myocardial Infarction via Collagen I/III Reorganisation. *Gels* **2024**, *10*, 568. https://doi.org/10.3390/gels10090568

Academic Editor: Maria Valentina Dinu

Received: 19 July 2024
Revised: 23 August 2024
Accepted: 29 August 2024
Published: 31 August 2024

1. Introduction

The mortality rate of cardiovascular diseases remains the highest [1]. Myocardial infarction (MI) is a disease with a poor prognosis that inevitably leads to heart failure. In the later stage of MI, collagen I stacks at the infarct tissue, the infarcted myocardial tissue develops fibrosis, and the compensatory hypertrophy of the residual myocardium causes the loss of contractility, irreversible ventricular remodelling, and heart failure [2]. The restricted regenerative capacity of the heart cannot restore or replace damaged tissue, causing local tissue death and ultimately culminating in mortality [3]. Various therapeutics, such as drug thrombolysis and interventional approaches, have been used in clinical scenarios, with varying degrees of success, concentrating on restoring blood perfusion; only a few methods have been proposed to improve the local microenvironment of MI [4,5].

Cell transplantation has shown great potential in enhancing heart function [6]. Several stem cell types, such as mesenchymal, embryonic, adipose mesenchymal, and induced pluripotent stem cells have been used to treat MI through myocardial regeneration [7,8].

Brown adipose-derived mesenchymal stem cells (BADSCs) have advantages including broad availability, low immunogenicity, notable myocardial differentiation capacity, and paracrine effects and are available for the cardiac tissue engineering of seed cells to restore damaged myocardium [9,10]. The infarct site is harsh and lacks a supportive matrix for cells. By delivering cells alone, it is difficult to achieve long-term cell retention and sufficient cell differentiation; therefore, other approaches are required.

Injectable cardiac tissue engineering is an appealing method to improve the harsh microenvironment of the infarcted myocardium, which can be regulated at the molecular, cellular, and individual levels and can directly deliver drugs to the MI site [11–13]. Applying carriers to load cells and cytokines can achieve proangiogenic and anti-fibrotic effects. Among the previously proposed carriers, hydrogels have the unique capacity to construct a microenvironment that can resist rigorous damage and avoid the degradation or rapid inhibition of loaded substances [14]. Vectors are crucial for protection, delivering loads, and improving therapeutic efficacy.

A hydrogel is one of the hot spots in the research field of biomaterials and regenerative medicine, and research progresses rapidly. Recently, scientists investigated the thermal performance and mechanical stability of methacrylic acid porous hydrogels in an aqueous medium at different initial temperatures and hydrogel volume fraction using the molecular dynamics simulation [15]. In addition, hydrogels have been developed as a promising vector carrier. Some injectable hydrogels are administered in a liquid state to the infarcted region and then undergo a rapid change from liquid to gel state owing to the unique microenvironment conditions. Materials such as collagen, dECM, polyethylene glycol (PEG), Polyvinyl Alcohol (PVA), and chitosan are used in injectable treatments [16]. Brown adipose stem cells that have been loaded in a pH-sensitive chitosan hydrogel used for myocardial regeneration showed a high cell survival rate [17]. In addition, alginate gels, specifically Algisyl LVRTM [18] and IK-5001 [19], and decellularized matrix hydrogels, known as VentriGel [20], have been approved and have progressed into clinical trials, demonstrating notable therapeutic effects. Studies have identified their beneficial properties, such as rich sources and low immunogenicity. However, reports on the mechanical variation of biomaterials used for treating MI are limited.

Myocardial tissue fibrosis is accompanied by extracellular matrix (ECM) replacement with denatured collagen. These detrimental matrix changes lead to poor mechanical properties and a less conducive environment for recovery (systolic dysfunction and arrhythmia) [21]. The mechanical clues of fibrotic microenvironment are essential to cell–ECM and cell–cell reactions [22]. Elastomeric hydrogels with favorable elasticity and extensibility can improve the MI microenvironment by providing mechanical support and replacing the ECM. Its intervention can improve local myocardial movement disorders in the infarcted tissue, alleviate stress on the ventricular wall, and facilitate damage repair [23,24]. A correlation between the mechanical properties of hydrogels and the stress of the infarcted ventricular wall has been identified, and a stiffness of less than 50 kPa can reduce muscle fibre stress [24]. The synthetic elastic hydrogel poly (octamethylene maleate (anhydride) citrate) (POMAC) was confirmed to achieve provascularisation in rat and swine models of MI [25]. The research on mechanical clues is essential for the growth of cardiomyocytes. When the substrates with a range of elastic moduli reach a similar stiffness to native tissue, the behaviour and contractile function is promoted [26].

Resilin, an excellent natural elastin, has been further developed. Natural resilin, a water-soluble elastic protein found in insect arthropods, consists of a structurally flexible and flowable array of randomly oriented polypeptide chains [27], which are interconnected through the covalent crosslinking of dityrosine and trityrosine to form a stable structure [28]. It has exceptional mechanical properties, including a low modulus of elasticity, a high-tensile strain capacity, and outstanding durability [29]. The tensile modulus of natural resilin is 600–2000 kPa [30], and its resilience can reach 92% [31]. The protein structure includes three exons: N-terminal elastic, chitin-binding, and C-terminal elastic domains [32]. Both exons 1 and 3 contain a large number of repetitive sequences that provide resilin elasticity. Thus, exon 1 may be a critical structural domain for providing resilience [33,34].

The resilin-like protein (RLP) has strong mechanical and biological functions obtained by repeating the core elastic sequence of resilin exon 1 (GGRPSDSYGAPGGGN) 12 times. RLP-based hydrogels obtained using trimethylolpropane (THP) as a crosslinking agent demonstrate remarkable resilience and facilitate the attachment and growth of human mesenchymal stem cells [35] while being associated with good biocompatibility [36]. RLP-derived biomaterials have been effectively applied in the fields of tissue engineering and drug delivery, particularly in tissue engineering research involving blood vessels [37], vocal cords [38], cartilage [39], and muscles [40].

In this study, an injectable RLP12 hydrogel with mechanical properties is synthesised, and whether the hydrogel could serve as a vehicle for stem cells to enhance the function of the infarcted myocardium is investigated. The study is designed as shown in Figure 1.

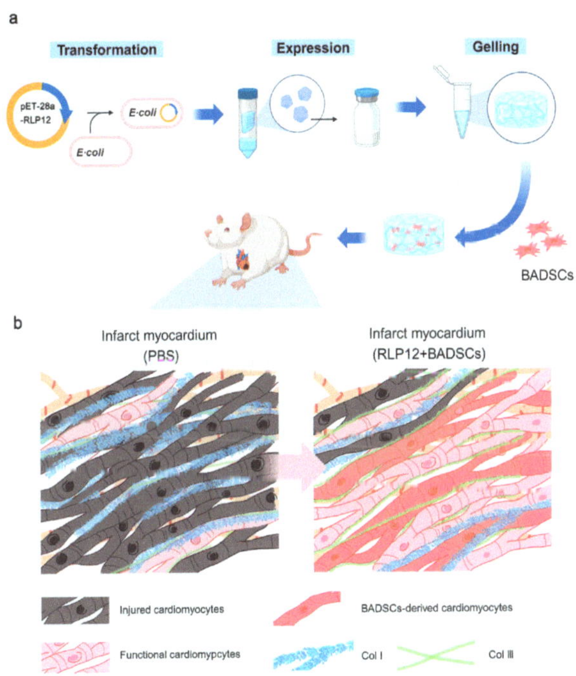

Figure 1. A schematic demonstrating the preparation and application of the RLP12-based delivery system. (**a**) Schematic image for experimental design. (**b**) Illustrations of infarcted myocardium after treatment.

2. Results and Discussion

2.1. Synthesis of RLP12 Expression Vector and Protein Expression

A schematic of the construction of the RLP12 expression vector is shown in Figure 2a. The RLP12 gene sequence was synthesised using the pUC57 plasmid as a vector. After transformation with the recombinant vector pET28a-RLP12, transformants were selected and identified. The results showed that the small fragment band of lane 1 was approximately 800 bp, which is consistent with RLP12, indicating that the expression strain *E. coli* BL21 (DE3)-pET-28a-RLP12 was successfully constructed (Figure 2b). The RLP12 protein eluate collected after Ni^{+}-column purification was subjected to sodium dodecyl sulphate-polyacrylamide gel electrophoresis (SDS-PAGE) (Figure 2c). There is a clear protein band between 25 and 34 kDa, which is that of the recombinant RLP12 protein.

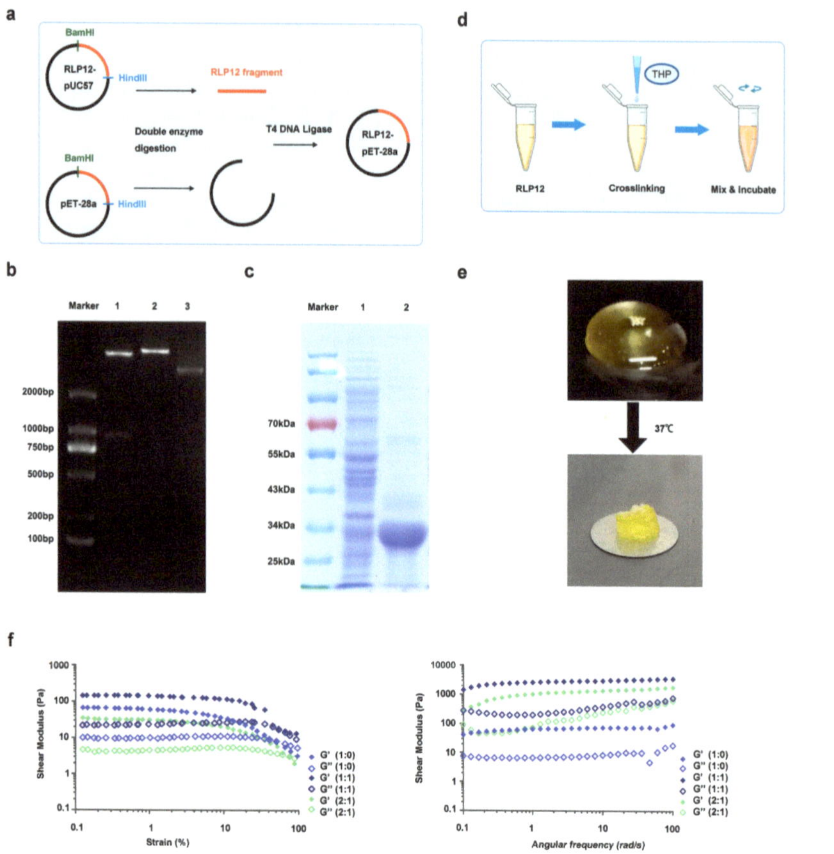

Figure 2. The preparation and characterisation of the RLP12 hydrogel. (**a**) A schematic of the construction of the RLP12 expression vector. (**b**) The identification of the pET-28a-RLP12 recombinant clone vector by dual enzyme digestion. (Lane M: DNA marker; Lane 1: pET-28a-RLP12 plasmid digested by BamH I and Hind III; Lane 2: pET-28a-RLP12 plasmid digested by Hind III; Lane 3: pET-28a-RLP12 plasmid). (**c**) The purification of the RLP12 protein using SDS-PAGE. (Lane M: Prestained protein marker; Lane 1: cell supernatant of bacterial disruption; Lane 2: eluate of target protein). (**d**) A schematic of the process of RLP12 solution gelation. (**e**) RLP 12 solution gelation. (**f**) The oscillatory shear rheological characterisation of resilin hydrogel materials based on RLP12. (Left: strain sweep measurements; right: frequency sweep measurements).

2.2. Preparation and Mechanical Properties Identification of RLP12 Hydrogel

Three kinds of RLP12 hydrogels were prepared according to the ratios of RLP12 to PEG: RLP12, PEG = 1:0; RLP12, PEG = 1:1; and RLP12, PEG = 2:1 (Figure 2d). After incubation at 37 °C for 15 min, all the solutions formed the light-yellow translucent gels and exhibited viscoelasticity. Figure 2e shows RLP12, PEG = 1:0 as an example. Tests involving stress, frequency, and time scanning were conducted to investigate the in situ oscillatory shear rheological mechanics of samples. Over time, the energy storage modulus of the three samples continued to increase and gradually and steadily reached approximately 10–15 kPa. The stability of the hydrogel was further confirmed using a frequency-scanning experiment. Over the entire frequency range of the experiment (0.1–100 rad/s), the elasticity was higher than the viscosity, and the storage modulus (G′) was always higher than the loss modulus (G″). This indicates that the three RLP12 hydrogel groups formed by crosslinking exhibited elastic solid behaviour (Figure 2f).

2.3. Cardiac Function

An ultrasound cardiogram was used to assess the heart function of rats with MI four weeks after treatment (Figure 3a). The left ventricular ejection fraction (LVEF) in the PBS-treated animals was 36.3 ± 2.3%, in the RLP12 hydrogel-treated animals was 45.1 ± 3.6%, in the BADSCs-treated animals was 43.9 ± 3.1%, and in the RLP12 hydrogel + BADSCs-treated animals was 59.2 ± 5.2%. Statistical analyses revealed that compared with those in the PBS group, the LVEF levels of the other three groups were significantly increased. Furthermore, the LVEF levels associated with the RLP12 hydrogel + BADSC group were considerably higher than those of the other three groups (Figure 3b).

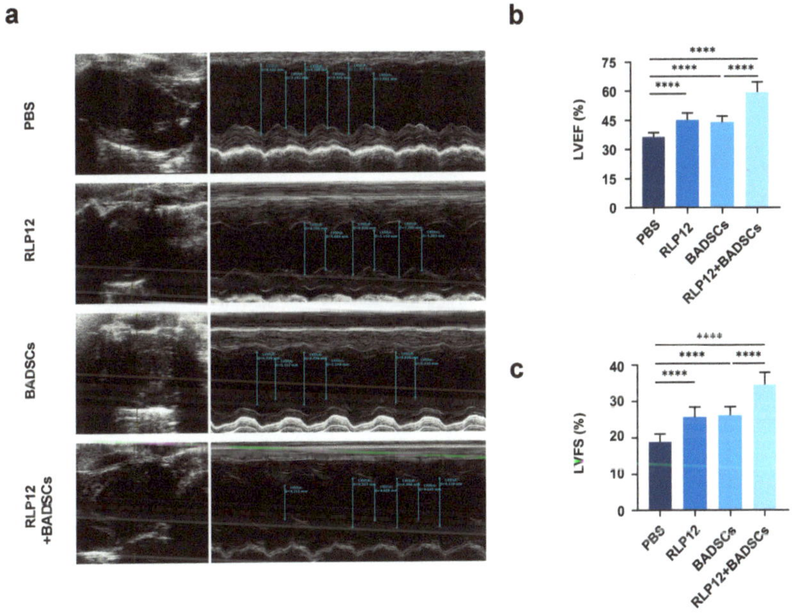

Figure 3. The evaluation of the cardiac function of rats with MI using cardiac ultrasound at 4 weeks after injection. (**a**) Representative echocardiographic images of rats from the different treatment groups. (**b**) LVEF; **** $p < 0.0001$. (**c**) LVFS. **** $p < 0.0001$.

Our results showed that compared with the PBS-administered group, the left ventricular short-axis shortening index (LVFS) was significantly increased in the other three groups. The LVFS values were as follows: 18.8 ± 2.2% in the PBS; 25.6 ± 2.8% in the RLP12 hydrogel; 26.1 ± 2.4% in the BADSC; and 34.5 ± 3.3% in the RLP12 hydrogel + BADSC groups. The LVFS in the RLP12 hydrogel + BADSC group was significantly higher than that in the other groups (Figure 3c). Thus, delivering RLP hydrogels, BADSCs, or RLP12 hydrogel + BADSCs to the infarcted myocardium of rats can substantially enhance cardiac function. The RLP12 hydrogel showed promising characteristics as a BADSC carrier for intramyocardial injection. Compared with BADSCs or RLP hydrogel alone, RLP12 hydrogel + BADSCs had the most pronounced effect in improving cardiac function.

2.4. MI Area Size and Fibrosis Level

After four weeks of treatment, cardiac tissue sections were sampled and subjected to Masson's trichrome staining (Figure 4a). The MI area was quantified by measuring the ratio of the endocardial length in the infarcted area to the left ventricular endocardial circumference. The results showed that the infarct area was 51.1 ± 5.6% in the PBS, 34.9 ± 4.4% in the RLP12 hydrogel, 35.4 ± 4.0% in the BADSCs, and 24.8 ± 2.5% in the RLP12 hydrogel + BADSC groups (Figure 4b). There was a significant reduction in the MI area of all experimental groups compared to that in the PBS group at 4 weeks. Furthermore, the MI area of the RLP12 hydrogel + BADSC group was significantly reduced compared to that of the other three groups, suggesting that injecting the RLP12 hydrogel with BADSCs was the most effective approach. Alternatively, the level of collagen deposition in the PBS group was 68.3 ± 9.5%, in the RLP12 hydrogel group was 52.4 ± 6.2%, in the BADSC group was 55.6 ± 6.4%, and in the RLP12 hydrogel + BADSC group was 35.9 ± 4.8% (Figure 4c). The collagen deposition level of the RLP12 hydrogel + BADSC group was lower than that of the RLP12 hydrogel and BADSC groups, indicating that delivering BADSCs with the RLP12 hydrogel was the most effective in alleviating the degree of fibrosis in rats with MI.

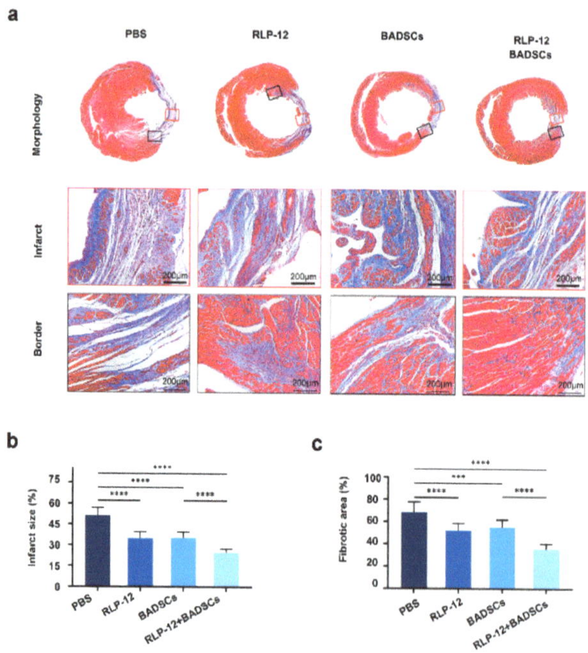

Figure 4. Fibrosis levels assessed using Masson's trichrome staining. (**a**) The Masson's trichrome staining of heart slices from a rat with myocardial infarction at four weeks of treatment (scale bar = 200 μm). (**b**) The statistical analysis of the myocardial infarction area, **** $p < 0.0001$. (**c**) The statistical analysis of the collagen deposition level, *** $p < 0.001$, **** $p < 0.0001$.

2.5. Distribution Levels of Type I Collagen in the Infarct Area and Border Area

Fibrosis is a major pathological change observed during the end stage of MI, in which type I collagen assembles into dense and rigid fibres, causing collagen deposition. Through immunohistochemistry, we discovered that the distribution level of type I collagen in the infarcted area was 85.1 ± 7.6% in the PBS, 64.7 ± 6.0% in the RLP12 hydrogel, 66.7 ± 5.3% in the BADSC, and 50.9 ± 4.4% in the RLP12 hydrogel + BADSC groups (Figure 5a). The statistical analysis showed that compared to the PBS group, the distribution level of type I collagen at the MI site in the three groups was significantly reduced. The proportion of type I collagen in the RLP12 hydrogel + BADSC group was considerably lower than that in the RLP12 hydrogel and BADSC groups (Figure 5b). In the border area of MI, the distribution level of collagen I was 42.4 ± 5.8% in the PBS, 32.1 ± 4.5% in the RLP12 hydrogel, 30.9 ± 4.9% in the BADSC, and 21.1 ± 3.0% in the RLP12 hydrogel + BADSC groups. Further analysis showed that compared with that in the PBS group, the distribution level of type I collagen in the border area of the MI was significantly decreased in the other groups, and the type I collagen in the RLP12 hydrogel + BADSC group was significantly reduced compared to that in the other three groups (Figure 5c). The distribution of type I collagen aligned with the collagen deposition level, suggesting that either the RLP12 hydrogel or BADSCs, alone or in combination, can effectively mitigate fibrosis in the infarcted tissue.

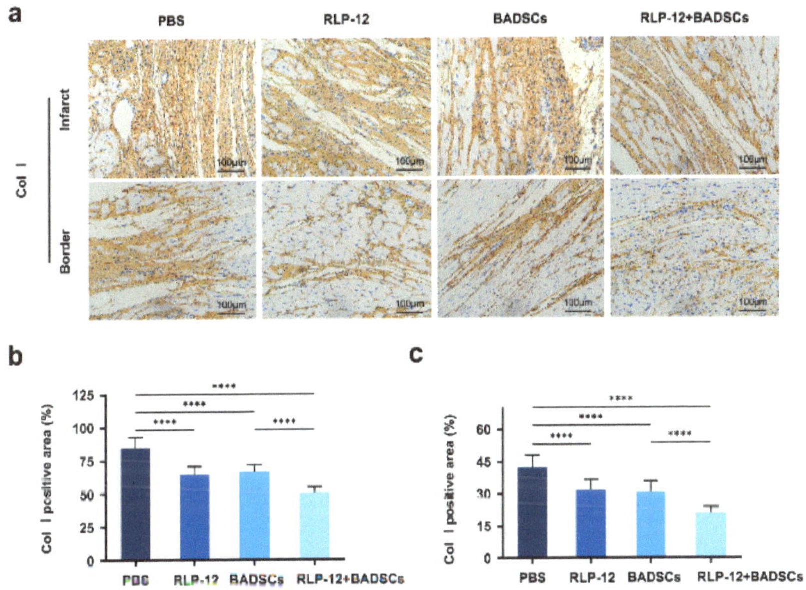

Figure 5. The distribution level of type I collagen in the myocardial infarction and border zone at 4 weeks after treatment. (**a**) The immunohistochemical staining of type I collagen in the myocardial infarction area. (**b**) The statistical analysis of type I collagen distribution in the myocardial infarction area. **** $p < 0.0001$. (**c**) The statistical analysis of type I collagen distribution in the border zone. **** $p < 0.0001$.

2.6. Distribution Levels of Type III Collagen in the Infarct Area and Border Area

In contrast, type III collagen forms fine and pliable fibres that are crucial for maintaining tissue elasticity (Figure 6a). The collagen III distribution was 10.5 ± 1.8% in the PBS, 22.9 ± 3.1% in the RLP12 hydrogel, 20.7 ± 2.1% in the BADSC, and 33.7 ± 3.5% in the RLP12 hydrogel + BADSC groups. This indicates a significant increase in the distribution level of type III collagen at the site of MI after four weeks of treatment with the RLP12 hydrogel alone, BADSCs alone, and BADSCs carried by the RLP12 hydrogel when compared with that in the PBS group. Furthermore, the type III collagen distribution level was significantly higher in the RLP12 hydrogel + BADSC group than in the RLP12 hydrogel and BADSC groups (Figure 6b). In the border area of MI, the distribution level of type III collagen in each group was 9.5 ± 2.5% in the PBS, 34.5 ± 4.1% in the RLP12 hydrogel, 31.8 ± 3.8% in the BADSC, and 51.5 ± 6.3% in the RLP12 hydrogel + BADSC groups. The distribution level of type III collagen at the border area of the MI was significantly increased in the RLP12 hydrogel, BADSC, and RLP12 hydrogel + BADSC groups after four weeks of treatment. The distribution of type III collagen in the RLP12 hydrogel + BADSC group was significantly higher than that in the other three groups (Figure 6c). The distribution of type III collagen showed that its content could be increased by injecting the RLP12 hydrogel and BADSCs alone or in combination, which indicated that this strategy was helpful in maintaining myocardium compliance after MI.

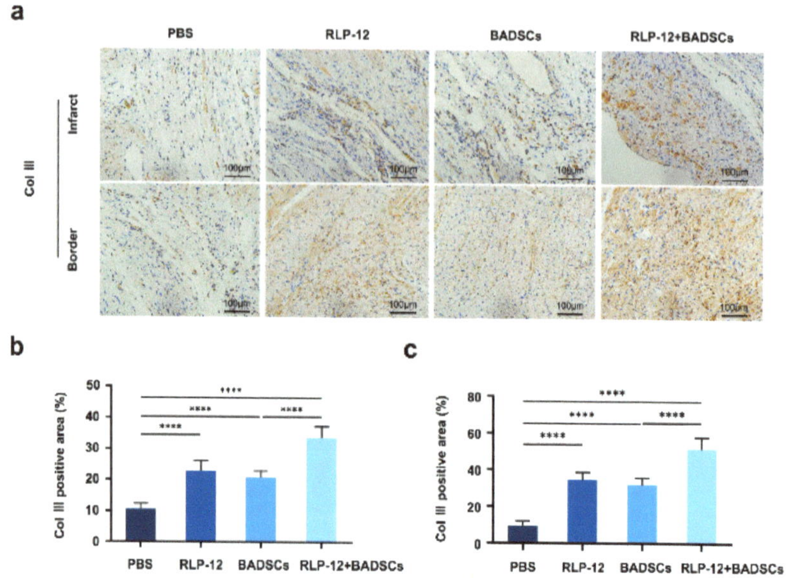

Figure 6. The distribution level of type III collagen in the myocardial infarction and border zone at 4 weeks after treatment. (**a**) The immunohistochemical staining of type III collagen in the myocardial infarction area. (**b**) The statistical analysis of type III collagen distribution in the myocardial infarction area. **** $p < 0.0001$. (**c**) The statistical analysis of type III collagen distribution in the border zone. **** $p < 0.0001$.

2.7. BADSC Survival Rate and In Vivo Differentiation after Transplantation

Representative images of surviving BADSCs are shown in Figure 7a. A week after transplantation, the mRFP-positive cells in the RLP12 hydrogel + BADSC group accounted for 16.5 ± 4.2% of DAPI-stained cells and mainly distributed in the myocardium. The mRFP-positive cells in the BADSC group accounted for 9.3 ± 2.7%, which were more dispersed than those in the RLP12 hydrogel + BADSC group (Figure 7b). After four weeks of transplantation, the RLP12 hydrogel + BADSC group exhibited a percentage of mRFP-positive cells in DAPI-stained cells at 12.4 ± 3.3%. This proportion is significantly higher than that of the BADSC group (7.8 ± 2.2%). In the RLP hydrogel + BADSC group, the BADSCs exhibited even dispersion in the infarct and border areas (Figure 7c). Conversely, in the group treated with BADSCs alone, the transplanted cells were primarily localised in the border area. These results indicate that the RLP hydrogel can lead to a substantial enhancement in the viability of BADSCs, both at one and four weeks after transplantation.

Four weeks after treatment, the specimens were collected, and the differentiation of BADSCs when injecting cells alone and alongside the RLP12 hydrogel in vivo was assessed through immunofluorescence staining (IF) (Figure 7d). The percentage of differentiated mRFP cells in the sections was calculated using Image Pro software (version 6.0). Our results showed that mRFP-positive and cTnT-positive cells matched in both the BADSC and RLP12 hydrogel + BADSC groups, indicating that BADSCs injected into the myocardium differentiated into cardiomyocytes (Figure 7e). The quantitative analysis showed that the number of cTnT-positive cardiomyocytes differentiated in the RLP12 hydrogel + BADSC group was significantly higher than that in the BADSC group (19.2 ± 4.7% vs. 9.5 ± 3.1%). In addition, the observed α-SMA-positive cells and mRFP-positive cells in the BADSC group and RLP12 hydrogel + BADSC groups matched, indicating that the BADSCs may differentiate into vascular smooth muscle cells (Figure 7f). Our quantitative analysis shows that the number of α-SA-positive BADSCs in the RLP12 hydrogel + BADSC group is significantly higher than that in the BADSC group (8.7 ± 2.3% vs. 3.9 ± 1.4%).

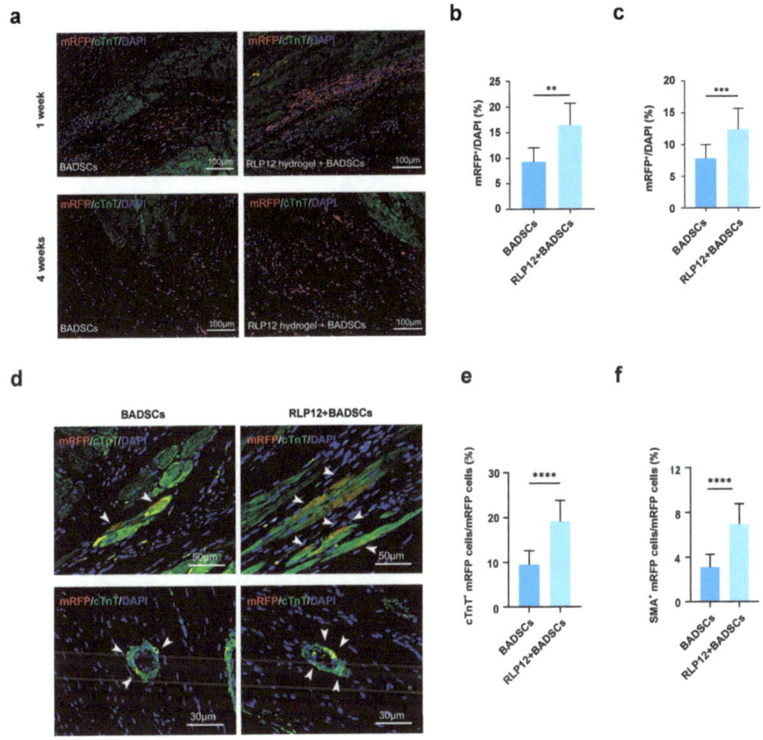

Figure 7. The cell survival and differentiation of BADSCs transplanted into infarcted myocardium. (**a**) Representative images of fluorescence microscopy at 1 and 4 weeks after the RLP12 hydrogel + BADSC treatment (scale bar = 100 μm). (**b**) The statistical analysis of the percentage of mRFP-positive cells to DAPI-stained cells at 1 week after treatment. ** $p < 0.01$. (**c**) The statistical analysis of the percentage of mRFP-positive cells to DAPI-stained cells at 4 weeks after treatment. *** $p < 0.001$. (**d**) Representative images of the immunofluorescence staining of cTnT-positive cardiomyocyte differentiation in the BADSC and RLP12 hydrogel + BADSC groups (myocardial infarction area, scale bar = 50 μm); representative immunofluorescence staining images of α SMA-positive vascular smooth muscle cell differentiation in the BADSC and RLP12 hydrogel + BADSC groups (myocardial infarction area, scale bar = 30 μm). (**e**) The quantitative analysis of myocardial differentiation in the BADSC and RLP12 hydrogel + BADSC groups. **** $p < 0.0001$. (**f**) The quantitative analysis of vascular smooth muscle cell differentiation in the BADSC and RLP12 hydrogel + BADSC groups. **** $p < 0.0001$.

2.8. Microvessel Density (MVD)

Capillaries are the smallest blood vessels, which are densely distributed in the cardiovascular system [41]. In order to evaluate the capillary status of an infarct heart after treatment, vWF immunochemistry staining was performed after 4 weeks of treatment (Figure 8a). The results show that the MVD at the infarct site was 96.3 ± 7.7/mm^2 in the PBS group, 191.5 ± 22.0/mm^2 in the RLP12 hydrogel group, 203.1 ± 18.3/mm^2 in the BADSC group, and 288.9 ± 30.7/mm^2 in the RLP12 hydrogel + BADSC group (Figure 8b). Statistical analyses showed that compared with that in the PBS group, the MVD was significantly increased in the other three groups. The MVD of the RLP12 hydrogel + BADSC group was significantly higher than that of the other three groups.

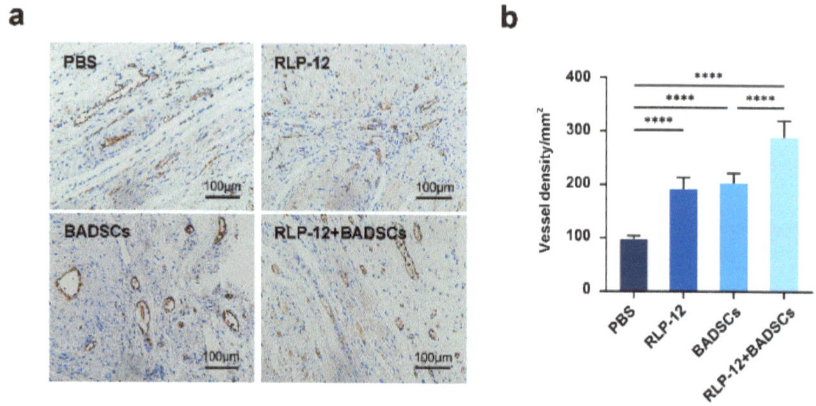

Figure 8. Vascularisation levels assessed using immunohistochemistry. (**a**) Representative images of MVD in the infarction area of MI rats after treatment. (**b**) The statistical analysis of the MVD in each group, ****$p < 0.0001$.

In this study, we prepared an RLP12 hydrogel based on resilin for injectable myocardial tissue engineering. First, an RLP12 protein expression vector was constructed, and the RLP12 protein was obtained after induction and purification. The molecular weight of the RLP12 protein was between 25 and 34 kDa, consistent with the reported 27.5 kDa [42].

THP was used as a crosslinking agent in the formation of an elastin-like polypeptide gel. It was discovered that the gel could adequately maintain the viability of mouse fibroblast cultures in vitro, indicating that THP has low cytotoxicity [43]. Moreover, the subcutaneous injection of the RLP12 hydrogel obtained by THP crosslinked into mice did not cause noticeable inflammatory responses, proving its good biocompatibility [36]. Importantly, the mechanical strength (10–15 kPa) of the *RLP12 hydrogel is consistent with the myocardial tissue (7.8 ± 4.1 kPa -26.2 ± 5.1 kPa) of rats in the state of natural pulsation [44,45]. Therefore, we selected the RLP12 hydrogel for further studies.

Studies have confirmed that MI forms a microenvironment of local ischemia, hypoxia, and extracellular matrix degradation [46]. Within this context, it is difficult for stem cells to survive and function when injected by themselves. Therefore, myocardial tissue engineering using hydrogels carrying stem cells for in situ injections has received increasing attention [47,48]. However, dealing with fibrosis and the local microenvironment following MI remains challenging. Uncontrollable negative ventricular remodelling after MI is the pathological basis for progression to heart failure. Both biological and mechanical malignant microenvironments are involved in negative ventricular remodelling [47]. Currently, most injectable materials are designed to protect biological microenvironments. However, a mechanical microenvironment is typically formed, which promotes the development of heart failure. Recently, studies have discovered that elastic hydrogels can also provide mechanical assistance for infarct tissues owing to their elasticity and extensibility. This was associated with improved local myocardial dyskinesia and reduced stress on the ventricular wall, which is conducive to the improvement of cardiac function [49]. A study on the inter-impact of the hydrogel and the stress and thickness of the infarcted ventricular wall showed that a hydrogel with a stiffness of less than 50 kPa can significantly reduce the stress on muscle fibres [24]. In 2019, Matsumura et al. injected a synthetic elastic hydrogel material poly(NIPAAm-co-HEMA-co-MAPLA) into the MI site of a miniature pig MI model and discovered that it could improve the mechanical strength of the infarcted tissue and reduce the formation of scar tissue, improving ventricular remodelling and cardiac function [50]. In addition, the mechanical signals of viscoelastic hydrogel could regulate the directed differentiation of tissues [51]. In this study, we used an RLP12 hydrogel as an injectable vehicle of BADSCs to the MI site of rats. We observed that the RLP12 hydrogel alone or with

BADSCs significantly decreased the fibrosis level in the infarct area and improved cardiac function, which may be related to the RLP12 hydrogel in the mechanical microenvironment of the infarct tissue. Studies have shown that resilin hydrogels have a significant repair effect on elastic tissues. In 2019, King et al. combined a resilin-like protein with hyaluronic acid (HA) to produce RLP/HA and RLP-acrylamide/mercaptanisation HA (RLP-AM/HA-SH) composite hydrogels with elastic shear moduli of ~600 and ~1500 Pa, respectively. These hydrogels have a similar elastic shear to that of human vocal cord tissue (400–2000 Pa). The hydrogels could maintain the viscoelasticity of the vocal cord when being injected into vocal cord tissue [52]. The shear modulus of the RLP12 hydrogel used in this study was 10–15 kPa, which is similar to the elastic modulus of rat myocardium in the state of natural pulsation, which ranges between 7.8 ± 4.1 and 26.2 ± 5.1 kPa. Therefore, the RLP12 hydrogel can satisfy the mechanical compliance requirements of the myocardial tissue.

The excessive deposition of denatured collagen leads to fibrosis, with type I collagen mainly deposited, whereas type III collagen is beneficial for maintaining tissue elasticity [53]. The ratio of Col I to Col III in the healthy myocardium is lower than that in the infarct area of the myocardium [54]. Research has shown that a high Col III/I ratio is commonly observed in young rats, and high expression of type III collagen is beneficial for reducing fibrosis [55]. Wen et al. suggested that an elastic Dex PCL-HEMA/PNIPAAm (DPHP) hydrogel could improve the distribution of type III collagen after transplantation to the infarct site [56]. In this study, the distribution of type I and III collagen in the heart slices of each group was analysed. The results show that the content of type III collagen in the infarct site and the border area increased, indicating that it can maintain the compliance of myocardial tissue after MI, further confirming that the RLP12 hydrogel improved the mechanical microenvironment. Although the content of type III collagen increases and can alleviate fibrosis to some extent, excessive deposition of type III collagen may cause myocardial hypertrophy, leading to cardiac dilation and heart failure [57].

We believe that RLP12 hydrogels, as injectable vehicles, can significantly improve the cardiac function of MI rats, primarily in the following aspects. First, the RLP12 hydrogel can form a gel in situ, improving the survival of BADSCs and promoting the role of BADSCs in myocardial repair. Second, the RLP12 hydrogel can alleviate fibrosis by improving the mechanical microenvironment of the MI and the compliance of the myocardium by increasing the distribution of type III collagen. Third, improving the mechanical microenvironment can optimise the biological microenvironment, thereby promoting the differentiation of BADSCs into cardiovascular lineage cells and increasing the MVD in the infarcted area. Next, in order to develop the functional materials based on the resilin hydrogel with superior physical/chemical properties and biological functions, we will carry out the studies related to the properties and degradation of RLP hydrogels with different protein concentrations and crosslinking degrees, the interaction between hydrogels and stem cells, and evaluate their function on repairing damaged myocardium in large animal MI models.

3. Conclusion

In this study, the RLP12 hydrogel combined with BADSCs significantly improved cardiac function by increasing microvessel density and reducing infarct area size and collagen deposition in MI rats four weeks after transplantation. The distribution ratio of collagen III to I increased in both the centre and edge areas of the MI, indicating improved compliance of the infarct heart. This study provides novel insights into the mechanical interactions between the hydrogel and the infarct microenvironment.

4. Materials and Methods

4.1. Expression and Purification of RLP12 Protein

The RLP12 protein sequence reported by Charati et al. was selected [41], with a total of 822 bp. BamH I and Hind III cleavage sites were added at both ends, synthesised by the GenScirpt Company (Nanjing, China), and inserted into the pUC57 cloning vector (Figure 1). Double enzyme digestion of the pET-28a vector and RLP12-pUC57 was performed to recover the target fragments. The T4 DNA ligase was used to link the pET-28a vector fragment (carrying 6xHis tag) with the RLP12 fragment, and the sample was incubated overnight at 16 °C. The recombinant expression vector was transfected into *Escherichia coli* BL21 (DE3) cells. The bacterial solution obtained was subjected to nucleic acid sequence analysis. The positive clone was identified as *E. coli* BL21 (DE3)-RLP12-pET-28a. The plasmid was extracted, and double restriction enzyme identification was performed. To analyse and compare existing methods for expressing the resilin protein, we selected the automatic induction method and referred to a previous study that introduced the use of the ZYP-5052 automatic induction medium (Sinopharm Chemical Reagent Co., Ltd., Shanghai, China) to express the RLP12 protein [36]. RLP12 protein was extensively expressed and extracted. The target protein was purified using a Ni^+ column (Qiagen, Germany), and SDS-PAGE was performed.

4.2. Preparation and Identification of the RLP12 Hydrogel

First, we desalinated and concentrated the RLP12 protein. We selected a centrifugal filter (Amicon® Ultra-15 10 K, Millipore, Burlington, MA, USA), added the protein sample to the filter (≤15 mL), and centrifuged the sample at 4 °C at 4000× *g* for 15–40 min. Deionised water was added to the concentrated solution up to the original volume of the sample, and this was repeated several times until the salt content decreased. Concentrated protein samples (150 µL to 1 mL) were freeze-dried overnight. The protein content in each tube was measured. The RLP12 hydrogel was prepared by dissolving lyophilised RLP12 protein in PBS, adjusting the concentration to 100 mg/mL, and preparing THP and PEG solutions. Before mixing, each solution was placed on ice to reduce the reaction rate and prevent excessive crosslinking. A 10 wt% hydrogel was prepared by vortexing the ingredients. Subsequently, the hydrogel was incubated at 37 °C for crosslinking for 1 h.

4.3. Examination of Mechanical and Rheometric Properties

The RLP12 hydrogel was analysed using a rheometer (Thermo Fisher, Waltham, MA, USA). At 37 °C, dynamic oscillation time, frequency, and strain scanning were assessed using a C35/1° cone-plate insulation sleeve. The total volume of each sample was 200 µL. The mixture was transferred to the bottom plate of the rheometer for in situ rheological analysis. Strain scanning measurements were performed on each sample from 0.01% strain to a maximum of 1000% strain to determine the linear viscoelastic region (LVE). Time scanning at a fixed strain amplitude of 1% in LVE (t: 0–8000 s, ω: 6.28 rad/s) and frequency scanning experiment (ω: 0.1–100 rad/s) were performed to assess the rheological properties of the hydrogel. The experiments were repeated thrice for each sample.

4.4. Establishment of a Rat MI Model and Injection

All animal experiments were approved by the Animal Ethics Committee of the College of Life Sciences and Bioengineering (approval number SS-HBW-2018-07). A rat model of MI was established as previously reported [58]. Rats were anaesthetised for surgery; the ligation was performed using a 6-0 Prolene suture (Lingqiao, Zhejiang, China) between the rat pulmonary artery cone and the left atrial appendage at 2–3 mm from the beginning of the coronary artery to 2 mm below the left atrial appendage and at a depth of 0.5–1 mm. The colour of the left ventricular anterior wall myocardium changed to cyan, followed by whitening, indicating success. MI rats were randomly divided into four groups: (1) control (PBS, n = 12); (2) RLP12 hydrogel (n = 12); (3) BADSCs (n = 18); and (4) RLP12 hydrogel +BADSCs (n = 18). Six animals from groups 3 and 4 were collected one week after surgery to evaluate the survival rate of the transplanted cells. The remaining rats underwent cardiac ultrasound examination four weeks after surgery, followed by sampling for histological and immunohistochemical testing.

4.5. Cardiac Function Testing

After four weeks of postoperative feeding, the cardiac function of the rats was tested using the Vevo 1100 ultrasound imaging system (VisualSonics, Toronto, Canada). After anaesthesia, the skin of the rats was prepared, and an appropriate amount of ultrasound gel was applied. When the heart rate stabilised, ultrasound images were collected, and the electrocardiogram and heart rate were monitored. Guided by two-dimensional images, M-shaped curves were measured to determine the left ventricular end-systolic diameter (LVESD), left ventricular end-diastolic diameter (LVEDD), left ventricular fractional shortening (LVFS), and left ventricular ejection fraction (LVEF). Subsequently, 3–6 cardiac cycle measurement data points were collected from each rat, and the average value was calculated. The formulas were as follows: LVFS (%) = [(EDD − ESD)/ EDD] × 100 and LVEF (%) = $[(LVIDd)^3 - (LVIDs)^3]/(LVIDd)^3 \times 100$.

EDD: End-Diastolic Diameter;
ESD: End-Systolic Dimension;
LVIDd: Left Ventricular Internal Diameter in Diastole;
LVIDs: Left Ventricular Internal Diameter in Systole.

4.6. Histological Testing

After the cardiac function assessment, rats were euthanised using a 2% potassium chloride solution, and the heart was collected. The ventricular wall below the trisection ligation, along the short axis of the heart, was immersed in 4% paraformaldehyde. Samples were cleaned, dehydrated with distilled water, and sliced in paraffin. Subsequently, Masson's trichrome staining was performed, and slice images were acquired. The fibrosis level and infarcted area were determined using Image Pro analysis software (version 6.0). The formula used is as follows: myocardial infarction size = (left ventricular endocardial length in the infarcted area/left ventricular intact endocardial circumference) × 100. Collagen deposition was evaluated using Image Pro software. We calculated the percentage of the left ventricular fibrosis area in the left ventricle area. Five cross-sections were selected from the bottom to the apex of each heart specimen for immunohistochemical staining of type I and III collagen. After image acquisition, we calculated the percentage of positive areas of the two types of collagens in the left ventricular MI area and the MI edge area using Image Pro analysis software. Using the same method for vWF immunohistochemistry staining, the number of microvessels with a diameter of 10–100 μm in the MI area was assessed, and the MVD of each group was calculated.

IF staining was performed using cardiac troponin T (cTnT) antibodies. The percentage of red fluorescent cells among the DAPI-stained cells was calculated using Image Pro analysis software. Using this information, we evaluated the survival of BADSCs at different time points. IF staining with α-SA antibodies was performed to detect the differentiation of BADSCs in vivo. We calculated the number of mRFP-positive myocardial cells (cTnT + mRFP) and the number of mRFP-positive vascular smooth muscle cells (α-SA + mRFP).

4.7. Statistical Analysis

The data were expressed as mean ± standard deviation and analysed using the SPSS software (version 17.0, IBM, Armonk, NY, USA). Statistical differences between two groups were analysed using a *t*-test; comparisons between two or more groups were analysed using a one-way ANOVA combined with the least significant difference (LSD) test. Values of $p < 0.05$ were considered significant.

Author Contributions: Conception: H.W., J.W.; Supervision: H.W., J.W., S.K.; Methodology; Validation; Writing—original draft: L.Z., H.L.; Investigation; Project administration: R.G., K.Z., Y.G., Y.C.; Writing—editing and revising: H.W., J.W., S.K.; Resources: H.W.; Funding acquisition: J.W. All authors have read and agreed to the published version of the manuscript.

Funding: This study was funded by the Natural Science Foundation of Beijing (No. 7222262) and the National Key Research and Development Program of China (No. 2023YFC2605002).

Institutional Review Board Statement: All animal experiments were approved by the Animal Ethics Committee of the College of Life Sciences and Bioengineering (approval number SS-HBW-2018-07).

Informed Consent Statement: Not applicable.

Data Availability Statement: The original contributions presented in the study are included in the article, further inquiries can be directed to the corresponding authors.

Acknowledgments: The authors would like to thank to J Shi from Tiantan hospital for her valuable help in animal studies.

Conflicts of Interest: The authors declare no conflicts of interest.

References

1. Perry, A.S.; Dooley, E.E.; Master, H.; Spartano, N.L.; Brittain, E.L.; Gabriel, K.P. Physical Activity Over the Lifecourse and Cardiovascular Disease. *Circ. Res.* **2023**, *132*, 1725–1740. [CrossRef] [PubMed]
2. Samsky, M.D.; Morrow, D.A.; Proudfoot, A.G.; Hochman, J.S.; Thiele, H.; Rao, S.V. Cardiogenic Shock After Acute Myocardial Infarction. *J. Am. Med. Assoc.* **2021**, *326*, 1840–1850. [CrossRef] [PubMed]
3. Kong, P.; Dong, J.; Li, W.; Li, Z.; Gao, R.; Liu, X.; Wang, J.; Su, Q.; Wen, B.; Ouyang, W.; et al. Extracellular Matrix/Glycopeptide Hybrid Hydrogel as an Immunomodulatory Niche for Endogenous Cardiac Repair after Myocardial Infarction. *Adv. Sci.* **2023**, *10*, e2301244. [CrossRef] [PubMed]
4. Shiraki, T.; Ichibori, Y.; Ohtani, T.; Mizote, I.; Kioka, H.; Tsukamoto, Y.; Nakamura, D.; Yokoi, K.; Ide, S.; Nakamoto, K.; et al. Pathophysiological evaluations of initial plaque development after heart transplantation via serial multimodality imaging and cytokine assessments. *J. Heart Lung Transplant.* **2022**, *41*, 877–885. [CrossRef]
5. Frantz, S.; Hundertmark, M.J.; Schulz-Menger, J.; Bengel, F.M.; Bauersachs, J. Left ventricular remodelling post-myocardial infarction: Pathophysiology, imaging, and novel therapies. *Eur. Heart J.* **2022**, *43*, 2549–2561. [CrossRef]
6. Carvalho, A.B.; Kasai-Brunswick, T.H.; de Carvalho, A.C.C. Advanced cell and gene therapies in cardiology. *eBioMedicine* **2024**, *103*, 105125. [CrossRef]
7. Bolli, R. Cell therapy for acute myocardial infarction: Requiescat in Pace. *Eur. Heart J.* **2020**, *41*, 3711–3714. [CrossRef]
8. Thomas, D.; Cunningham, N.J.; Shenoy, S.; Wu, J.C. Human-induced pluripotent stem cells in cardiovascular research: Current approaches in cardiac differentiation, maturation strategies, and scalable production. *Cardiovasc Res.* **2022**, *118*, 20–36. [CrossRef]
9. Qiao, L.; Kong, Y.; Shi, Y.; Sun, A.; Ji, R.; Huang, C.; Li, Y.; Yang, X. Synergistic effects of adipose-derived stem cells combined with decellularized myocardial matrix on the treatment of myocardial infarction in rats. *Life Sci.* **2019**, *239*, 116891. [CrossRef]
10. Lu, K.Y.; Primus Dass, K.T.; Lin, S.Z.; Harn, H.J.; Liu, S.P. The application of stem cell therapy and brown adipose tissue transplantation in metabolic disorders. *Cytotherapy* **2020**, *22*, 521–528. [CrossRef]
11. Omidian, H.; Chowdhury, S.D. Advancements and Applications of Injectable Hydrogel Composites in Biomedical Research and Therapy. *Gels* **2023**, *9*, 533. [CrossRef] [PubMed]
12. Hao, T.; Qian, M.; Zhang, Y.; Liu, Q.; Midgley, A.C.; Liu, Y.; Che, Y.; Hou, J.; Zhao, Q. An Injectable Dual-Function Hydrogel Protects Against Myocardial Ischemia/Reperfusion Injury by Modulating ROS/NO Disequilibrium. *Adv. Sci.* **2022**, *9*, e2105408. [CrossRef] [PubMed]
13. Jalilinejad, N.; Rabiee, M.; Baheiraei, N.; Ghahremanzadeh, R.; Salarian, R.; Rabiee, N.; Akhavan, O.; Zarrintaj, P.; Hejna, A.; Saeb, M.R.; et al. Electrically conductive carbon-based (bio)-nanomaterials for cardiac tissue engineering. *Bioeng. Transl. Med.* **2022**, *8*, e10347. [CrossRef] [PubMed]
14. Bai, R.; Tian, L.; Li, Y.; Zhang, J.; Wei, Y.; Jin, Z.; Liu, Z.; Liu, H. Combining ECM Hydrogels of Cardiac Bioactivity with Stem Cells of High Cardiomyogenic Potential for Myocardial Repair. *Stem Cells Int.* **2019**, *2019*, 6708435. [CrossRef] [PubMed]
15. Shahshahani, S.; Shahgholi, M.; Karimipour, A. The thermal performance and mechanical stability of methacrylic acid porous hydrogels in an aqueous medium at different initial temperatures and hydrogel volume fraction using the molecular dynamics simulation. *J. Mol. Liq.* **2023**, *382*, 122001. [CrossRef]
16. He, S.; Zhang, Z.; Luo, R.; Jiang, Q.; Yang, L.; Wang, Y. Advances in Injectable Hydrogel Strategies for Heart Failure Treatment. *Adv. Healthc. Mater.* **2023**, *12*, e2300029. [CrossRef]
17. Alimirzaei, F.; Vasheghani-Farahani, E.; Ghiaseddin, A.; Soleimani, M.; Pouri; Najafi-Gharavi, Z. pH-Sensitive Chitosan Hydrogel with Instant Gelation for Myocardial Re generation. *J. Tissue Sci. Eng.* **2017**, *8*, 3. [CrossRef]
18. Mann, D.L.; Lee, R.J.; Coats, A.J.-S.; Neagoe, G.; Dragomir, D.; Pusineri, E.; Piredda, M.; Bettari, L.; Kirwan, B.A.; Dowling, R.; et al. One-year follow-up results from AUGMENT-HF: A multicentre randomized controlled clinical trial of the efficacy of left ventricular augmentation with Algisyl in the treatment of heart failure. *Eur. J. Heart Fail.* **2015**, *18*, 314–325. [CrossRef]
19. Rao, S.V.; Zeymer, U.; Douglas, P.S.; Al-Khalidi, H.; White, J.A.; Liu, J.; Levy, H.; Guetta, V.; Gibson, C.M.; Tanguay, J.F.; et al. Bioabsorbable Intracoronary Matrix for Prevention of Ventricular Remodeling After Myocardial Infarction. *J. Am. Coll. Cardiol.* **2016**, *68*, 715–723. [CrossRef]
20. Seif-Naraghi, S.B.; Singelyn, J.M.; Salvatore, M.A.; Osborn, K.G.; Wang, J.J.; Sampat, U.; Kwan, O.L.; Strachan, G.M.; Wong, J.; Schup-Magoffin, P.J.; et al. Safety and Efficacy of an Injectable Extracellular Matrix Hydrogel for Treating Myocardial Infarction. *Sci. Transl. Med.* **2013**, *5*, 173ra25. [CrossRef]

21. Maruyama, K.; Imanaka-Yoshida, K. The Pathogenesis of Cardiac Fibrosis: A Review of Recent Progress. *Int. J. Mol. Sci.* **2022**, *23*, 2617. [CrossRef] [PubMed]
22. Long, Y.; Niu, Y.; Liang, K.; Du, Y. Mechanical communication in fibrosis progression. *Trends Cell Biol.* **2022**, *32*, 70–90. [CrossRef] [PubMed]
23. Vettori, L.; Tran, H.A.; Mahmodi, H.; Filipe, E.C.; Wyllie, K.; Liu Chung Ming, C.; Cox, T.R.; Tipper, J.; Kabakova, I.V.; Rnjak-Kovacina, J.; et al. Silk fibroin increases the elasticity of alginate-gelatin hydrogels and regulates cardiac cell contractile function in cardiac bioinks. *Biofabrication* **2024**, *16*, 035025. [CrossRef] [PubMed]
24. Wang, H.; Rodell, C.B.; Lee, M.E.; Dusaj, N.N.; Gorman, J.H., 3rd; Burdick, J.A.; Gorman, R.C.; Wenk, J.F. Computational sensitivity investigation of hydrogel injection characteristics for myocardial support. *J. Biomech.* **2017**, *64*, 231–235. [CrossRef]
25. Montgomery, M.; Ahadian, S.; Huyer, L.D.; Lo Rito, M.; Civitarese, R.A.; Vanderlaan, R.D.; Wu, J.; Reis, L.A.; Momen, A.; Akbari, S.; et al. Flexible shape-memory scaffold for minimally invasive delivery of functional tissues. *Nat. Mater.* **2017**, *16*, 1038–1046. [CrossRef]
26. Tallawi, M.; Rai, R.; Boccaccini, A.R.; Aifantis, K.E. Effect of substrate mechanics on cardiomyocyte maturation and growth. *Tissue Eng. Part B Rev.* **2015**, *21*, 157–165. [CrossRef]
27. Su, R.S.-C.; Kim, Y.; Liu, J.C. Resilin: Protein-based elastomeric biomaterials. *Acta Biomater.* **2014**, *10*, 1601–1611. [CrossRef]
28. van Eldijk, M.B.; McGann, C.L.; Kiick, K.L.; van Hest, J.C. Elastomeric polypeptides. *Top. Curr. Chem.* **2012**, *310*, 71–116.
29. Li, L.; Kiick, K.L. Transient dynamic mechanical properties of resilin-based elastomeric hydrogels. *Front. Chem.* **2014**, *2*, 21. [CrossRef]
30. Gosline, J.M. The elastic properties of rubber-like proteins and highly extensible tissues. *Symp. Soc. Exp. Biol.* **1980**, *34*, 332–357.
31. Gosline, J.; Bailey, A.J.; Macmillan, J.; Guerette, P.; Ortlepp, C.; Savage, K. Elastic proteins: Biological roles and mechanical properties. *Philos. Trans. R. Soc. London. Ser. B Biol. Sci.* **2002**, *357*, 121–132. [CrossRef] [PubMed]
32. Ardell, D.H.; Andersen, S.O. Tentative identification of a resilin gene in Drosophila melanogaster. *Insect Biochem. Mol. Biol.* **2001**, *31*, 965–970. [CrossRef] [PubMed]
33. Li, L.; Charati, M.B.; KiickK, L. Elastomeric polypeptide-based biomaterials. *Polym. Chem.* **2010**, *1*, 1160–1170. [CrossRef]
34. Qin, G.; Rivkin, A.; Lapidot, S.; Hu, X.; Preis, I.; Arinus, S.B.; Dgany, O.; Shoseyov, O.; Kaplan, D.L. Recombinant exon-encoded resilins for elastomeric biomaterials. *Biomaterials* **2011**, *32*, 9231–9243. [CrossRef] [PubMed]
35. Li, L.; Mahara, A.; Tong, Z.; Levenson, E.A.; McGann, C.L.; Jia, X.; Yamaoka, T.; Kiick, K.L. Recombinant Resilin-Based Bioelastomers for Regenerative Medicine Applications. *Adv. Healthc. Mater.* **2015**, *5*, 266–275. [CrossRef]
36. Li, L.; Stiadle, J.M.; Levendoski, E.E.; Lau, H.K.; Thibeault, S.L.; Kiick, K.L. Biocompatibility of injectable resilin-based hydrogels. *J. Biomed. Mater. Res. Part A* **2018**, *106*, 2229–2242. [CrossRef] [PubMed]
37. McGann, C.L.; Levenson, E.A.; Kiick, K.L. Resilin-Based Hybrid Hydrogels for Cardiovascular Tissue Engineering. *Macromol. Chem. Phys.* **2012**, *214*, 203–213. [CrossRef]
38. Li, L.; Teller, S.; Clifton, R.J.; Jia, X.; Kiick, K.L. Tunable Mechanical Stability and Deformation Response of a Resilin-Based Elastomer. *Biomacromolecules* **2011**, *12*, 2302–2310. [CrossRef]
39. Renner, J.N.; Cherry, K.M.; Su, R.S.; Liu, J.C. Characterization of resilin-based materials for tissue engineering applications. *Biomacromolecules* **2012**, *13*, 3678–3685. [CrossRef]
40. Lv, S.; Dudek, D.M.; Cao, Y.; Balamurali, M.M.; Gosline, J.; Li, H. Designed biomaterials to mimic the mechanical properties of muscles. *Nature* **2010**, *465*, 69–73. [CrossRef]
41. Rahimnejad, M.; Boroujeni, N.N.; Jahangiri, S.; Rabiee, N.; Rabiee, M.; Makvandi, P.; Akhavan, O.; Varma, R.S. Prevascularized Micro-/Nano-Sized Spheroid/Bead Aggregates for Vascular Tissue Engineering. *Nano-Micro Lett.* **2021**, *13*, 182. [CrossRef]
42. Charati, M.B.; Ifkovits, J.L.; Burdick, J.A.; Linhardt, J.G.; Kiick, K.L. Hydrophilic elastomeric biomaterials based on resilin-like polypeptides. *Soft Matter.* **2009**, *5*, 3412–3416. [CrossRef] [PubMed]
43. Lim, D.W.; Nettles, D.L.; Setton, L.A.; Chilkoti, A. Rapid Cross-Linking of Elastin-like Polypeptides with (Hydroxymethyl)phosphines in Aqueous Solution. *Biomacromolecules* **2007**, *8*, 1463–1470. [CrossRef]
44. Su, R.S.-C.; Gill, E.E.; Kim, Y.; Liu, J.C. Characterization of resilin-like proteins with tunable mechanical properties. *J. Mech. Behav. Biomed. Mater.* **2019**, *91*, 68–75. [CrossRef]
45. Zhu, Y.; Zhang, Y.; Chen, M. Research Status of the Measurement for Young's Modulus of Myocardial Cells Using Atomic Force Microscopy. *Chin. J. Biomed. Eng.* **2014**, *33*, 93–97.
46. Soni, S.S.; D'Elia, A.M.; Rodell, C.B. Control of the post-infarct immune microenvironment through biotherapeutic and biomaterial-based approaches. *Drug Deliv. Transl. Res.* **2023**, *13*, 1983–2014. [CrossRef]
47. Frangogiannis, N.G. Cardiac fibrosis: Cell biological mechanisms, molecular pathways and therapeutic opportunities. *Mol. Asp. Med.* **2019**, *65*, 70–99. [CrossRef] [PubMed]
48. Hao, T.; Li, J.; Yao, F.; Dong, D.; Wang, Y.; Yang, B.; Wang, C. Injectable Fullerenol/Alginate Hydrogel for Suppression of Oxidative Stress Damage in Brown Adipose-Derived Stem Cells and Cardiac Repair. *ACS Nano* **2017**, *11*, 5474–5488. [CrossRef]
49. Hernandez, M.J.; Christman, K.L. Designing Acellular Injectable Biomaterial Therapeutics for Treating Myocardial Infarction and Peripheral Artery Disease. *JACC Basic Transl. Sci.* **2017**, *2*, 212–226. [CrossRef]
50. Matsumura, Y.; Zhu, Y.; Jiang, H.; D'Amore, A.; Luketich, S.K.; Charwat, V.; Yoshizumi, T.; Sato, H.; Yang, B.; Uchibori, T.; et al. Intramyocardial injection of a fully synthetic hydrogel attenuates left ventricular remodeling post myocardial infarction. *Biomaterials* **2019**, *217*, 119289. [CrossRef]

51. Nerger, B.A.; Sinha, S.; Lee, N.N.; Cheriyan, M.; Bertsch, P.; Johnson, C.P.; Mahadevan, L.; Bonventre, J.V.; Mooney, D.J. 3D Hydrogel Encapsulation Regulates Nephrogenesis in Kidney Organoids. *Adv. Mater.* **2024**, *36*, 2308325. [CrossRef]
52. King, R.E.; Lau, H.K.; Zhang, H.; Sidhu, I.; Christensen, M.B.; Fowler, E.W.; Li, L.; Jia, X.; Kiick, K.L.; Thibeault, S.L. Biocompatibility and Viscoelastic Properties of Injectable Resilin-Like Polypeptide and Hyaluronan Hybrid Hydrogels in Rabbit Vocal Folds. *Regen. Eng. Transl. Med.* **2019**, *5*, 373–386. [CrossRef]
53. Carver, W.; Nagpal, M.; Nachtigal, M.; Borg, T.K.; Terracio, L. Collagen Expression in MechanicallyStimulated Cardiac Fibroblasts. *Circ. Res.* **1991**, *69*, 116–122. [CrossRef]
54. Roman, B.; Kumar, S.A.; Allen, S.C.; Delgado, M.; Moncayo, S.; Reyes, A.M.; Suggs, L.J.; Chintalapalle, R.; Li, C.; Joddar, B. A Model for Studying the Biomechanical Effects of Varying Ratios of Collagen Types I and III on Cardiomyocytes. *Cardiovasc. Eng. Technol.* **2021**, *12*, 311–324. [CrossRef]
55. Ivan, M.; Ruthard, J. Characterisation of left ventricular collagen in the rat. *Cardiovasc. Res.* **1983**, *17*, 15–21.
56. Wen, Y.; Li, X.; Li, Z.; Wang, M.L.; Chen, P.P.; Liu, Y.; Zhang, X.Z.; Jiang, X.J. Intra-myocardial Delivery of a Novel Thermosensitive Hydrogel Inhibits Post-infarct Heart Failure After Degradation in Rat. *J. Cardiovasc. Transl. Res.* **2020**, *13*, 677–685. [CrossRef] [PubMed]
57. Mukherjee, D.; Sen, S. Alternation of Cardiac Collagen Phenotypes in Hypertensive Hypertrophy: Role of Blood Pressure. *J. Mol. Cell. Cardiol.* **1993**, *25*, 185–196. [CrossRef] [PubMed]
58. Wei, M.; Xin, P.; Li, S.; Tao, J.; Li, Y.; Li, J.; Liu, M.; Li, J.; Zhu, W.; Redington, A.N. Repeated remote ischemic postconditioning protects against adverse left ventricular remodeling and improves survival in a rat model of myocardial infarction. *Circ. Res.* **2011**, *108*, 1220–1225. [CrossRef] [PubMed]

gels

Review

The Clinical Application of Gel-Based Composite Scaffolds in Rotator Cuff Repair

Shebin Tharakan [1,2], Michael Hadjiargyrou [3] and Azhar Ilyas [1,4,*]

[1] Bio-Nanotechnology and Biomaterials (BNB) Laboratory, New York Institute of Technology, Old Westbury, NY 11568, USA
[2] College of Osteopathic Medicine, New York Institute of Technology, Old Westbury, NY 11568, USA
[3] Department of Biological & Chemical Sciences, New York Institute of Technology, Old Westbury, NY 11568, USA; mhadji@nyit.edu
[4] Department Electrical and Computer Engineering, New York Institute of Technology, Old Westbury, NY 11568, USA
* Correspondence: ailyas@nyit.edu; Tel.: +1-516-686-7455; Fax: +1-516-686-7933

Abstract: Rotator cuff tears are a common injury that can be treated with or without surgical intervention. Gel-based scaffolds have gained significant attention in the field of tissue engineering, particularly for applications like rotator cuff repair. Scaffolds can be biological, synthetic, or a mixture of both materials. Collagen, a primary constituent of the extracellular matrix (ECM) in musculoskeletal tissues, is one of the most widely used materials for gel-based scaffolds in rotator cuff repair, but other ECM-based and synthetic-based composite scaffolds have also been utilized. These composite scaffolds can be engineered to mimic the biomechanical and biological properties of natural tissues, supporting the healing process and promoting regeneration. Various clinical studies examined the effectiveness of these composite scaffolds with collagen, ECM and synthetic polymers and provided outstanding results with remarkable improvements in range of motion (ROM), strength, and pain. This review explores the material composition, manufacturing process and material properties of gel-based composite scaffolds as well as their clinical outcomes for the treatment of rotator cuff injuries.

Keywords: rotator cuff; gel-based composite scaffolds; grafts; clinical outcomes; augmentation; orthopedic surgery

Academic Editor: Esmaiel Jabbari

Received: 15 November 2024
Revised: 17 December 2024
Accepted: 22 December 2024
Published: 24 December 2024

Citation: Tharakan, S.; Hadjiargyrou, M.; Ilyas, A. The Clinical Application of Gel-Based Composite Scaffolds in Rotator Cuff Repair. *Gels* **2025**, *11*, 2. https://doi.org/10.3390/gels11010002

1. Introduction

The human shoulder is very complex and dynamic with multiple anatomical structures that stabilize the glenohumeral synovial joint. Compared to the acetabulum, the glenoid fossa is shallow conferring greater mobility in exchange for stability making the shoulder prone to luxation [1]. The four muscles stabilizing the glenohumeral joint are collectively known as the rotator cuff. The rotator cuff consists of the subscapularis, infraspinatus, supraspinatus, and teres minor muscles anchoring the humeral head to the glenoid cavity [2]. In addition to the rotator cuff, accessory structures such as the bursae and the fibrocartilage labrum enhance joint stability [3]. Injuries of the shoulder may arise due to overuse, sports injuries, or potential genetic predisposition which result in muscle impingement or decreased range of motion (ROM) [4–6]. Age also plays a factor in decreased ROM as a recent study in a community-based cohort indicated that flexion, abduction, and external rotation tend to decrease with age [7].

A rotator cuff tear entails damage to the tendons surrounding the glenohumeral joint. The most common tear involves the supraspinatus tendon which presents with a decreased

ROM, a positive Jobe's test, and a subacromial grind test [8–10]. Two types of tears are a partial-thickness tear (PTT) or a full-thickness tear (FTT). PTTs are incomplete tears that arise due to trauma but can propagate into an FTT over time developing into a cuff tear arthropathy. FTTs involve tendon discontinuance due to trauma which does not heal without aid and often require surgical intervention [11]. Currently, rotator cuff tears are prevalent in athletes and older individuals. Shoulder overuse by overhead motion as seen in athletes pose the threat of impingement and rotator cuff pathology. In particular, athletes older than 40 years are at significant risk for rotator cuff tears after an anterior dislocation [12]. Previous studies have determined rotator cuff tears (asymptomatic and symptomatic) are in tandem with age affecting individuals 60 years and older [13,14]. Bilateral cuff tears correlate with age as one study indicated a 50% likelihood of tearing after the age of 66 [15].

Clinical management of rotator cuff tears may be nonoperative or operative dependent upon the pathology. Conservative measures for management include NSAID treatments, muscle massages, and physiotherapy for the initial treatment of PTTs [16]. Operative measures for PTT repair include surgical debridement or arthroscopic tendon repair [17]. Subacromial decompression may also be used for Ellman Grade I or II without the need for rotator cuff repair for long-term outcomes [18]. FTTs may be managed with analgesics but operative procedures are necessary to prevent further degeneration [19]. Surgical intervention for FTTs showed to have a positive long-term response indicating that 80% of rotator cuff repairs were excellent or satisfactory after an average of 13 years [20]. Traditional repair methods, such as suturing, may fail to restore the normal structure and function of the tendon, leading to re-tears. Surgical management also showed promising results with the use of biological or synthetic 3D tissue engineered scaffolds in the repair of rotator cuff tears [21]. This has prompted the development of advanced biomaterials like gel-based composite scaffolds to improve tissue regeneration and healing. Below, we describe the various types of gel-based scaffolds used for the repair of rotator cuff tears.

2. Material Composition, Manufacturing Process and Properties of Gel-Based Composite Scaffolds

2.1. Materials for Gel-Based Composite Scaffolds

Gel-based composite scaffolds for rotator cuff repair are an evolving area of research, with promising potential to improve healing outcomes. The combination of natural and synthetic materials (composite scaffolds) offers a balance between biological compatibility and mechanical strength [22]. For example, collagen, hyaluronic acid, and fibrin are highly beneficial for promoting cellular interactions and tissue regeneration, while Poly(lactic-co-glycolic acid) (PLGA) and Polyethylene glycol (PEG) provide superior mechanical strength [23,24]. Table 1 provides an overview of the key gel-based composite scaffold materials used for rotator cuff repair, comparing their material and biological properties.

Table 1. Materials of comparative characteristics for gel-based composite scaffolds.

Material	Mechanical Strength	Biological Activity	Degradation Rate	Clinical Application
Collagen-based [25]	Low (enhanced with crosslinking)	High (cell adhesion and migration)	Moderate (enzymatic)	Ideal for tendon regeneration
Hyaluronic Acid [26,27]	Low (soft and hydrophilic)	High (anti-inflammatory)	Fast (enzymatic)	Supports healing and reduces inflammation
Alginate-based [28]	Moderate (crosslinking enhances)	Low (requires modification)	Moderate (enzyme-controlled)	Good for cell encapsulation
Fibrin-based [29]	Moderate (initial strength)	High (promotes healing and angiogenesis)	Moderate (fibrinolysis)	Wound healing and tendon-to-bone integration

Table 1. *Cont.*

Material	Mechanical Strength	Biological Activity	Degradation Rate	Clinical Application
PEG-based [30]	High (tunable)	Low (requires functionalization)	Slow (stable, controlled)	Ideal for mechanical scaffold designs
PLGA-based [23,31]	High (strong and load bearing)	Low (requires biofunctionalization)	Moderate (hydrolytic)	Load-bearing, controlled degradation
Composite (e.g., Collagen + PEG) [25,32]	High (balanced strength)	High (synergistic effect)	Controlled (depends on components)	Combines strength and biological support

2.2. Manufacturing Processes of Gel-Based Composite Scaffolds

The manufacturing processes and technological innovations of gel-based composite scaffolds are critical for their successful application in tissue engineering, particularly for complex repairs like rotator cuff injuries. These processes aim to produce scaffolds with the ideal mechanical properties, biocompatibility, and degradation rates, while promoting cell adhesion, migration, and tissue regeneration. Table 2 provides an overview of each manufacturing process, its technological innovations, and how they compare in terms of benefits, challenges, and applications.

Table 2. Manufacturing processes used for the production of gel-based composite scaffolds.

Manufacturing Process	Advantages	Challenges	Ideal Applications
Freeze-Drying [33]	High porosity, retains material structure, customizable pore size	Mechanical weakness, complex process	Soft tissue scaffolds, natural polymer scaffolds (collagen, alginate)
Electrospinning [34,35]	High surface area, fiber alignment, ECM mimicry	Mechanical weakness, scalability issues	Tendon, ligament tissue engineering, rotator cuff repairs
3D Bioprinting [36–39] (Biofabrication)	High precision, patient-specific, complex geometry	High cost, slow production speed, bioink challenges	Patient-specific scaffolds for rotator cuff repair, bone, cartilage
Solvent Casting and Particulate Leaching [40,41]	Simple, cost-effective, porosity control	Mechanical strength, solvent residue	Soft tissue regeneration, muscle and tendon scaffolds
Self-Assembly and Crosslinking [42,43]	Mimics natural processes, high biocompatibility	Lack of reproducibility, weak strength	Collagen-based scaffolds, tendon and ligament repair
Gelation [44]	Minimally invasive, customizable shape, ease of use	Weak mechanical properties, gelation consistency issues	Injectable scaffolds for rotator cuff and soft tissue repair
Smart Scaffolds [45]	Controlled release of growth factors, tailored response	Complexity in design and fabrication	Rotator cuff and tendon regeneration
Hybrid Scaffolds (Natural + Synthetic) [46,47]	Combines the best of both materials (biocompatibility + strength)	Complexity in material blending and manufacturing	Used for load-bearing rotator cuff repair
Injectable Hydrogels [48,49]	Minimally invasive delivery and customization	Mechanical strength may not be sufficient	Ideal for soft tissue repair and rotator cuff

2.3. Mechanical and Biological Propoerties of Gel-Based Composite Scaffolds

Gel-based composite scaffolds for rotator cuff repair integrate exceptional mechanical and biological properties, making them highly effective for tissue regeneration. Biologically, these scaffolds offer high biocompatibility, supporting cell adhesion, proliferation, and migration [21]. Their porous structure allows for nutrient and oxygen diffusion, cell migration, fostering angiogenesis and new tissue formation. Bioactive molecules such as growth factors (e.g., TGF-β, BMPs) and other types of therapeutic agents can be integrated into the scaffold to promote tendon-bone integration and control inflammation [50,51]. The controlled degradability of the materials ensures the scaffold provides temporary support while gradually being replaced by native ECM without generating harmful toxic by-products that can elicit a local or even a systemic immune response.

Mechanically, these scaffolds are engineered to match the elasticity and tensile strength of native tendons, providing resistance to physiological loads [52]. Their stiffness gradient [53], and bone interface, enable efficient load transfer and reduced mechanical discrepancies [54]. Fatigue resistance and durability, enhanced through the incorporation of various types of nanomaterials, ensure the scaffold's structural integrity during repetitive movements [55,56]. These combined properties make gel-based composite scaffolds a promising solution for improving outcomes in clinical treatment of rotator cuff repair. Table 3 provides an overview of the mechanical properties of gel-based composite scaffolds.

Table 3. Mechanical properties of gel-based composite scaffold.

Property	Description	Purpose/Benefit
Elasticity	Matches the flexibility of native tendons. Elastic moduli for human tendons range from 1.2 to 1.8 GPa [57].	Reduces stress concentrations and supports biomechanical compatibility.
Tensile Strength	Enhanced through reinforcement with polymers or ceramics. Ranges from 45 to 150 MPa for human tendons [57,58].	Supports the mechanical loads during tendon movement.
Stiffness Gradient	Gradual transition from soft (gel-based) to stiff (ceramic/polymer-based) zones.	Mimics the natural tendon-to-bone interface, ensuring smooth load transfer.
Load-Bearing Capability	Improved with composite materials like hydroxyapatite or nanofibers.	Withstands physiological loads, ensuring durability during the healing process.
Compression Resistance	Enhanced resistance to compressive forces. Compressive strength of cortical bone is typically 100–200 MPa [59].	Useful for distributing loads at the tendon-bone junction.
Porosity	Desired mechanical properties with controlled porosity, and optimal pore size.	Porosity impacts the material's strength and facilitate better cell and tissue integration.
Degradation Rate	Designed to degrade gradually, allowing for tissue formation and integration.	Degradation process can be influenced by the gel matrix (e.g., natural versus synthetic polymers) and the crosslinking agents used to stabilize the scaffold.
Fatigue Resistance	Improved by adding nanoscale fillers (e.g., graphene oxide).	Ensures performance under repetitive mechanical stresses over extended periods.
Durability	Maintains structural integrity over time with reinforced hydrogels.	Reduces risk of scaffold failure during tissue regeneration.

3. Scaffold Augmentations

As aforementioned, gel-based and synthetic composite scaffolds can be biodegradable and encourage cell proliferation for tendon or muscle repair [60]. They can consist of different biomaterials forming composites or one material either biological or synthetic [61,62]. The bioengineered constructs may contain collagen, ECM, decellularized ECM, human/porcine dermis, or biosynthetic material which can be implanted into the rotator cuff in a process known as scaffold augmentation [63]. Figure 1 demonstrates one example using a synthetic Poly-L-lactic acid (PLLA)-based scaffold augmentation for rotator cuff repair. This type of scaffold augmentation for shoulder treatment has been tested in different in vivo models which showed positive results [64,65].

Gel-based scaffolds typically consist of biocompatible and biodegradable materials that mimic the natural ECM, providing a 3D structure for cellular growth and tissue development. They are typically designed to be injectable or moldable, allowing them to conform to the shape of the damaged area. These scaffolds support cell attachment, proliferation, and differentiation, and can be loaded with bioactive factors to enhance healing. These scaffolds can be inserted and fixed arthroscopically onto the rotator cuff [67]. Collagen, ECM, cells, and growth factors can constitute the composite scaffold to enhance regeneration and recovery [68–70]. Figure 2 provides an example of some materials that can constitute composite scaffolds. Previous in vivo studies conducted on rats, sheep, and canine models with synthetic or biological scaffolds have demonstrated positive outcomes for augmentation. Tenocyte growth was observed with little immune or inflammatory response present and the scaffolds were also biodegradable [71–74]. Unfortunately, that does not necessitate the same outcome clinically. Scaffolds have been used clinically for spinal fusion, fractures and defects, sinus augmentation, hand surgery, and cranioplasty with varying clinical outcomes [75]. In the case of shoulder augmentation, different clinical outcomes have been reported from previous scaffold evaluations [68,76,77] and can be grouped into three major categories: (i) Collagen-based (ii) ECM-based and (iii) Synthetic-based augmentation. A summary of several studies dealing with scaffold augmentation is represented in Table 4 below.

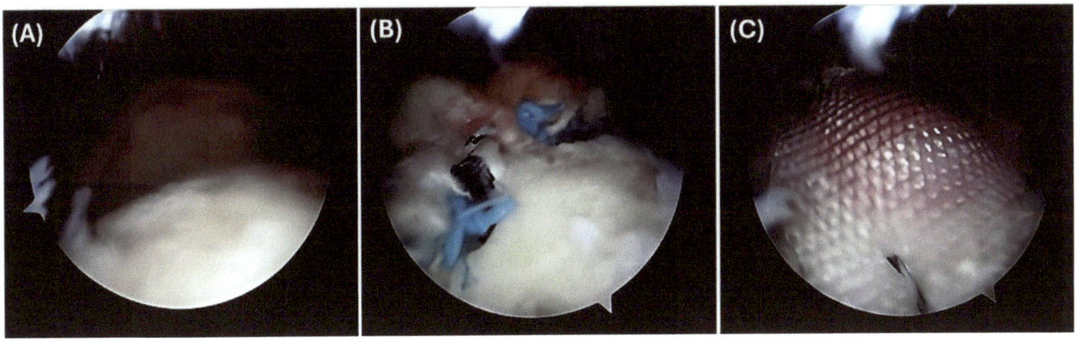

Figure 1. Right shoulder observed from the lateral portal. (**A**) Demonstrates a large tear at the level of the glenoid. (**B**) Rotator cuff repair prior to the addition of the synthetic PLLA scaffold. (**C**) Placement of the scaffold with medial sutures and anchors. Reprinted with permissions from [66].

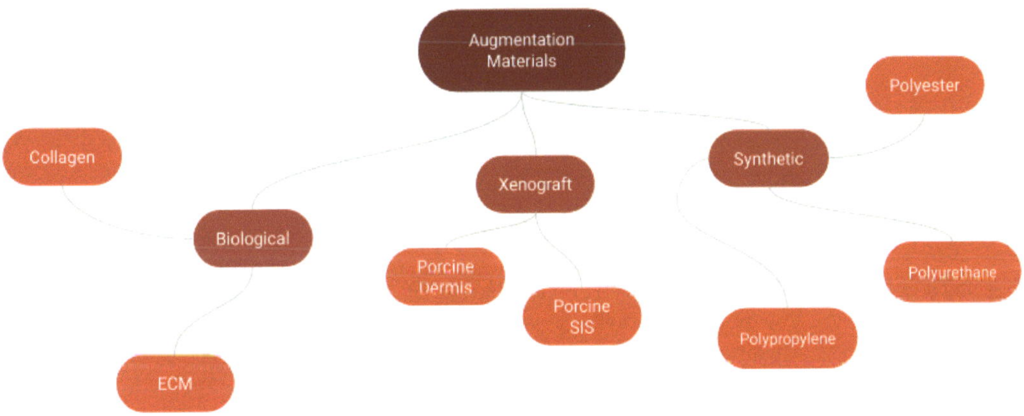

Figure 2. An example of some biological and synthetic materials that can constitute composite scaffold for shoulder augmentation.

Table 4. Summary of studies.

Study	Material	Summary of Findings
		Collagen-based Augmentation
Bokor [78–80]	Collagen	↑ Tendon thickness; reduced defect size
	Collagen *	No degeneration observed; tendon thickness better than pre-operative
	Collagen	↑ Tendon thickness and integration
Thon [81]	Collagen	↑ Tendon thickness and vascularity; 96% of patients healed
Schlegel [82]	Collagen	↓ Defect size for 70% of patients; ↑ Tendon thickness
Dai [83]	Collagen **	↑ Tendon thickness
Bushnell [84]	Collagen	Integration present and lower re-tear rates
		ECM-based Augmentation
Iannotti [85]	SIS **	Less healing in massive tears; no clear clinical benefit
Bryant [86]	SIS **	52.9% augmentation failure; no clear clinical benefit
Castagna [87]	DDCM **	Fewer re-tears, determined to be safe
Avanzi [88]	Dermal Patch **	97.6% healing with patch; ↑ Tendon thickness and strength recovery
Maillot [89]	Dermal Patch **	No superior benefit of use; slower anterior elevation, and external rotation recovery
Gouk [90]	Acellular Dermal Patch ***	86% of grafts failed; failed biocompatibility
Johnson [91]	Dermal Patch ***	Intact grafts by 35 months; functional ROM

Table 4. *Cont.*

Study	Material	Summary of Findings
		Synthetic-based Augmentation
Encalada-Diaz [92]	Polycarbonate Polyurethane	90% of tears healed; ROM recovered
Petriccioli [93]	Polyurethane	Greater tendon thickness in repaired shoulder compared to healthy shoulder
Proctor [66]	PLLA	83% repair rate (12 months); 78% (48 months), ↑ Tendon integrity
Ciampi [94]	Propylene	Fewer re-tears than the collagen group; stronger abduction strength and elevation
Vitali [95]	Propylene	Greater abduction and scapular plane elevation, low retears; ↑ Tendon integrity
Smolen [96]	Polyester	86% of repairs intact, 14% re-tear; significant improvement in internal rotation
Cowling [97]	Polyester	Improved shoulder functionality; no difference in Goutallier classification

* 5-year follow-up study on Bokor et al. [78]; ** porcine xenograft; *** human allograft; DDCM, dermal-derived collagen membrane; SIS, small intestine submucosa; PLLA, poly-L-lactic acid.

3.1. Collagen-Based Augmentation

Collagen scaffolds have been used pre-clinically and clinically, and demonstrated tendon regeneration and remodeling [74,98]. Bokor et al. [79] utilized porous type I collagen scaffolds in regenerating PTTs of the supraspinatus tendon. The study involved 13 patients evaluated by MRI over 2 years to monitor tendon regeneration. The scaffolds were placed on the bursal side arthroscopically and were observed to increase tendon thickness throughout the study. A total of 92% of the patients experienced a positive result in decreasing tear size. Tears regressed from high-grade to low-grade or none as the scaffold demonstrated potent healing capabilities in reducing the defect size. The collagen-based scaffold provided inductive capabilities to assist in the formation of new connective tissue [74,79]. A 5-year follow-up on 11 patients from this study was conducted by the same authors. Tendon thickness decreased at the 5-year MRI but was better than the pre-operative thickness. Tendon integrity was established in 72.7% of the patients indicating no degenerative effects in this sample [78]. A separate but similar study conducted by Bokor et al. involved type I collagen scaffolds in FTT repair of the supraspinatus tendon. The study recruited nine patients, with 89% having satisfactory results after 24 months. Similar to the first study, the collagen-based scaffolds were placed on the bursal side and an MRI was used to evaluate patient status post-operatively. Tendon thickness increased for up to 6-months with new tendon growth indistinguishable from native tendon [80]. These results support the notion of utilizing collagen-based tissue engineered scaffolds for PTT and FTT repair, albeit in samples of 13 and 9 patients, respectively. Similarly, Thon et al. [81] arthroscopically implanted collagen scaffolds in patients with large or massive Cofield classified cuff tears. Ultrasound and MRI were used to determine the progress of tendon healing in 23 patients. A total of 96% of patients had tendon healing with a 6% failure rate due to two patients having either progressive arthritis or a lack of tendon healing. Overall tendon thickness and vascularity increased within 12 months as determined by imaging [81]. Schlegel et al. [82] recruited 33 patients for augmentation with a collagen patch for PTT repair. MRI assessment of the defects indicated that 70% of patients showed a decrease of at least 1 grade in tear severity 3 months post-augmentation. Six patients had their tears completely heal, with one individual having no reduction in tear size. Supraspinatus tendon thickness improved after 3 months and persisted until 12 months in 94% of the patients post-operatively. A 1-year follow-up indicated that all tear defect sizes decreased except for one [82]. A recent study conducted by Dai et al. used bovine-collagen-derived scaffolds from Smith and Nephew Regeneten for PTT repair in 30 patients with 24 available for follow-up. Tendon thickness increased post-operatively from 0.8 mm to 6.5 mm at the average follow-up.

Additionally, the implant reduced strain on the supraspinatus tendon for bursal, articular, and intrasubstance tears due to its inductive abilities [83]. Furthermore, Bushnell et al. performed a larger study involving 115 patients with the Smith and Nephew Regeneten collagen scaffold for repairing FTTs and determined re-tear rates. Patients involved in the study had medium or large Cofield classified tears with the scaffold arthroscopically inserted. These collagen scaffolds were resorbed in 100% of patients at the 1-year follow up but a boundary was detected in 8.9% of patients at the 3-month checkup. This indicates good biocompatibility as the scaffold was able to encourage tissue growth [84]. There was a 16.5% re-tear rate within 1 year, including re-torn and failed to heal rotator cuffs which is lower than other reported studies [99–103]. Although these studies lack a control shoulder, improvements were observed with the use of scaffolds.

3.2. Extracellular Matrix-Based Augmentation

ECM-based hydrogels are typically injected into the injury site to fill gaps or defects, improving tendon healing by promoting cell infiltration, thus representing the predominant approach for augmentation [104,105]. Porcine products (acellular matrix or collagen membrane) were reported to have varying efficacy from older studies [85,106–109]. Iannotti et al. [85] studied 30 patients with large or massive supraspinatus or infraspinatus tendon tears with porcine small intestine submucosa (SIS) implants. A total of 89% of the cohort patients with large tears had fully healed, however, with massive tears only 24% were fully healed. Overall, 40% of the tears did not heal due to varying factors. There was no significant difference between groups and no clear clinical benefit was observed in the 30 shoulders evaluated [85]. A similar study was conducted by Bryant et al. [86] and included 62 randomized patients with porcine SIS augmentations. The primary outcome evaluated the success of the implants with 52.9% failure in the augmentation group and 65.4% failure in the control. Although the sample size was greater than Iannotti et al. [53], the risk ratio (0.81) was in favor of the augmentation group, but again, the results were not statistically significant between groups. Similar to Iannotti et al. [53], the authors concluded there was no apparent clinical benefit from using the SIS implant [86].

Castagna et al. [87] tested porcine dermis-derived collagen membranes in 35 patients for large or massive rotator cuff tears. The treatment group included 7 shoulders as compared to 13 in the control which experienced re-tears as determined by MRI evaluation at 24 months. Although tendon integrity was not fully restored by the implant, there was no evidence of tissue rejection due to the xenograft. As such, the authors concluded that the porcine dermis-derived collagen to be safe for human use [87]. A separate study evaluated a porcine dermal patch in 96 patients randomized into two arms. Avanzi et al. [88] followed up with 38 patients with augmentation and 30 without. Significant healing was present in the augmented group with a healing rate of 97.6% as compared to 59.6% in the control. Tendon thickness and footprint coverage were significantly higher in the augmented group throughout the 24 months. Density scores were similar with 87.3 and 88.4 for the augmented and control group, respectively. Additionally, the augmented group displayed a quicker tendon strength recovery. On the contrary, Maillot et al. [89] provide contradictory results with the use of porcine dermal patches. Thirty-two patients were placed into either an arthroscopic repair, patch, or debridement group and no difference was observed between the three groups. Recovery for anterior elevation and external rotation progressed slower as compared to the debridement and arthroscopic repair groups. The analysis of clinical outcomes indicated no superior benefit to utilizing the porcine dermal patch but is limited by the sample size of 11 patients [89].

ECM based scaffolds using human acellular dermal allograft have also been tested. Gouk et al. [110] used a human acellular dermal allograft from Graft Jacket® in eight patients as an interposition graft. MRI was used post-operatively and indicated that 14% of grafts were intact within 6 months. The other 86% had failed grafts within 6 months, consistent with previous reports that indicated failure as well [110]. Additionally, the grafts displayed integration deficits suggesting failed compatibility with the rotator cuff [90]. Alternatively, Johnson et al. [91] had positive outcomes with the same Graft Jacket® patch on 14 shoulders. Ultrasound evaluation indicated intact grafts within 35 months with three grafts undergoing attenuation or heterogeneous change (Figure 3). Also, ROM was functional at a 2-year follow-up. The graft was viable at the final follow-up of 41 months and indicated healing of the rotator cuff as seen by shoulder functionality [91].

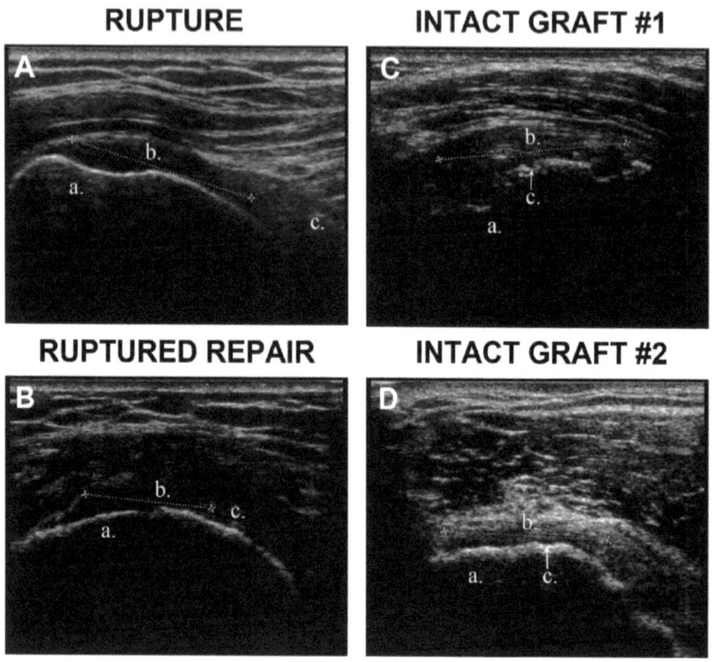

Figure 3. Ultrasound image of the rotator cuff tears and repairs from representative patients. (**A**) Rotator cuff tear. a. humeral head; b. defect; c. retracted cuff. (**B**) Ruptured repair. a. humeral head; b. defect; c. retracted cuff and graft. (**C**,**D**) Intact GraftJacket® and cuff repair. a. humeral head; b. intact graft; c. suture. Adapted with permission from [91].

3.3. Synthetic-Based Augmentation

Synthetic hydrogels are a group of synthetic polymers that can be photo-crosslinked, allowing for precise control over their physical properties, including stiffness and degradation rate. These scaffolds are versatile and can be tailored to provide the appropriate biomaterial properties of the target tissue. Synthetic composite scaffolds can incorporate polymers such as polyester, PLLA, polypropylene, polyacrylamide, Dacron, carbon, silicone, and nylon [111–113]. These synthetic composite materials are considered foreign bodies and are associated with complications due to poor biocompatibility [114]. Many studies were conducted on synthetic materials for rotator cuff repair with promising results due to their strong mechanical characteristics. Figure 4 shows tendon-bone junction extracts from sheep demonstrating proper healing in groups with PLGA scaffolds [73]. Encalada-Diaz et al. [92] used a non-resorbable polycarbonate polyurethane patch in 10 patients with FTTs to evaluate post-operative outcomes. Healing was present in 90% of patients as determined by MRI and ultrasound, while one patient experienced a re-tear. The low re-tear rate, progressive ROM, and high healing rate demonstrated efficacy but, the study was limited by the small sample size and sex (cohort of 10 females) [92]. A degradable polyurethane urea scaffold was employed by Petriccioli et al. [93] in 10 patients for the repair of the subscapularis. Ultrasound and MRI were also used to identify FTTs of the subscapularis in 1 patient who experienced a re-tear within 23 months. The remaining nine patients had intact repairs with all patients having greater tendon thickness in the repaired tendon compared to the healthy control shoulder. The scaffold provided a low re-tear rate in this small cohort [93]. Another study involved the use of a PLLA-based resorbable scaffold for massive rotator cuff tears in 18 patients. Results indicated an 83% success in repair rate 12 months post-operatively with long-term success at 78% by 42 months. Three repairs failed due to factors unrelated to the scaffold. The integrity of the rotator cuff was strengthened due to the stability of the scaffold [66]. Despite being the first clinical synthetic study, no control group was present, but the long-term results are indeed promising.

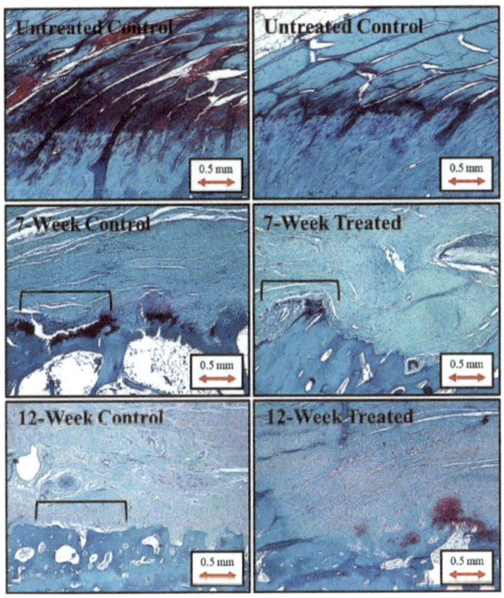

Figure 4. Histology specimens extracted from sheep infraspinatus tenocytes at the tendon-bone junction were stained with Safranin O, demonstrating healing with a PLGA anchored scaffold. Black brackets indicate interfaces where tissue and bone are not properly integrated. Magnification of 40×. Reprinted with permissions from [73].

Polypropylene is another biosynthetic material that can be used in composite scaffolds. Ciampi et al. [94] separated 152 patients into three groups receiving a collagen patch, polypropylene patch, or no patch (control). Elevation and abduction strength at 36 months with the propylene patch was significantly higher than those with the collagen and control group, probably because propylene provides greater long-term strength since it is not resorbed, unlike collagen. The re-tear rate was also significantly lower in the propylene group, at 17% compared to 41% and 51% in the control and collagen group, respectively [94]. Similar results were reported by Vitali et al. [95] with the use of a propylene scaffold in 60 patients. Greater abduction strength and scapular plane elevation were observed in the propylene group. Re-tear rates were also lower at 15% in the propylene group as compared to 40% in the control group, demonstrating greater tendon integrity with the scaffold [95]. The improvements presented with the propylene scaffold shows clinical potential for reducing re-tears and increasing rotator cuff integrity. Polyester is another synthetic material that was tested by Smolen et al. [96] in 50 patients. There was a significant improvement in internal rotation within 22 months. Flexion, abduction, external rotation, and strength did improve post-operatively but were not significant. As for re-tears, 86% of patients had an intact repair with 14% experiencing a re-tear [96]. A recent study by Cowling et al. treated 29 patients with a polyester patch and 39 patients without. Comparable results with those of Smolen et al. [96] were observed since the augmentation with polyester improved shoulder functionality. However, the Goutallier classification of both groups showed little difference at 6 months. Although Smolen et al. [96] conducted a longer study, the authors acknowledged that 6 months may be too early to identify changes in the Goutallier classification [97]. Despite this, the results indicated beneficial short-term results with the polyester patch.

4. Patient Outcomes

The clinical implications for various types of gel-based scaffolds used for rotator cuff repair are described in Table 5.

Table 5. Clinical implications in treatment groups.

Study	Material	Patients Treated (n)	AE (n)	Functionality	Level
		Collagen-based Augmentation			
Bokor [78–80]	Collagen	13	4	↑ CM	-
	Collagen *	11	0	↑ CM, ASES	IV
	Collagen	9	2	↑ CM, ASES, ↓ Pain	-
Thon [81]	Collagen	23	9	High ASES scores (82.87)	IV
Schlegel [82]	Collagen	33	5	↑ CM, ASES, ↓ Pain	IV
Dai [83]	Collagen **	30	1	↑ ASES, VAS	-
Bushnell [84]	Collagen	115	7, 2 †	↑ CM, ASES	IV
		ECM-based Augmentation			
Iannotti [85]	SIS **	15	3	↓ PENN, No clinical benefit	II
Bryant [86]	SIS **	34	4	52.9% risk of failure, No clinical benefit	I
Castagna [87]	DDCM **	35	0	↑ CM, More functionality in treatment	III
Avanzi [88]	Dermal Patch **	46	2	↑ EQ-VAS, DASH, SST, CM post-operative	II
Maillot [89]	Dermal Patch **	11	4, 1 †	↑ CM post-operative, No difference in VAS/mobility	II
Gouk [90]	Acellular Dermal Patch ***	8	1	No difference in Oxford/DASH, No clinical benefit	-
Johnson [91]	Dermal Patch ***	14	0	No difference in Oxford/DASH	IV
		Synthetic-based Augmentation			
Encalada-Diaz [92]	Polycarbonate Polyurethane	10	0	↑ VAS, SST, ASES, CADL, UCLA	IV
Petricciolі [93]	Polyurethane	10	0	↑ VAS, DASH, CM, subscapular function	IV
Proctor [66]	PLLA	18	0	↑ ASES	IV
Ciampi [94]	Propylene	52	0	↑ VAS, UCLA	III
Vitali [95]	Propylene	60	0	↑ VAS, UCLA	-
Smolen [96]	Polyester	50	0	↑ CM, SSV, ↓ Pain,	IV
Cowling [97]	Polyester	29	0	↑ Oxford, SPADI, EQ-VAS, CM similar across groups	-

* 5-year follow-up study on Bokor et al.; ** porcine xenograft; *** human allograft; † complications may be attributed to the implant, otherwise unlinked; AE, adverse events; DDCM, dermal-derived collagen membrane; SIS, small intestine submucosa; PLLA, poly-L-lactic acid; CM, Constant–Murley Score; ASES, American Shoulder and Elbow Society Shoulder Scale; VAS, visual analog scale; PENN, PENN Shoulder Score; EQ-VAS, EuroQol-Visual Analog Scale; DASH, disabilities of the arm, shoulder, and hand; SST, simple shoulder test; Oxford, Oxford Shoulder Score; CADL, constant activities of daily living; UCLA, UCLA Shoulder Score; SSV, subjective shoulder value; SPADI, shoulder pain and disability index.

4.1. Collagen-Based Scaffolds

The clinical assessment of Bokor et al. [39] in repairing supraspinatus PTTs indicated improvement and functionality. The Constant–Murley and American Shoulder and Elbow

Society (ASES) shoulder scale indicated progress as scores improved. Pain scales involving both scores demonstrated improvement as well. One patient out of thirteen reported persistent pain despite improvement in the Constant–Murley score and successful MRIs. Four patients experienced complications such as shoulder inflammation during arthroscopy, bursitis, rupture of the bicep's long head, and adhesive capsulitis. These complications were resolved along with satisfactory healing of the PTT with no evidence of the scaffold causing these [79]. The 5-year follow-up showed improvement in the Constant–Murley and ASES scales. Apart from the complications from the initial study, no further adverse effects were observed due to the scaffold. Degenerative rotator cuff tears progressed in two patients as identified by MRI. The collagen implant induced healing of the original defect but was unable to prevent progressive underlying disease [78,81].

Evaluating complications and outcomes in FTT repair by Bokor et al. [80] indicate an overall positive outcome. The rotator cuffs were intact with comprehensive tendon repair at the implantation sites as demonstrated by patient MRIs. Constant–Murley and ASES scores increased significantly over the 24 months along with a significant decrease on the pain scale post-operatively. However, one patient presented with pain at 26 months with no evidence of a faulty repair. Additionally, a patient had capsulitis pre-operatively which persisted for three months with improvement. The authors noted no adverse effects as a result of the collagen scaffold implantation [80].

Clinical assessment was similar in Thon et al. [80], however, ASES scores were not taken pre-and post-operatively. The overall ASES shoulder score across all groups was 82.87 with no difference between large and massive tears. One patient had a healing failure which was determined by MRI at 3 months in the supraspinatus tendon due to lack of repair. Another patient had progressive arthritis and atrophy of the rotator cuff muscles which required reverse shoulder arthroplasty 25 months post-operatively. However, it was noted by the authors that the patient's rotator cuff tear healed but the scaffold had no impact on her osteoarthritis and atrophy. Eight patients had post operative scapular dyskinesia, but this was not attributed to the scaffold. Overall, the collagen scaffold implanted by Thon et al. is indeed biocompatible and regenerative for shoulder augmentation [81].

In the study by Schlegel et al. [80], ASES pain and shoulder scores improved post-operatively by 1 year. Constant–Murley scores improved to 81.4 by 1 year post-operatively. The 1-year evaluation indicated that 94% of the patients were satisfied with the results of the augmentation. Regardless, complications such as subacromial fluid buildup, pain, dermatological reactions, and a case involving cardiac ablation were reported. The patients were treated accordingly, and the complaints were resolved with medication or surgically. These reactions were not linked to the presence of the collagen implant and no device-related adverse effects were reported [82].

Dai et al. [83] and Bushnell et al. [83] used the Smith and Nephew Regeneten scaffold for PTT and FTT repair, respectively. Dai et al. utilized ASES and Visual Analog Scale (VAS) scores for functional assessment pre- and post-operatively. ASES and VAS scores improved significantly at the 1-year follow up, indicating greater shoulder function and less pain. ASES improved from 45.6 to 68.1, and VAS improved from 8.3 to 3.8 post-operatively. Patients rated satisfaction at 7.5 out of 10 post-operatively. One complication emerged when a patient encountered a traumatic re-tear four months post-operatively unrelated to implantation [83]. Bushnell's larger sample had 19 re-tears at the 1-year follow-up. Variance in surgical technique produced a greater number of re-tears with single-row repair having greater rates compared to double-row repair. ASES and Constant–Murley scores improved within one year as patients met the minimal clinically important difference. Likewise, 96.5% of patients were satisfied with the results of the surgery. Seven reoperations were performed for revision rotator cuff repair due to re-tearing but were not related to the implant. Two reoperations may have been linked to the collagen scaffold, procedure, or anchors as the first patient experienced swelling and a possible infection post-operatively and thus required debridement. The second patient experienced inflammation and osteopenia in the greater tuberosity leading to future debridement and aspiration [84]. Despite the complications, the overall result of the Regeneten collagen implant is positive and demonstrates good tendon healing in the cohort.

4.2. ECM-Based Scaffolds

Iannotti et al. [85] demonstrated no clinical benefit from porcine SIS scaffolds. Specifically, the median post-operative PENN shoulder score was greater in the control group with 91 points. Patient satisfaction was at 10 and shoulder functionality was at 59 with fully healed rotator cuffs. Shoulder functionality was decreased in the augmentation group, but the median PENN score was not significant between both groups. Adverse effects were present in three patients in the augmentation group. One patient experienced erythema with spontaneous drainage which required debridement with tissue loss present at the posterior edge of the infraspinatus. Another patient experienced erythema and greater temperature near the wound which resolved without intervention. The third patient experienced swelling and pain requiring aspiration of the subacromial space. All adverse events were resolved with tendon healing by the 1-year checkup [85].

The clinical implications reported by Bryant et al. [86] with porcine SIS scaffolds indicate a slight effect bias towards the control group. Patients reported a similar range of motion and strength in both groups along with progressing Western Ontario Rotator Cuff Index scores (WORC) over 24 months. Two patients in the experimental group required surgical intervention as one had a deep infection, and another experienced a biceps tendon rupture post-operatively. Warmth near the surgical site and fever were also present in two patients resolving without intervention. The control group with 3 patients experienced adverse events. One patient experienced an infection that was treated by antibiotics, and the others required surgical repair or shoulder manipulation. These results are concurrent with Iannotti et al. [86] as it is unlikely to have clinical benefit from SIS scaffolds [86].

Castagna et al. [87] utilized the Constant–Murley evaluation to determine shoulder functionality in both the control and experimental groups. Scores taken before surgery were lower than post-operative scores, but at the 24-month evaluation were significantly higher in the treatment group with the mean being 71.4 compared to 63.9 in the control. Patients with re-tearing in the experimental group had significant functionality compared to patients with re-tears in the control. No adverse events were noted and there was no indication of tissue rejection indicating that the porcine dermal-derived ECM scaffolds are safe and effective for improving shoulder functionality [87]. This, however, is limited to the small sample size which requires studies with larger patient population for determining the exact efficacy.

A study by Avanzi et al. [88] evaluating the effectiveness of a porcine dermal patch produced excellent clinical results. The authors employed the EuroQol-Visual Analog Scale (EQ-VAS), Disabilities of the Arm, Shoulder, and Hand (DASH), Simple Shoulder Test (SST), and Constant–Murley scores to determine the functionality of the augmentation. The EQ-VAS, DASH, and SST scores reflected no significant difference between groups but were significant in improvement from pre- to post-operative scores. Specifically, Constant–Murley scores improved post-operatively in both groups with scores higher in the augmentation group at 95.5 at 24 months, indicating that shoulder functionality improved with the porcine dermal patch. Four patients were classified as "not healed" and two patients experienced a re-tear. Furthermore, two patients experienced complications unrelated to the porcine dermal patch [88]. The clinical assessment of this clinical trial provides promising results, despite its contradictory results when compared to those of Maillot et al. [89]. The clinical outcomes indicate no difference between Constant–Murley scores in each group, but scores increased post-operatively. There was a significant difference between (1) patch and debridement and (2) repair and debridement groups at 12 and 24 months. Similarly, no difference was noted for VAS and active mobility between groups. Complications arose in the augmentation and arthroscopic repair group. One patient experienced a deep infection that required removal of the patch and four others experienced post-operative stiffness which improved within 12 months in the augmentation group. One patient in the arthroscopic repair group experienced a superficial wound infection that cleared after antibiotic treatment [89].

Shoulder assessment conducted by Gouk et al. [91] in the use of human acellular dermal patches were measured by DASH, Constant–Murley, and Oxford shoulder scores. Average scores were 48.2, 37.16, and 24.94 for Constant–Murley, Oxford, and DASH, respectively. The majority of patients were also satisfied with surgery, however, MRI results by 6 months indicated graft failure. Also, one patient experienced an infection that required the removal of the graft [90]. The patients evaluated by Johnson et al. [91] had greater Constant–Murley scores in the non-operated shoulder, but function in the augmented shoulder was satisfactory. There was no difference between Oxford and DASH scores between operated and non-operated shoulders. No complications were recorded from the scaffold [91]. The satisfactory results and intact grafts detected by ultrasound from Johnson et al. are in contrast with the graft failure and complications reported by Gouk et al. [91].

4.3. *Synthetic-Based Scaffolds*

FTT repair by Encalada-Diaz et al. [92] with the polycarbonate polyurethane scaffold showed promising clinical results with VAS (2.6), SST (7.7), and ASES (73.3) scores all significantly higher post-operatively at the 1-year follow-up. Also, the Constant Activities of Daily Living (CADL) and UCLA scores improved post-operatively from 6 to 12 months. No complications were found associated with the synthetic patch and no inflammatory response was present [92]. This is a significant and promising finding because no adverse reaction was present despite the presence of the synthetic patch, thus, indicating its clinical efficacy.

The use of the polyurethane urea scaffold by Petriccioli et al. [93] resulted in improved VAS, DASH, and Constant–Murley scores. Patients experienced improvements post-operatively in shoulder function by 23 months. Additionally, functional use significant improvement in the belly-press, lift-off, and bear-hug test indicating subscapular function was re-emerging. The re-tear rate was at 10% with no complications present in any patient. No infection or inflammation was present as a result of the exposure to the polyurethane urea scaffold [93], again indicating promising results in subscapularis repairs.

Another study utilized ASES scores to evaluate shoulder functionality post-operatively. There was a significant improvement in ASES scores 6 to 42 months after surgery. The final score was 82 at 42 months. Moreover, ASES scores were significantly higher for patients with an intact repair compared to a re-tear by 42 months. No adverse events or tissue rejections were reported due to exposure to the poly-L-lactic acid scaffold [66], indicating satisfactory clinical results in shoulder functionality.

The results of Ciampi et al. [94] indicate an excellent clinical outcome with the use of polypropylene scaffolds. The re-tear rate was low in addition to improvement in shoulder functionality. VAS scores improved overall over the 36 months. UCLA scores were significantly better in the propylene group at 36 months with the mean being 24.61 as compared to the control and collagen groups. No adverse events or inflammatory reactions were detected from the augmentation [94]. The overall results are more favorable toward the usage of polypropylene compared to the biological collagen for augmentation. Furthermore, the clinical assessment of Vitali et al. [94] supports the study of Ciampi et al. [94]. Similarly, this study also reported improvements in the VAS and UCLA over 36 months. The post-operative UCLA score for the propylene group was 24.6 compared to 14.73 in the control. Patient satisfaction was recorded in 52 patients with no foreign-body reactions or adverse events recorded [95].

Constant–Murley and Subjective Shoulder Value (SSV) scores improved significantly by the final 22-month follow-up in a study conducted by Smolen et al. [94] Reports of pain improved by the final follow-up. There was high patient satisfaction since shoulder functionality was also restored. In the cases of re-tearing, one involved detachment of the polyester while it remained attached in the other six cases. One patient experienced crepitus requiring the scaffold to be removed. There was no evidence of a foreign body reaction with no other adverse events attributed to the scaffold [96]. The clinical implications of the polyester patch indicate substantial functionality and lower re-tear rates. This is reiterated by another study where polyester scaffolds were tested in a smaller sample size (n = 29). Greater improvements were present in the group with the polyester as compared to the control by 6 months. Both the Oxford score and the Shoulder Pain and Disability Index (SPADI) improved significantly in the patch group. Constant–Murley scores were similar across groups while EQ-VAS was improved with the polyester patch. No adverse events were reported due to the polyester [97]. The concurrent results of these studies indicate favorability for polyester patches in both small and medium sample sizes for augmentation.

5. Discussion and Future Prospective

Gel-based composite scaffolds represent a promising avenue in tissue engineering, combining mechanical and biological innovative formulations and designs to address the complex requirements of rotator cuff repair. Ongoing research aims to optimize these scaffolds for clinical use, ensuring effective and successful tissue regeneration. Recent progress involves the development of composite scaffolds that combine hydrogels with reinforcing materials like ceramics or synthetic polymers [24,115]. This approach addresses the mechanical challenges of rotator cuff repair by providing a stiffness gradient that mimics the natural transition from tendon to bone [54,55]. Nanofibers and nanoparticles are also being integrated into gel-based scaffolds to enhance their mechanical strength and fatigue resistance while maintaining elasticity [56,116].

Advanced manufacturing technology like Electrospinning and 3D printing has further advanced the fabrication of scaffolds with custom designed and highly organized structures and controlled porosity, allowing for better nutrient diffusion, cell infiltration and angiogenesis [34,39]. These scaffolds often include biodegradable materials that degrade at controlled rates, ensuring temporary mechanical support as cells begin to produce native ECM and new tissue forms. Moreover, researchers are exploring drug delivery capabilities within these scaffolds to release anti-inflammatory or antimicrobial agents, reducing post-operative complications and improving healing outcomes [27].

However, these scaffolds also have limitations. Hydrogels alone lack sufficient mechanical strength and often require reinforcement with polymers or ceramics [24], which can complicate their design. Achieving an optimal degradation rate remains challenging since premature or delayed scaffold breakdown can hinder healing [39,117]. Also, complex manufacturing processes and high production costs limit scalability and accessibility. Furthermore, despite promising preclinical results, clinical data on large-scale efficacy and safety are still limited [118]. Some advanced components, like nanoparticles, may pose a risk of localized inflammation or immune responses. Addressing these challenges is essential in order to make these tissue engineered scaffolds viable for widespread clinical use.

Overall, gel-based scaffolds represent a promising approach to enhancing rotator cuff repair. Gel-based scaffolds with bioactive molecules such as genes, growth factors or small molecules can further enhance their healing potential. For example, incorporating tendon-specific growth factors like tendon-derived growth factor (TDGF) could improve the regeneration of tendon tissues [119]. Combining gel scaffolds with stem cells (e.g., mesenchymal stem cells or tendon-derived stem cells) holds significant potential for improving healing and regeneration [120]. The scaffold provides biomechanical support while the stem cells differentiate into tenocytes, accelerating tissue repair. In the future, greater advances in tissue engineering might lead to the development of personalized scaffolds tailored to an individual's specific injury, improving the chances of successful rotator cuff repair.

6. Conclusions

Advances in orthopedic surgery have allowed individuals to regain shoulder functionality after rotator cuff tears. More recently, gel-based composite scaffolds offer significant promise in improving the clinical outcomes of rotator cuff repair. Their ability to promote tendon regeneration, support cell migration, prevent immune reaction, and reduce complications makes them a valuable adjunct to surgical repair, especially for large or difficult-to-heal tears. Although some biomaterials such as porcine SIS grafts may provoke an immune reaction, a wide array of others such as collagen-based, ECM-based and synthetic-based composites can be used as alternatives. The clinical implications of using such scaffolds/patches are beneficial as shoulder functionality, ROM, and pain improves following augmentation. Moreover, adverse events and re-tear rates appear to be low which further encourages the use of such hydrogel-based composite scaffolds. Furthermore, the incorporation of stem cells and bioactive molecules such as growth factors, hormones, and drugs into the scaffolds represents a significant technological advancement, enhancing scaffold performance in rotator cuff repair and other tissue engineering applications. The future of shoulder rotator cuff repair looks very promising with the use of these various types of tissue-engineered composite scaffolds. But, searching clinicaltrials.gov with the "term rotator cuff tears" and "scaffold" indicates only two active and two recruiting clinical trials using tissue-engineered scaffolds. Ten other studies that were also listed are either completed or withdrawn. Clearly, given this small number of active clinical trials, many more are warranted. Regardless, this review serves as a quick guide for clinical evaluation research of novel surgical or regenerative material options for rotator cuff treatment.

Author Contributions: Conceptualization, S.T., M.H. and A.I.; methodology, S.T.; writing—original draft preparation, S.T., M.H. and A.I.; writing—review and editing, S.T., M.H. and A.I.; supervision, M.H. and A.I.; project administration, M.H. and A.I. All authors have read and agreed to the published version of the manuscript.

Funding: This research received no external funding.

Data Availability Statement: No new data were created or analyzed in this study. Data sharing is not applicable to this article.

Conflicts of Interest: The authors declare no conflict of interest.

References

1. May, T.; Garmel, G.M. Rotator Cuff Injury. In *StatPearls [Internet]*; StatPearls Publishing: St. Petersburg, FL, USA, 2019.
2. Papatheodorou, A.; Ellinas, P.; Takis, F.; Tsanis, A.; Maris, I.; Batakis, N. US of the shoulder: Rotator cuff and non–rotator cuff disorders. *Radiographics* **2006**, *26*, e23. [CrossRef] [PubMed]

3. Lugo, R.; Kung, P.; Ma, C.B. Shoulder biomechanics. *Eur. J. Radiol.* **2008**, *68*, 16–24. [CrossRef] [PubMed]
4. Dabija, D.I.; Gao, C.; Edwards, T.L.; Kuhn, J.E.; Jain, N.B. Genetic and familial predisposition to rotator cuff disease: A systematic review. *J. Shoulder Elb. Surg.* **2017**, *26*, 1103–1112. [CrossRef] [PubMed]
5. Minagawa, H.; Yamamoto, N.; Abe, H.; Fukuda, M.; Seki, N.; Kikuchi, K.; Kijima, H.; Itoi, E. Prevalence of symptomatic and asymptomatic rotator cuff tears in the general population: From mass-screening in one village. *J. Orthop.* **2013**, *10*, 8–12. [CrossRef]
6. Tashjian, R.Z.; Farnham, J.M.; Albright, F.S.; Teerlink, C.C.; Cannon-Albright, L.A. Evidence for an inherited predisposition contributing to the risk for rotator cuff disease. *J. Bone Jt. Surg. Am. Vol.* **2009**, *91*, 1136. [CrossRef] [PubMed]
7. Gill, T.K.; Shanahan, E.M.; Tucker, G.R.; Buchbinder, R.; Hill, C.L. Shoulder range of movement in the general population: Age and gender stratified normative data using a community-based cohort. *BMC Musculoskelet. Disord.* **2020**, *21*, 676. [CrossRef] [PubMed]
8. Gillooly, J.J.; Chidambaram, R.; Mok, D. The lateral Jobe test: A more reliable method of diagnosing rotator cuff tears. *Int. J. Shoulder Surg.* **2010**, *4*, 41.
9. Hughes, P.C.; Taylor, N.F.; Green, R.A. Most clinical tests cannot accurately diagnose rotator cuff pathology: A systematic review. *Aust. J. Physiother.* **2008**, *54*, 159–170. [CrossRef] [PubMed]
10. Sawalha, S.; Fischer, J. The accuracy of "subacromial grind test" in diagnosis of supraspinatus rotator cuff tears. *Int. J. Shoulder Surg.* **2015**, *9*, 43. [CrossRef] [PubMed]
11. Sambandam, S.N.; Khanna, V.; Gul, A.; Mounasamy, V. Rotator cuff tears: An evidence based approach. *World J. Orthop.* **2015**, *6*, 902. [CrossRef]
12. Blevins, F.T. Rotator cuff pathology in athletes. *Sports Med.* **1997**, *24*, 205–220. [CrossRef]
13. Teunis, T.; Lubberts, B.; Reilly, B.T.; Ring, D. A systematic review and pooled analysis of the prevalence of rotator cuff disease with increasing age. *J. Shoulder Elb. Surg.* **2014**, *23*, 1913–1921. [CrossRef]
14. Kim, H.M.; Teefey, S.A.; Zelig, A.; Galatz, L.M.; Keener, J.D.; Yamaguchi, K. Shoulder strength in asymptomatic individuals with intact compared with torn rotator cuffs. *J. Bone Jt. Surg. Am. Vol.* **2009**, *91*, 289. [CrossRef] [PubMed]
15. Yamaguchi, K.; Ditsios, K.; Middleton, W.D.; Hildebolt, C.F.; Galatz, L.M.; Teefey, S.A. The demographic and morphological features of rotator cuff disease: A comparison of asymptomatic and symptomatic shoulders. *J. Bone Jt. Surg. Am.* **2006**, *88*, 1699–1704. [CrossRef] [PubMed]
16. Fukuda, H. The management of partial-thickness tears of the rotator cuff. *J. Bone Jt. Surg. Br. Vol.* **2003**, *85*, 3–11. [CrossRef] [PubMed]
17. Clement, N.D.; Nie, Y.X.; McBirnie, J.M. Management of degenerative rotator cuff tears: A review and treatment strategy. *Sports Med. Arthrosc. Rehabil. Ther. Technol.* **2012**, *4*, 48. [CrossRef] [PubMed]
18. Liem, D.; Alci, S.; Dedy, N.; Steinbeck, J.; Marquardt, B.; Möllenhoff, G. Clinical and structural results of partial supraspinatus tears treated by subacromial decompression without repair. *Knee Surg. Sports Traumatol. Arthrosc.* **2008**, *16*, 967–972. [CrossRef] [PubMed]
19. Franceschi, F.; Papalia, R.; Palumbo, A.; Del Buono, A.; Maffulli, N.; Denaro, V. Operative management of partial-and full-thickness rotator cuff tears. In *Rotator Cuff Tear*; Karger Publishers: Basel, Switzerland, 2012; Volume 57, pp. 100–113.
20. Cofield, R.H.; Parvizi, J.; Hoffmeyer, P.J.; Lanzer, W.L.; Ilstrup, D.M.; Rowland, C.M. Surgical repair of chronic rotator cuff tears: A prospective long-term study. *J. Bone Jt. Surg. Am.* **2001**, *83*, 71. [CrossRef] [PubMed]
21. Patel, S.; Gualtieri, A.P.; Lu, H.H.; Levine, W.N. Advances in biologic augmentation for rotator cuff repair. *Ann. N. Y. Acad. Sci.* **2016**, *1383*, 97–114. [CrossRef]
22. Maisani, M.; Ziane, S.; Ehret, C.; Levesque, L.; Siadous, R.; Le Meins, J.F.; Chevallier, P.; Barthélémy, P.; De Oliveira, H.; Amédée, J. A new composite hydrogel combining the biological properties of collagen with the mechanical properties of a supramolecular scaffold for bone tissue engineering. *J. Tissue Eng. Regen. Med.* **2018**, *12*, e1489–e1500. [CrossRef] [PubMed]
23. Rahman, C.V.; Kuhn, G.; White, L.J.; Kirby, G.T.; Varghese, O.P.; McLaren, J.S.; Cox, H.C.; Rose, F.R.; Müller, R.; Hilborn, J. PLGA/PEG-hydrogel composite scaffolds with controllable mechanical properties. *J. Biomed. Mater. Res. Part B Appl. Biomater.* **2013**, *101*, 648–655. [CrossRef] [PubMed]
24. Suvarnapathaki, S.; Wu, X.; Lantigua, D.; Nguyen, M.A.; Camci-Unal, G. Hydroxyapatite-incorporated composite gels improve mechanical properties and bioactivity of bone scaffolds. *Macromol. Biosci.* **2020**, *20*, 2000176. [CrossRef]
25. Zhang, D.; Wu, X.; Chen, J.; Lin, K. The development of collagen based composite scaffolds for bone regeneration. *Bioact. Mater.* **2018**, *3*, 129–138. [CrossRef]
26. Collins, M.N.; Birkinshaw, C. Hyaluronic acid based scaffolds for tissue engineering—A review. *Carbohydr. Polym.* **2013**, *92*, 1262–1279. [CrossRef]
27. Wang, S.; Lei, H.; Mi, Y.; Ma, P.; Fan, D. Chitosan and hyaluronic acid based injectable dual network hydrogels-Mediating antimicrobial and inflammatory modulation to promote healing of infected bone defects. *Int. J. Biol. Macromol.* **2024**, *274*, 133124. [CrossRef]

28. Venkatesan, J.; Bhatnagar, I.; Manivasagan, P.; Kang, K.-H.; Kim, S.-K. Alginate composites for bone tissue engineering: A review. *Int. J. Biol. Macromol.* **2015**, *72*, 269–281. [CrossRef] [PubMed]
29. Noori, A.; Ashrafi, S.J.; Vaez-Ghaemi, R.; Hatamian-Zaremi, A.; Webster, T.J. A review of fibrin and fibrin composites for bone tissue engineering. *Int. J. Nanomed.* **2017**, *12*, 4937–4961. [CrossRef]
30. Kutikov, A.B.; Song, J. Biodegradable PEG-based amphiphilic block copolymers for tissue engineering applications. *ACS Biomater. Sci. Eng.* **2015**, *1*, 463–480. [CrossRef] [PubMed]
31. Chen, P.; Cui, L.; Chen, G.; You, T.; Li, W.; Zuo, J.; Wang, C.; Zhang, W.; Jiang, C. The application of BMP-12-overexpressing mesenchymal stem cells loaded 3D-printed PLGA scaffolds in rabbit rotator cuff repair. *Int. J. Biol. Macromol.* **2019**, *138*, 79–88. [CrossRef] [PubMed]
32. Pazarçeviren, E.; Erdemli, Ö.; Keskin, D.; Tezcaner, A. Clinoptilolite/PCL–PEG–PCL composite scaffolds for bone tissue engineering applications. *J. Biomater. Appl.* **2017**, *31*, 1148–1168. [CrossRef] [PubMed]
33. Teimouri, A.; Azadi, M.; Emadi, R.; Lari, J.; Chermahini, A.N. Preparation, characterization, degradation and biocompatibility of different silk fibroin based composite scaffolds prepared by freeze-drying method for tissue engineering application. *Polym. Degrad. Stab.* **2015**, *121*, 18–29. [CrossRef]
34. Lim, T.K.; Dorthé, E.; Williams, A.; D D'Lima, D. Nanofiber scaffolds by electrospinning for rotator cuff tissue engineering. *Chonnam Med. J.* **2021**, *57*, 13. [CrossRef] [PubMed]
35. Orr, S.B.; Chainani, A.; Hippensteel, K.J.; Kishan, A.; Gilchrist, C.; Garrigues, N.W.; Ruch, D.S.; Guilak, F.; Little, D. Aligned multilayered electrospun scaffolds for rotator cuff tendon tissue engineering. *Acta Biomater.* **2015**, *24*, 117–126. [CrossRef]
36. Singh, S.; Ramakrishna, S.; Berto, F. 3D Printing of polymer composites: A short review. *Mater. Des. Process. Commun.* **2020**, *2*, e97. [CrossRef]
37. Chen, X.; Anvari-Yazdi, A.F.; Duan, X.; Zimmerling, A.; Gharraei, R.; Sharma, N.; Sweilem, S.; Ning, L. Biomaterials/bioinks and extrusion bioprinting. *Bioact. Mater.* **2023**, *28*, 511–536. [CrossRef] [PubMed]
38. Jiang, X.; Wu, S.; Kuss, M.; Kong, Y.; Shi, W.; Streubel, P.N.; Li, T.; Duan, B. 3D printing of multilayered scaffolds for rotator cuff tendon regeneration. *Bioact. Mater.* **2020**, *5*, 636–643. [CrossRef] [PubMed]
39. Pahlevanzadeh, F.; Emadi, R.; Valiani, A.; Kharaziha, M.; Poursamar, S.A.; Bakhsheshi-Rad, H.R.; Ismail, A.F.; RamaKrishna, S.; Berto, F. Three-dimensional printing constructs based on the chitosan for tissue regeneration: State of the art, developing directions and prospect trends. *Materials* **2020**, *13*, 2663. [CrossRef] [PubMed]
40. Deng, Y.; Zhang, M.; Chen, X.; Pu, X.; Liao, X.; Huang, Z.; Yin, G. A novel akermanite/poly (lactic-co-glycolic acid) porous composite scaffold fabricated via a solvent casting-particulate leaching method improved by solvent self-proliferating process. *Regen. Biomater.* **2017**, *4*, 233–242. [CrossRef]
41. Thadavirul, N.; Pavasant, P.; Supaphol, P. Development of polycaprolactone porous scaffolds by combining solvent casting, particulate leaching, and polymer leaching techniques for bone tissue engineering. *J. Biomed. Mater. Res. Part A* **2014**, *102*, 3379–3392. [CrossRef]
42. Yu, P.; Bao, R.-Y.; Shi, X.-J.; Yang, W.; Yang, M. B. Self-assembled high-strength hydroxyapatite/graphene oxide/chitosan composite hydrogel for bone tissue engineering. *Carbohydr. Polym.* **2017**, *155*, 507–515. [CrossRef] [PubMed]
43. Turnbull, G.; Clarke, J.; Picard, F.; Riches, P.; Jia, L.; Han, F.; Li, B.; Shu, W. 3D bioactive composite scaffolds for bone tissue engineering. *Bioact. Mater.* **2018**, *3*, 278–314. [CrossRef]
44. Liu, X.; Smith, L.A.; Hu, J.; Ma, P.X. Biomimetic nanofibrous gelatin/apatite composite scaffolds for bone tissue engineering. *Biomaterials* **2009**, *30*, 2252–2258. [CrossRef] [PubMed]
45. Yuan, X.; Zhu, W.; Yang, Z.; He, N.; Chen, F.; Han, X.; Zhou, K. Recent advances in 3D printing of smart scaffolds for bone tissue engineering and regeneration. *Adv. Mater.* **2024**, *36*, 2403641. [CrossRef] [PubMed]
46. Zou, L.; Zhang, Y.; Liu, X.; Chen, J.; Zhang, Q. Biomimetic mineralization on natural and synthetic polymers to prepare hybrid scaffolds for bone tissue engineering. *Colloids Surf. B Biointerfaces* **2019**, *178*, 222–229. [CrossRef]
47. Funakoshi, T.; Majima, T.; Iwasaki, N.; Suenaga, N.; Sawaguchi, N.; Shimode, K.; Minami, A.; Harada, K.; Nishimura, S.-I. Application of tissue engineering techniques for rotator cuff regeneration using a chitosan-based hyaluronan hybrid fiber scaffold. *Am. J. Sports Med.* **2005**, *33*, 1193–1201. [CrossRef] [PubMed]
48. Han, B.; Jones, I.A.; Yang, Z.; Fang, W.; Vangsness, C.T., Jr. Repair of rotator cuff tendon defects in aged rats using a growth factor injectable gel scaffold. *Arthrosc. J. Arthrosc. Relat. Surg.* **2020**, *36*, 629–637. [CrossRef] [PubMed]
49. Chen, C.-H.; Chang, C.-H.; Wang, K.-C.; Su, C.-I.; Liu, H.-T.; Yu, C.-M.; Wong, C.-B.; Wang, I.-C.; Whu, S.W.; Liu, H.-W. Enhancement of rotator cuff tendon–bone healing with injectable periosteum progenitor cells-BMP-2 hydrogel in vivo. *Knee Surg. Sports Traumatol. Arthrosc.* **2011**, *19*, 1597–1607. [CrossRef] [PubMed]
50. Narayanan, G.; Nair, L.S.; Laurencin, C.T. Regenerative engineering of the rotator cuff of the shoulder. *ACS Biomater. Sci. Eng.* **2018**, *4*, 751–786. [CrossRef] [PubMed]
51. Tsekes, D.; Konstantopoulos, G.; Khan, W.S.; Rossouw, D.; Elvey, M.; Singh, J. Use of stem cells and growth factors in rotator cuff tendon repair. *Eur. J. Orthop. Surg. Traumatol.* **2019**, *29*, 747–757. [CrossRef]

52. Stace, E.T.; Nagra, N.S.; Tiberwel, S.; Khan, W.; Carr, A.J. The use of electrospun scaffolds in musculoskeletal tissue engineering: A focus on tendon and the rotator cuff. *Curr. Stem Cell Res. Ther.* **2018**, *13*, 619–631. [CrossRef]
53. Moffat, K.L.; Levine, W.N.; Lu, H.H. In vitro evaluation of rotator cuff tendon fibroblasts on aligned composite scaffold of polymer nanofibers and hydroxyapatite nanoparticles. In Proceedings of the Transactions of the 54th Orthopaedic Research Society, San Francisco, CA, USA, 2–5 March 2008.
54. Chen, Y.; Li, Y.; Zhu, W.; Liu, Q. Biomimetic gradient scaffolds for the tissue engineering and regeneration of rotator cuff enthesis. *Biofabrication* **2024**, *16*, 032005. [CrossRef] [PubMed]
55. Shang, P.; Xiang, Y.; Du, J.; Chen, S.; Cheng, B.; Li, J.; Yuan, F. Gradient Bipolar Nanofiber Scaffolds with a Structure of Biomimetic Tendon-Bone Interface as Rotator Cuff Patches. *ACS Appl. Polym. Mater.* **2023**, *5*, 6107–6116. [CrossRef]
56. Liu, J.; Lin, S.; Liu, X.; Qin, Z.; Yang, Y.; Zang, J.; Zhao, X. Fatigue-resistant adhesion of hydrogels. *Nat. Commun.* **2020**, *11*, 1071. [CrossRef] [PubMed]
57. Kaya, D.Ö. Architecture of tendon and ligament and their adaptation to pathological conditions. In *Comparative Kinesiology of the Human Body*; Elsevier: Amsterdam, The Netherlands, 2020; pp. 115–147.
58. Martin, R.B.; Burr, D.B.; Sharkey, N.A.; Fyhrie, D.P.; Martin, R.B.; Burr, D.B.; Sharkey, N.A.; Fyhrie, D.P. Mechanical properties of ligament and tendon. In *Skeletal Tissue Mechanics*; Springer: New York, NY, USA, 2015; pp. 175–225.
59. Ryan, E.; Yin, S. Compressive strength of β-TCP scaffolds fabricated via lithography-based manufacturing for bone tissue engineering. *Ceram. Int.* **2022**, *48*, 15516–15524. [CrossRef]
60. Smith, R.; Carr, A.; Dakin, S.; Snelling, S.; Yapp, C.; Hakimi, O. The response of tenocytes to commercial scaffolds used for rotator cuff repair. *eCells Mater. J.* **2016**, *30*, 107–118. [CrossRef]
61. Yokoya, S.; Mochizuki, Y.; Natsu, K.; Omae, H.; Nagata, Y.; Ochi, M. Rotator cuff regeneration using a bioabsorbable material with bone marrow–derived mesenchymal stem cells in a rabbit model. *Am. J. Sports Med.* **2012**, *40*, 1259–1268. [CrossRef] [PubMed]
62. Zhao, S.; Zhao, J.; Dong, S.; Huangfu, X.; Li, B.; Yang, H.; Zhao, J.; Cui, W. Biological augmentation of rotator cuff repair using bFGF-loaded electrospun poly (lactide-co-glycolide) fibrous membranes. *Int. J. Nanomed.* **2014**, *9*, 2373. [CrossRef]
63. Longo, U.G.; Lamberti, A.; Maffulli, N.; Denaro, V. Tendon augmentation grafts: A systematic review. *Br. Med. Bull.* **2010**, *94*, 165–188. [CrossRef]
64. Adams, J.E.; Zobitz, M.E.; Reach, J.S., Jr.; An, K.-N.; Steinmann, S.P. Rotator cuff repair using an acellular dermal matrix graft: An in vivo study in a canine model. *Arthrosc. J. Arthrosc. Relat. Surg.* **2006**, *22*, 700–709. [CrossRef]
65. Schlegel, T.F.; Hawkins, R.J.; Lewis, C.W.; Motta, T.; Turner, A.S. The effects of augmentation with Swine small intestine submucosa on tendon healing under tension: Histologic and mechanical evaluations in sheep. *Am. J. Sports Med.* **2006**, *34*, 275–280. [CrossRef] [PubMed]
66. Proctor, C.S. Long-term successful arthroscopic repair of large and massive rotator cuff tears with a functional and degradable reinforcement device. *J. Shoulder Elb. Surg.* **2014**, *23*, 1508–1513. [CrossRef] [PubMed]
67. Washburn III, R.; Anderson, T.M.; Tokish, J.M. Arthroscopic rotator cuff augmentation: Surgical technique using bovine collagen bioinductive implant. *Arthrosc. Tech.* **2017**, *6*, e297–e301. [CrossRef]
68. Cheung, E.V.; Silverio, L.; Sperling, J.W. Strategies in biologic augmentation of rotator cuff repair: A review. *Clin. Orthop. Relat. Res.* **2010**, *468*, 1476–1484. [CrossRef]
69. Montgomery, S.R.; Petrigliano, F.A.; Gamradt, S.C. Biologic augmentation of rotator cuff repair. *Curr. Rev. Musculoskelet. Med.* **2011**, *4*, 221. [CrossRef]
70. Voss, A.; McCarthy, M.B.; Allen, D.; Cote, M.P.; Beitzel, K.; Imhoff, A.B.; Mazzocca, A.D. Fibrin scaffold as a carrier for mesenchymal stem cells and growth factors in shoulder rotator cuff repair. *Arthrosc. Tech.* **2016**, *5*, e447–e451. [CrossRef] [PubMed]
71. Street, M.; Thambyah, A.; Dray, M.; Amirapu, S.; Tuari, D.; Callon, K.E.; McIntosh, J.D.; Burkert, K.; Dunbar, P.R.; Coleman, B. Augmentation with an ovine forestomach matrix scaffold improves histological outcomes of rotator cuff repair in a rat model. *J. Orthop. Surg. Res.* **2015**, *10*, 165. [CrossRef] [PubMed]
72. Derwin, K.A.; Codsi, M.J.; Milks, R.A.; Baker, A.R.; McCarron, J.A.; Iannotti, J.P. Rotator cuff repair augmentation in a canine model with use of a woven poly-L-lactide device. *J. Bone Jt. Surg. Am. Vol.* **2009**, *91*, 1159. [CrossRef] [PubMed]
73. Easley, J.; Puttlitz, C.; Hackett, E.; Broomfield, C.; Nakamura, L.; Hawes, M.; Getz, C.; Frankle, M.; Pierre, P.S.; Tashjian, R. A prospective study comparing tendon-to-bone interface healing using an interposition bioresorbable scaffold with a vented anchor for primary rotator cuff repair in sheep. *J. Shoulder Elb. Surg.* **2020**, *29*, 157–166. [CrossRef] [PubMed]
74. Van Kampen, C.; Arnoczky, S.; Parks, P.; Hackett, E.; Ruehlman, D.; Turner, A.; Schlegel, T. Tissue-engineered augmentation of a rotator cuff tendon using a reconstituted collagen scaffold: A histological evaluation in sheep. *Muscles Ligaments Tendons J.* **2013**, 3, 229. [CrossRef]
75. Fernandez de Grado, G.; Keller, L.; Idoux-Gillet, Y.; Wagner, Q.; Musset, A.-M.; Benkirane-Jessel, N.; Bornert, F.; Offner, D. Bone substitutes: A review of their characteristics, clinical use, and perspectives for large bone defects management. *J. Tissue Eng.* **2018**, *9*, 2041731418776819. [CrossRef] [PubMed]

76. Baldwin, M.; Nagra, N.; Greenall, G.; Carr, A.J.; Beard, D.; Rees, J.; Rangan, A.; Merritt, N.; Dritsaki, M.; Hopewell, S. Use of implantable meshes for augmented rotator cuff repair: A systematic review and meta-analysis. *BMJ Open* **2020**, *10*, e039552. [CrossRef] [PubMed]
77. Thangarajah, T.; Pendegrass, C.J.; Shahbazi, S.; Lambert, S.; Alexander, S.; Blunn, G.W. Augmentation of rotator cuff repair with soft tissue scaffolds. *Orthop. J. Sports Med.* **2015**, *3*, 2325967115587495. [CrossRef]
78. Bokor, D.; Sonnabend, D.; Deady, L.; Cass, B.; Young, A.; Van Kampen, C.; Arnoczky, S. Healing of partial-thickness rotator cuff tears following arthroscopic augmentation with a highly-porous collagen implant: A 5-year clinical and MRI follow-up. *Muscles Ligaments Tendons J. MLTJ* **2019**, *9*, 16–25. [CrossRef]
79. Bokor, D.J.; Sonnabend, D.; Deady, L.; Cass, B.; Young, A.; Van Kampen, C.; Arnoczky, S. Evidence of healing of partial-thickness rotator cuff tears following arthroscopic augmentation with a collagen implant: A 2-year MRI follow-up. *Muscles Ligaments Tendons J.* **2016**, *6*, 16. [CrossRef] [PubMed]
80. Bokor, D.J.; Sonnabend, D.; Deady, L.; Cass, B.; Young, A.; Van Kampen, C.; Arnoczky, S. Preliminary investigation of a biological augmentation of rotator cuff repairs using a collagen implant: A 2-year MRI follow-up. *Muscles Ligaments Tendons J.* **2015**, *5*, 144. [CrossRef] [PubMed]
81. Thon, S.G.; O'Malley, L.; O'Brien, M.J.; Savoie III, F.H. Evaluation of healing rates and safety with a bioinductive collagen patch for large and massive rotator cuff tears: 2-year safety and clinical outcomes. *Am. J. Sports Med.* **2019**, *47*, 1901–1908. [CrossRef] [PubMed]
82. Schlegel, T.F.; Abrams, J.S.; Bushnell, B.D.; Brock, J.L.; Ho, C.P. Radiologic and clinical evaluation of a bioabsorbable collagen implant to treat partial-thickness tears: A prospective multicenter study. *J. Shoulder Elb. Surg.* **2018**, *27*, 242–251. [CrossRef]
83. Dai, A.Z.; Campbell, A.L.; Bloom, D.A.; Baron, S.L.; Begly, J.P.; Meislin, R.J. Collagen-Based Bioinductive Implant for Treatment of Partial Thickness Rotator Cuff Tears. *Bull. NYU Hosp. Jt. Dis.* **2020**, *78*, 195–201.
84. Bushnell, B.D.; Connor, P.M.; Harris, H.W.; Ho, C.P.; Trenhaile, S.W.; Abrams, J.S. Re-tear rates and clinical outcomes at 1 year following repair of full-thickness rotator cuff tears augmented with a bioinductive collagen implant: A prospective multicenter study. *JSES Int.* **2021**, *5*, 228–237. [CrossRef] [PubMed]
85. Iannotti, J.P.; Codsi, M.J.; Kwon, Y.W.; Derwin, K.; Ciccone, J.; Brems, J.J. Porcine small intestine submucosa augmentation of surgical repair of chronic two-tendon rotator cuff tears: A randomized, controlled trial. *J. Bone Jt. Surg. Am.* **2006**, *88*, 1238–1244. [CrossRef] [PubMed]
86. Bryant, D.; Holtby, R.; Willits, K.; Litchfield, R.; Drosdowech, D.; Spouge, A.; White, D.; Guyatt, G. A randomized clinical trial to compare the effectiveness of rotator cuff repair with or without augmentation using porcine small intestine submucosa for patients with moderate to large rotator cuff tears: A pilot study. *J. Shoulder Elb. Surg.* **2016**, *25*, 1623–1633. [CrossRef] [PubMed]
87. Castagna, A.; Cesari, E.; Di Matteo, B.; Osimani, M.; Garofalo, R.; Kon, E.; Marcacci, M.; Chillemi, C. Porcine dermal xenograft as augmentation in the treatment of large rotator cuff tears: Clinical and magnetic resonance results at 2-year follow-up. *Joints* **2018**, *6*, 135. [CrossRef] [PubMed]
88. Avanzi, P.; Dei Giudici, L.; Capone, A.; Cardoni, G.; Lunardi, G., Foti, G.; Zorzi, C. Prospective randomized controlled trial for patch augmentation in rotator cuff repair: 24-month outcomes. *J. Shoulder Elb. Surg.* **2019**, *28*, 1918–1927. [CrossRef]
89. Maillot, C.; Harly, E.; Demezon, H.; Le Huec, J.-C. Surgical repair of large-to-massive rotator cuff tears seems to be a better option than patch augmentation or debridement and biceps tenotomy: A prospective comparative study. *J. Shoulder Elb. Surg.* **2018**, *27*, 1545–1552. [CrossRef]
90. Gouk, C.J.C.; Shulman, R.M.; Buchan, C.; Thomas, M.J.E.; Taylor, F.J. Failure of dermal allograft repair of massive rotator cuff tears in magnetic resonance imaging and clinical assessment. *Clin. Orthop. Surg.* **2019**, *11*, 200–207. [CrossRef] [PubMed]
91. Johnson, S.M.; Cherry, J.V.; Thomas, N.; Jafri, M.; Jariwala, A.; McLeod, G.G. Clinical outcomes and ultrasonographic viability of GraftJacket® augmented rotator cuff repair: A prospective follow-up study with mean follow-up of forty-one months. *J. Clin. Orthop. Trauma* **2020**, *11*, S372–S377. [CrossRef] [PubMed]
92. Encalada-Diaz, I.; Cole, B.J.; MacGillivray, J.D.; Ruiz-Suarez, M.; Kercher, J.S.; Friel, N.A.; Valero-Gonzalez, F. Rotator cuff repair augmentation using a novel polycarbonate polyurethane patch: Preliminary results at 12 months' follow-up. *J. Shoulder Elb. Surg.* **2011**, *20*, 788–794. [CrossRef] [PubMed]
93. Petriccioli, D.; Bertone, C.; Marchi, G.; Mujahed, I. Open repair of isolated traumatic subscapularis tendon tears with a synthetic soft tissue reinforcement. *Musculoskelet. Surg.* **2013**, *97*, 63–68. [CrossRef] [PubMed]
94. Ciampi, P.; Scotti, C.; Nonis, A.; Vitali, M.; Di Serio, C.; Peretti, G.M.; Fraschini, G. The benefit of synthetic versus biological patch augmentation in the repair of posterosuperior massive rotator cuff tears: A 3-year follow-up study. *Am. J. Sports Med.* **2014**, *42*, 1169–1175. [CrossRef] [PubMed]
95. Vitali, M.; Cusumano, A.; Pedretti, A.; Rodriguez, N.N.; Fraschini, G. Employment of synthetic patch with augmentation of the long head of the biceps tendon in irreparable lesions of the rotator cuff: Our technique applied to 60 patients. *Tech. Hand Up. Extrem. Surg.* **2015**, *19*, 32–39. [CrossRef]

96. Smolen, D.; Haffner, N.; Mittermayr, R.; Hess, F.; Sternberg, C.; Leuzinger, J. Application of a new polyester patch in arthroscopic massive rotator cuff repair—A prospective cohort study. *J. Shoulder Elb. Surg.* **2020**, *29*, e11–e21. [CrossRef]
97. Cowling, P.; Hackney, R.; Dube, B.; Grainger, A.; Biglands, J.; Stanley, M.; Song, D.; Conaghan, P.; Kingsbury, S. The use of a synthetic shoulder patch for large and massive rotator cuff tears–a feasibility study. *BMC Musculoskelet. Disord.* **2020**, *21*, 213. [CrossRef] [PubMed]
98. Smith, M.J.; Bozynski, C.C.; Kuroki, K.; Cook, C.R.; Stoker, A.M.; Cook, J.L. Comparison of biologic scaffolds for augmentation of partial rotator cuff tears in a canine model. *J. Shoulder Elb. Surg.* **2020**, *29*, 1573–1583. [CrossRef] [PubMed]
99. Bishop, J.; Klepps, S.; Lo, I.K.; Bird, J.; Gladstone, J.N.; Flatow, E.L. Cuff integrity after arthroscopic versus open rotator cuff repair: A prospective study. *J. Shoulder Elb. Surg.* **2006**, *15*, 290–299. [CrossRef]
100. Jo, C.H.; Park, J.W.; Shin, J.S. Changes of muscle atrophy according to the immediate postoperative time point in magnetic resonance imaging after arthroscopic rotator cuff repair. *Arthrosc. J. Arthrosc. Relat. Surg.* **2016**, *32*, 2477–2487. [CrossRef] [PubMed]
101. Rashid, M.S.; Cooper, C.; Cook, J.; Cooper, D.; Dakin, S.G.; Snelling, S.; Carr, A.J. Increasing age and tear size reduce rotator cuff repair healing rate at 1 year: Data from a large randomized controlled trial. *Acta Orthop.* **2017**, *88*, 606–611. [CrossRef]
102. Sugaya, H.; Maeda, K.; Matsuki, K.; Moriishi, J. Repair integrity and functional outcome after arthroscopic double-row rotator cuff repair: A prospective outcome study. *J. Bone Jt. Surg.* **2007**, *89*, 953–960. [CrossRef] [PubMed]
103. Taniguchi, N.; Suenaga, N.; Oizumi, N.; Miyoshi, N.; Yamaguchi, H.; Inoue, K.; Chosa, E. Bone marrow stimulation at the footprint of arthroscopic surface-holding repair advances cuff repair integrity. *J. Shoulder Elb. Surg.* **2015**, *24*, 860–866. [CrossRef] [PubMed]
104. Derwin, K.A.; Badylak, S.F.; Steinmann, S.P.; Iannotti, J.P. Extracellular matrix scaffold devices for rotator cuff repair. *J. Shoulder Elb. Surg.* **2010**, *19*, 467–476. [CrossRef]
105. Goldenberg, B.T.; Lacheta, L.; Dekker, T.J.; Spratt, J.D.; Nolte, P.C.; Millett, P.J. Biologics to Improve Healing in Large and Massive Rotator Cuff Tears: A Critical Review. *Orthop. Res. Rev.* **2020**, *12*, 151. [CrossRef] [PubMed]
106. Malcarney, H.L.; Bonar, F.; Murrell, G.A. Early inflammatory reaction after rotator cuff repair with a porcine small intestine submucosal implant: A report of 4 cases. *Am. J. Sports Med.* **2005**, *33*, 907–911. [CrossRef]
107. Phipatanakul, W.P.; Petersen, S.A. Porcine small intestine submucosa xenograft augmentation in repair of massive rotator cuff tears. *Am. J. Orthop.* **2009**, *38*, 572–575. [PubMed]
108. Walton, J.R.; Bowman, N.K.; Khatib, Y.; Linklater, J.; Murrell, G.A. Restore orthobiologic implant: Not recommended for augmentation of rotator cuff repairs. *J. Bone Jt. Surg. Am.* **2007**, *89*, 786–791.
109. Zheng, M.H.; Chen, J.; Kirilak, Y.; Willers, C.; Xu, J.; Wood, D. Porcine small intestine submucosa (SIS) is not an acellular collagenous matrix and contains porcine DNA: Possible implications in human implantation. *J. Biomed. Mater. Res. Part B Appl. Biomater.* **2005**, *73*, 61–67. [CrossRef] [PubMed]
110. Kokkalis, Z.T.; Mavrogenis, A.F.; Scarlat, M.; Christodoulou, M.; Vottis, C.; Papagelopoulos, P.J.; Sotereanos, D.G. Human dermal allograft for massive rotator cuff tears. *Orthopedics* **2014**, *37*, e1108–e1116. [CrossRef]
111. Ahmed, J.; Varshney, S.K. Polylactides—Chemistry, properties and green packaging technology: A review. *Int. J. Food Prop.* **2011**, *14*, 37–58. [CrossRef]
112. Chen, J.; Xu, J.; Wang, A.; Zheng, M. Scaffolds for tendon and ligament repair: Review of the efficacy of commercial products. *Expert Rev. Med. Dev.* **2009**, *6*, 61–73. [CrossRef] [PubMed]
113. Longo, U.G.; Lamberti, A.; Khan, W.S.; Maffulli, N.; Denaro, V. Synthetic augmentation for massive rotator cuff tears. *Sports Med. Arthrosc. Rev.* **2011**, *19*, 360–365. [CrossRef]
114. Gillespie, R.J.; Knapik, D.M.; Akkus, O. Biologic and synthetic grafts in the reconstruction of large to massive rotator cuff tears. *JAAOS-J. Am. Acad. Orthop. Surg.* **2016**, *24*, 823–828. [CrossRef]
115. Emadi, H.; Karevan, M.; Masoudi Rad, M.; Sadeghzade, S.; Pahlevanzadeh, F.; Khodaei, M.; Khayatzadeh, S.; Lotfian, S. Bioactive and biodegradable polycaprolactone-based nanocomposite for bone repair applications. *Polymers* **2023**, *15*, 3617. [CrossRef] [PubMed]
116. Chen, C.; Zhu, J.; Chen, J.; Yu, F.; Huang, K.; Jiang, J.; Zhu, T.; Mo, X.; Zhao, J. A reinforced nanofibrous patch with biomimetic mechanical properties and chondroinductive effect for rotator cuff tissue engineering. *Mater. Today Chem.* **2022**, *23*, 100655. [CrossRef]
117. Ozbolat, I.T.; Hospodiuk, M. Current advances and future perspectives in extrusion-based bioprinting. *Biomaterials* **2016**, *76*, 321–343. [CrossRef]
118. Mandalia, K.; Mousad, A.; Welborn, B.; Bono, O.; Le Breton, S.; MacAskill, M.; Forlizzi, J.; Ives, K.; Ross, G.; Shah, S. Scaffold-and graft-based biological augmentation of rotator cuff repair: An updated systematic review and meta-analysis of preclinical and clinical studies for 2010–2022. *J. Shoulder Elb. Surg.* **2023**, *32*, 1784–1800. [CrossRef] [PubMed]

119. Ogueri, K.S.; Laurencin, C.T. Nanofiber technology for regenerative engineering. *ACS Nano* **2020**, *14*, 9347–9363. [CrossRef] [PubMed]
120. Qi, Y.; Niu, L.; Zhao, T.; Shi, Z.; Di, T.; Feng, G.; Li, J.; Huang, Z. Combining mesenchymal stem cell sheets with platelet-rich plasma gel/calcium phosphate particles: A novel strategy to promote bone regeneration. *Stem Cell Res. Ther.* **2015**, *6*, 256. [CrossRef] [PubMed]

gels

Review

Hydrogel-Based Scaffolds: Advancing Bone Regeneration Through Tissue Engineering

Juan Luis Cota Quintero [1], Rosalío Ramos-Payán [2], José Geovanni Romero-Quintana [2], Alfredo Ayala-Ham [3], Mercedes Bermúdez [4] and Elsa Maribel Aguilar-Medina [2,*]

1 Faculty of Biology, Autonomous University of Sinaloa, Josefa Ortiz de Domínguez s/n y Avenida de las Américas, Culiacan 80010, Sinaloa, Mexico; luis@uas.edu.mx
2 Faculty of Biological and Chemical Sciences, Autonomous University of Sinaloa, Josefa Ortiz de Domínguez s/n y Avenida de las Américas, Culiacan 80010, Sinaloa, Mexico; rosaliorp@uas.edu.mx (R.R.-P.); geovanniromero@uas.edu.mx (J.G.R.-Q.)
3 Faculty of Odontology, Autonomous University of Sinaloa, Josefa Ortiz de Domínguez s/n y Avenida de las Américas, Culiacan 80010, Sinaloa, Mexico; endoalfredo@uas.edu.mx
4 Faculty of Odontology, Autonomous University of Chihuahua, Circuito Universitario Campus I, Chihuahua 31000, Chihuahua, Mexico; mbermudez@uach.mx
* Correspondence: maribelaguilar@uas.edu.mx; Tel.: +52-667-713-7860

Abstract: Bone tissue engineering has emerged as a promising approach to addressing the limitations of traditional bone grafts for repairing bone defects. This regenerative medicine strategy leverages biomaterials, growth factors, and cells to create a favorable environment for bone regeneration, mimicking the body's natural healing process. Among the various biomaterials explored, hydrogels (HGs), a class of three-dimensional, hydrophilic polymer networks, have gained significant attention as scaffolds for bone tissue engineering. Thus, this review aimed to investigate the potential of natural and synthetic HGs, and the molecules used for its functionalization, for enhanced bone tissue engineering applications. HGs offer several advantages such as scaffolds, including biocompatibility, biodegradability, tunable mechanical properties, and the ability to encapsulate and deliver bioactive molecules. These properties make them ideal candidates for supporting cell attachment, proliferation, and differentiation, ultimately guiding the formation of new bone tissue. The design and optimization of HG-based scaffolds involve adapting their composition, structure, and mechanical properties to meet the specific requirements of bone regeneration. Current research focuses on incorporating bioactive molecules, such as growth factors and cytokines, into HG scaffolds to further enhance their osteoinductive and osteoconductive properties. Additionally, strategies to improve the mechanical strength and degradation kinetics of HGs are being explored to ensure long-term stability and support for new bone formation. The development of advanced HG-based scaffolds holds great potential for revolutionizing bone tissue engineering and providing effective treatment options for patients with bone defects.

Keywords: engineering tissue; scaffolds; hydrogel (HG); bone regeneration

Academic Editors: Daihua Fu and Jieyu Zhang

Received: 29 January 2025
Revised: 21 February 2025
Accepted: 25 February 2025
Published: 27 February 2025

Citation: Cota Quintero, J.L.; Ramos-Payán, R.; Romero-Quintana, J.G.; Ayala-Ham, A.; Bermúdez, M.; Aguilar-Medina, E.M. Hydrogel-Based Scaffolds: Advancing Bone Regeneration Through Tissue Engineering. *Gels* **2025**, *11*, 175. https://doi.org/10.3390/gels11030175

1. Introduction

Tissue engineering is a multidisciplinary biomedical field focused on developing technologies and methodologies to promote tissue renovation, replacement, and regeneration [1,2]. The primary goal of tissue engineering is to create biological substitutes using cellular and extracellular components to restore the functionality of injured or impaired organs and tissues [3]. This emerging area of research holds vast potential and diverse biomedical applications, mostly due to the incorporation of HGs. These biomaterials possess a highly flexible structure resembling the extracellular matrix (ECM), making them suitable for use as grafts [4]. Additionally, the integration of bioactive materials in tissue engineering facilitates bone repair by filling defects and accelerating the healing of damaged tissues [5,6].

Biomedical techniques for tissue replacement traditionally rely on autologous grafts, which, despite their advantages, are associated with certain drawbacks, such as the risk of infection [7]. The selection of the most appropriate material for such procedures is contingent upon several factors, including tissue viability, as well as the size, shape, and volume of the defect [8]. In response to these challenges, regenerative medicine emerges as an innovative approach by integrating tissue engineering with the body's intrinsic self-healing capabilities to treat tissues that are otherwise incapable of repairing themselves [9,10]. Central to this approach are biomaterials, which serve as adaptable platforms for tissue engineering applications, tailored based on their biocompatibility, degradation kinetics, and mechanical properties [11]. Biomaterial-based scaffolds are categorized into bio-polymers, bio-metals, bio-ceramics, and bio-composites [12,13]. Another classification considers their consistency: for instance, certain modifiable materials exhibit strong compression and tensile resistance-like bone, yet are hydrophobic, which limits cell encapsulation, and their low degradation rate delays effective healing. For optimal outcomes, biomaterials must degrade at a rate proportional to new bone formation [14].

Bone regenerative therapies have earned significant attention in recent years. These therapies require the synergistic use of three critical components: (1) scaffolds such as HGs made from biomaterials to provide structural support, (2) growth factors to stimulate cellular activity, and (3) local cell populations to facilitate tissue regeneration. This triad is designed to mimic natural biological processes that promote fracture healing or the repair of critical-sized defects, ultimately restoring damaged biological functions; furthermore, the incorporation of exogenous cells can enhance the osteogenic potential of these therapies [15,16].

Three key features are also essential: the capability of a matrix to retain and transport cells to a specific site, to support the repair process by native cells capable of forming a functional matrix, and to promote the synthesis of bioactive molecules such as cytokines and growth factors of the host, as well as to transport molecules added in the HG to promote bone repair [17]. In other words, these scaffolds are employed to facilitate cellular adhesion and the subsequent formation of new tissues, providing an ideal environment for cell adhesion and proliferation; this environment often mimics the ECM within a highly porous three-dimensional structure [12,18]. Such biomaterials offer additional advantages, including the possibility of administration via minimally invasive surgical procedures or as injectable systems, involving self-setting materials that solidify under physiological conditions or soft gels designed to adapt to the defect site, thereby improving their clinical applicability and patient outcomes [19,20]. In this context, HG-based scaffolds, with a viscous, gelatin-like consistency, offer excellent degradation rates and permeability, allowing the efficient exchange of oxygen, nutrients, and metabolites, despite their mechanical properties [14,21]. Natural polymers such as chitosan, collagen, agarose, and alginate are the most commonly used in tissue engineering [21], and stand out due to their unique properties and high biocompatibility [22].

Synthetic polymer-based scaffolds, such as polylactic acid (PLA), are extensively studied for their applications in load-bearing implants, including fracture fixation devices, due to their biodegradable nature and favorable mechanical strength [23]. Also, polyethylene glycol (PEG) is valued for its excellent solubility in water and organic solvents, making it an effective modifier for starch/PLA composites, as it combines the biodegradability capacity of starch with the toughness of PLA [24]. Furthermore, polycaprolactone (PCL), an FDA-approved biocompatible and bioresorbable polyester, has shown potential in medical applications. Incorporating PCL into brittle PLA matrices enhances their toughness, creating a polymeric matrix more suitable for bone-related applications [25]. All these synthetic polymers stand out for their adjustable physicochemical properties, enabling the replication of physiological environments. Thus, adaptability highlights the complementary roles of synthetic and natural polymers in advancing tissue engineering applications for bone regeneration. Thus, this review aimed to investigate the potential of natural and synthetic HGs, and the molecules used for its functionalization, for enhanced bone tissue engineering applications.

2. Bone: A Highly Specialized and Dynamic Tissue

Bone, among other characteristics, is a dense, calcified, and porous connective tissue [26], and one of the hardest tissues in the human body, just below the enamel of the teeth [27]. The human skeleton comprises cortical and trabecular bone, which differ in structure despite identical chemical composition. Cortical bone, comprising 80% of the skeleton, is dense, compact, and slow-renewing, providing mechanical strength. Trabecular bone, accounting for 20% of bone mass, is less dense, more elastic, and metabolically active, with a higher turnover rate [28].

Bone is a mineralized, dense connective tissue primarily composed of a mineral component and an organic matrix that mainly consists of type I collagen (COL-I), with smaller amounts of type III collagen (COL-III). Hydroxyapatite, represented as $Ca_3(PO_4)_2 \cdot (OH)_2$, forms nanocrystals that are embedded between individual collagen molecules, significantly increasing the rigidity of bone. In addition to collagen, non-collagenous proteins such as fibronectin, osteocalcin (OCN), osteopontin (OPN), and osteonectin contribute to a scaffold for mineral deposition [29]. Bone is a highly specialized and dynamic tissue that undergoes constant regeneration. The process of bone remodeling is regulated and coordinated by various cell types. Maintaining bone remodeling and overall mineral homeostasis relies on a delicate balance between bone resorption and formation [30]. The bone remodeling cycle is defined by five stages: activation of osteoclast precursors, resorption carried out by differentiated osteoclasts, the reversal phase lasting approximately 4 to 5 weeks, the bone formation phase regulated by osteoblasts, and finally, the termination phase, where osteoblasts differentiate into osteocytes [31].

Bone Healing

Bone possesses a remarkable ability to repair itself following both, microtraumas and macrostructural injuries, such as fractures. This repair process is mediated through neurogenesis, angiogenesis, osteogenic differentiation, and biomineralization [32]. Bone healing is a naturally occurring phenomenon that can be categorized into two distinct mechanisms: primary and secondary repair. Primary repair occurs when the fracture gap is smaller than 0.1 mm, enabling direct ossification across the gap without the formation of an intermediary callus. In contrast, secondary repair, which is the predominant mechanism, takes place when the fracture gap exceeds 0.1 mm but is less than twice the diameter of the injured bone. This process involves a multistage sequence of biological events, including the formation of a callus that eventually undergoes ossification [33].

The physiological mechanisms underlying bone repair can further be divided into two primary pathways: direct and indirect healing. Direct healing, also known as intramembranous ossification, is characterized by the differentiation of mesenchymal progenitor cells directly into osteoblasts, resulting in the immediate deposition of bone matrix. Indirect healing, in contrast, is mediated by an intermediate phase of cartilage formation, known as endochondral ossification. This process relies on the differentiation of chondrocytes, which play a pivotal role in forming a cartilage template that is subsequently replaced by mineralized bone [34].

Bone repair is orchestrated by a highly complex cascade of cellular and molecular events involving various cytokines (e.g., IL-1, IL-6, TNF-α) and growth factors. These bioactive molecules are essential for the recruitment of progenitor and inflammatory cells, including lymphocytes, macrophages, eosinophils, and neutrophils, to the site of injury. These cells coordinate the initial inflammatory response, establish the regenerative microenvironment, and initiate downstream signaling pathways critical for tissue regeneration and repair (Figure 1) [34]. The repair of bone is a meticulously regulated process that unfolds in distinct sequential phases, including hematoma formation, an acute inflammatory response, the development of granulation tissue, bone regeneration, and subsequent remodeling [35].

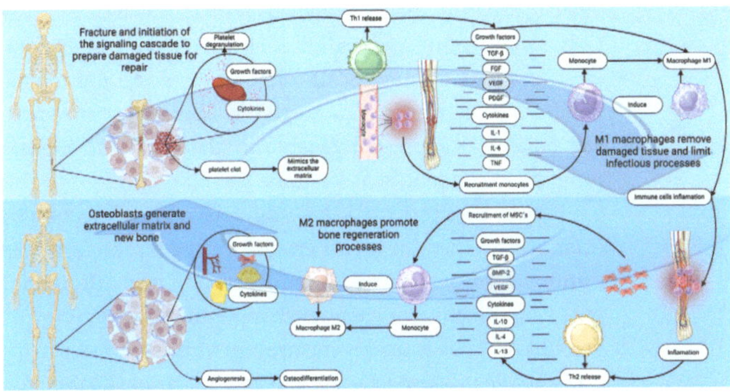

Figure 1. Fracture repair is a process that involves several physiological phenomena in a specific order: fracture of the periosteum and bone, immediately after which the formation of a clot begins to imitate the environment of a matrix, then the onset of inflammation thanks to platelet degranulation that releases growth factors like vascular endothelial growth factor (VEGF), fibroblast growth factor (FGF), platelet-derived growth factor (PDGF), and transforming growth factor β (TGF-β), and then cytokines like interleukin 1 (IL1), interleukin 6 (IL6), and tumor necrosis factor (TNF) induce macrophage polarization into the M1 phenotype. The expansion and proliferation of mesenchymal stem cells (MSCs) are initiated when macrophages undergo M2 polarization and release tissue repair signals, including interleukin 10 (IL-10), transforming growth factor-beta (TGF-β), bone morphogenetic protein-2 (BMP-2), and vascular endothelial growth factor (VEGF), thereby promoting angiogenesis. It is important to note that the application of MSCs has been extensively studied in vitro, in vivo using animal biomodels, and in clinical trials. The rationale for MSC utilization lies in their dual role: they can autonomously differentiate into bone cells (autocrine function) while also enhancing bone regeneration and repair through the secretion of various growth factors (paracrine function). Under these conditions, the process of bone formation (endochondral and intramembranous) begins, with fibroblasts penetrating the defect area to build a fibrous callus, and finally the recruitment of osteoclasts that will eliminate the primary bone and remodel the bone structure.

For successful bone repair, the injury must not exceed a critical size; otherwise, natural recovery through the immune system becomes unlikely, requiring external interventions such as bone grafting to expedite the healing process [36,37]. Approximately 5% of the millions of skeletal fractures that occur each year result in nonunion, where the bone defect fails to heal effectively with the initial treatment and remains mechanically unstable [38]. According to the Global Burden of Disease organization, 178 million new bone fractures were reported worldwide in 2019, representing a 33.4% increase since 1990. Lower limb fractures, including those of the patella, tibia, fibula, and ankle, were the most common, with an age-standardized rate of 419.9 cases per 100,000 population [39]. In addition, degenerative bone diseases affect more than 1.7 billion people worldwide; thus, bone is one of the most frequently transplanted tissues worldwide, with more than 4 million medical procedures performed annually to address bone-related conditions [40]. Furthermore, bone void-filling treatments serve more than half a million patients annually, representing a market valued at USD 2.8 billion [41].

In the United States, more than 450,000 patients are affected by bone and cartilage defects annually, requiring numerous bone grafting procedures and a substantial number of autologous transplants each year [10,16], with autologous bone grafts, particularly those from the iliac crest [9], due to their three key regenerative properties: osteoconductive, osteoinductive, and osteogenic capacity [42,43].

3. Hydrogel-Based Scaffolds and Their Role in Tissue Engineering

HG-based scaffolds have been extensively used across various domains, including agriculture, pharmaceuticals, cosmetics, and biomedicine, with significant relevance in tissue engineering applications [44]. The development of tissues in artificial regenerative environments requires scaffolds that replicate the microstructure of the ECM, thereby facilitating the repair of damaged or diseased tissues and organs [26,45]. Both synthetic and natural HGs present a viable and less invasive alternative, addressing the limitations inherent to conventional scaffolds [1].

These HGs provide a three-dimensional architecture that supports tissue regeneration, establishing them as a cornerstone in regenerative therapies. This structure creates an optimal environment for the proliferation and differentiation of endogenous cells while ensuring effective nutrient exchange and waste removal [46]. Moreover, the structural stability of HGs is achieved through various cross-linking methods, including physical and chemical approaches (Table 1), which also enable the controlled release of bioactive agents, such as drugs and growth factors, encapsulated within their matrix [47]. Critical factors influencing physical cross-linking include temperature, pH, and the electrostatic interactions inherent to the polymeric components, highlighting their importance in the functional optimization of these materials [47].

Regarding osteoinductivity, the use of growth factors is essential during the synthesis of various HGs. In addition to growth factors, peptides such as parathyroid hormone-related peptide (PTHrP), arginine–glycine–aspartic acid (RGD), and LL37 are also used. However, bone morphogenetic proteins (BMPs) have proven to possess the greatest osteoinductive potential. These growth factors significantly increase the expression of Runx-2 and osterix (OSX), mediating both the Smad and MAPK pathways. It is crucial that during the synthesis of HGs, they possess the ability to maintain and sustain the controlled release of these growth factors and peptides, thereby contributing to tissue regeneration and repair [48]. A key factor is the method of growth factor binding to the HG. Covalent binding of the BMP-2 growth factor via amino groups has been shown to enhance cell adhesion and differentiation in culture. However, this synthesis method is both time-consuming and costly. Alternatively, non-covalent interactions between the HG and the growth factor such as electrostatic interactions, hydrophobic forces, hydrogen bonds, and van der Waals forces facilitate a simpler binding approach but present the disadvantage of an uncontrolled initial release, which may lead to ectopic bone formation [49].

Table 1. Types of HGs for scaffolds design.

Physical-Responsive HG		Chemical-Responsive HG	
Temperature-responsive HG	Characterized by the presence of hydrophobic groups in the monomers such as methyl, ethyl, and propyl	pH-responsive HG	They work especially in the administration and incorporation of bioactive agents.

Table 1. *Cont.*

Physical-Responsive HG		Chemical-Responsive HG	
Photo/light-responsive HG	They contain a photoreceptor fraction, where photoirradiation induces a reaction involving isomerization (cis–trans, open–close), cleavage, and dimerization, which is then transferred to the functional group	Glucose-responsive HG	Developing insulin delivery system. They promote water absorption capacity and an increase in the size of the matrix.
Electro and magnetic-responsive HG	They change their properties in response to small changes in electrical current stimuli or external fields	Biological/biochemical-responsive HG	They contain biological functionalities/instructive fragments designed to interact with the biological environment. They use biomolecules such as peptide sequences that act as cross-linkers.

[50].

4. Key Issues in HG Selection

One of the key factors for achieving satisfactory outcomes in tissue engineering, specifically in bone regeneration, is the selection of the HG. In recent years, the use of aggregated MSCs has gained attention due to their pluripotent properties, self-renewal capacity, and multispectral differentiation potential [51]. Some contraindications of using mesenchymal stem cells (MSCs) include the potential formation of heterotopic bone, particularly in joints, which can impair their normal function. Therefore, it is crucial to regulate homeostasis and control the regenerative process through strategies that prevent ectopic bone formation. The use of HGs with nanoparticles containing curcumin-loaded zeolitic imidazolate and cerium oxide has been shown to modulate immune activity, promoting inflammatory balance and enhancing efferocytosis, thereby reducing the likelihood of heterotopic bone regeneration [52].

Other materials used to regulate immune functions during bone regeneration include the incorporation of peptides such as MP196, which is rich in arginine and tryptophan, into HGs. These peptides have been shown to upregulate the expression of genes associated with neutrophil apoptosis, macrophage polarization towards the M2 phenotype, and the promotion of efferocytosis, thereby contributing to the establishment of a favorable microenvironment for bone regeneration [53].

Another important factor in choosing HGs is their capacity to cause minimal or no inflammation. As mentioned earlier, macrophage polarization to M1 and M2 phenotypes plays a critical role in bone regeneration. In vitro studies using bone marrow-derived macrophages (BMDMs) have demonstrated that calcium alginate/amelogenin-based HGs promote macrophage polarization towards the M2 phenotype, as evidenced by the expression of CD206. These findings also show an inhibition of the inflammatory response collaborating in regenerative processes [54].

Finally, an important aspect in material selection is that it should not trigger a foreign body reaction, such as fibrotic encapsulation. While commonly used materials, such as polyethylene glycol (PEG), exhibit excellent biocompatibility, studies have shown that they can still induce fibrotic encapsulation. Research on HGs with alternating sequences of glutamic acid and lysine has reported minimal inflammation two weeks after subcutaneous implantation, as well as no evident collagen encapsulation after six months in mice. These findings open new possibilities for the use of such materials in the field of tissue engineering [55].

5. Physical Properties of Hydrogels

5.1. Cross-Linking

One way that HGs are classified is based on their cross-linking method, which can be physical through non-covalent cross-links or chemical through intra- or interpolymeric covalent bonds. Some studies on gelatin-based HGs report that chemical cross-linking enhances the network strength of the hydrogel and increases its water absorption capacity. Fourier transform infrared (FTIR) spectroscopy evaluations show a significant increase in the strength of triple helix structures, which resemble collagen [56]. The cross-linking method plays a crucial role in modifying the physical structure of HGs. Some materials that serve as cross-linking agents while also enhancing the physical properties of HGs include resveratrol and tannic acid. In the case of polyvinyl alcohol (PVA), it lacks the ability to form hydrogels independently at temperatures of 25 °C or higher. Therefore, it requires the addition of cross-linking agents such as resveratrol, which facilitates HG formation through strong hydrogen bonds between both materials, as confirmed by attenuated total reflection FTIR spectroscopy [57].

5.2. Stiffness

Stiffness refers to the ability of a material to withstand forces or stress, measured by the elastic modulus or Young's modulus. This is a critical property of HGs, particularly those intended for load-bearing sites, as it depends on polymer molecular weight and cross-linking density. For MSC-based applications, the stiffness of HGs ranges from 0.1 kPa to over 100 kPa [58]. This property plays a key role in MSC differentiation, as these cells sense hydrogel stiffness by exerting forces on the substrate to which they adhere [59].

The combination of different polymers significantly influences HG stiffness. A formulation consisting of gelatin/polyethylene glycol diacrylate/2-(dimethylamino)ethyl methacrylate demonstrated stiffness values exceeding 170 kPa and compressive strength above 700 kPa, making it a promising material for bone regeneration [60].

The impact of stiffness is further supported by studies on N-carboxyethyl chitosan hydrogels mixed with oxidized hyaluronic acid at different concentrations (5%, 7.5%, and 10% wt). The resulting stiffness values were 576.36 Pa, 3842.662 Pa, and 7368.937 Pa, respectively. Notably, higher stiffness was correlated with greater cell proliferation rates of BMSCs at days 1, 7, and 14, highlighting its importance in cellular responses [61].

5.3. Porosity

Another important characteristic of HGs is their porosity, as it plays a crucial role in cell proliferation and migration. The porous structure of HGs influences nutrient diffusion, waste removal, and cellular interactions, all of which are essential for effective tissue regeneration. In humans, the porosity of both trabecular and cortical bone varies between 5% and 90%, while the pore diameter ranges approximately from 0.1 to 2000 μm [62]. Research reports that HGs composed of calcium carbonate microcapsules, nanohydroxyapatite, chitosan, and collagen with varying hydroxyapatite concentrations (0%, 1%, 2%, and 5%) exhibit good porosity, with pore diameters below 100 μm when synthesized with 2% nanohydroxyapatite. However, their mechanical properties, including strength, viscoelasticity, and physical stability, decrease compared to HGs with higher nanohydroxyapatite concentrations [63].

5.4. Water Absorption

Another property is the water absorption capacity of HGs, which is a key property for promoting cell adhesion and migration, essential for tissue regeneration. This feature is often evaluated using the contact angle test to assess swelling behavior. Hydrogels composed of chitosan/β-glycerophosphate sodium/polyvinyl alcohol with varying antler powder deer concentrations (6%, 8%, and 10%) exhibit different contact angles, with the 10% formulation showing the most favorable results. However, no direct correlation has been established with water absorption capacity, as the 6% formulation shows no significant difference from the others [64].

6. Bioactive Properties of Hydrogels

Finally, it is important to highlight the bioactive properties that HGs should possess, such as controlled drug or metal ion (e.g., Cu) release, antibacterial activity, reactive oxygen species (ROS) scavenging, inflammation inhibition, and promotion of M2 macrophage polarization. These properties contribute to the establishment of a microenvironment that supports osteoblast and endothelial cell adhesion, proliferation, migration, and differentiation while minimizing osteoclast recruitment [65].

6.1. *Drug Delivery*

Bone tissue damage is not always solely the result of trauma; underlying diseases can also impede bone regeneration under homeostatic conditions. Therefore, HGs designed for bone regeneration must possess the capacity to be loaded with pharmacological agents that actively promote the regenerative process [66].

HGs composed of GelMA and chitosan loaded with acetylsalicylic acid (aspirin) have demonstrated satisfactory drug release capabilities in promoting bone regeneration. Viability and proliferation assays conducted on adipose tissue-derived stromal cells (ADSCs) after 24 h of exposure to the HGs showed high proliferation levels. Additionally, the long-term sustained release of aspirin contributes to anti-inflammatory effects and enhances osteoinductive properties [67].

6.2. *Osteogenic Potential*

Various HGs exhibit biocompatibility, degradability, and low cytotoxicity, among other properties. However, their osteogenic potential is often weak or even absent. Therefore, it is necessary to incorporate bioactive components or biological factors into HGs to enhance or maximize their osteogenic capabilities.

One example is the use of amine-modified bioactive glass (BAG) nanoparticles in polyethylene glycol (PEG) HG, which exhibit strong osteogenic properties, as evidenced by the upregulation of osteogenesis-related genes such as RUNX2, ALP, OCN, and OSTERIX in bone marrow-derived mesenchymal stem cell (BMSC) cultures [68].

On the other hand, the use of osteocyte cultures, which have been less explored, has demonstrated remarkable effects on the osteogenic potential of HG. The interstitial fluid velocity within the bone matrix and vertical channels ranges between 0.04–0.2 and 0.06–0.9 m/s, respectively. By modifying the interstitial fluid flow velocity in MLO-Y4 osteocyte cultures, it is possible to stimulate anabolic responses in osteocytes, leading to a downregulation of osteoclastogenesis-related genes and the production of soluble mediators that induce osteoblastic differentiation, ultimately contributing to new bone formation [69].

6.3. *Proliferation Potential*

Another crucial bioactive property of HGs is their ability to promote the proliferation of resident cells at the injury site as well as the cells incorporated into the HG, particularly mesenchymal stem cells (MSCs). Bioactive molecules such as peptides are commonly used to enhance this proliferative capacity. The E7 peptide, a short peptide with the amino acid sequence EPLQLKM, has been shown to influence MSC proliferation. In gelatin methacrylate Wharton's jelly hydrogels functionalized with E7, results indicate that cell proliferation within E7-modified HGs is significantly higher compared to control groups, confirming that this peptide can effectively enhance proliferative properties [70].

Another material that enhances the proliferative bioactivity of HGs is graphene oxide. In vitro experiments involving the encapsulation of single MSCs within alginate micro-hydrogels containing graphene oxide nanolayers have demonstrated the ability to maintain MSC viability and proliferation, as assessed through live/dead assays and fluorescence microscopy. These findings highlight the potential application of graphene oxide-functionalized HGs in minimally invasive tissue regeneration therapies [71].

7. Hydrogel-Based Scaffolds for Bone Regeneration

One of the primary characteristics required for HGs is biocompatibility, i.e., the ability of a material to avoid provoking a harmful immune response in the body [47], while performing its intended function, such as supporting bone regeneration [21]. HGs require structural and design properties that protect active molecules, ensure nutrient permeability, and mimic the native bone environment. Their porous structures are essential for nutrient uptake, waste diffusion, and intercellular communication (Figure 2) [17,72]. These hydrophilic polymeric products [73], with a great capacity to absorb and retain significant amounts of water [74], form a three-dimensional structural network due to the weak interactions between the polymeric chains [75]. This network is stabilized by cross-linking within the polymer structure, which may involve covalent bonds, ionic bonds, and physical interactions such as hydrogen bonding, van der Waals forces, and physical entanglements [21,74]. Typically, these HGs contain between 1% and 20% dry mass [15] and can be directly transplanted to the injury site after being molded into a specific form in vitro through controlled cross-linking [16].

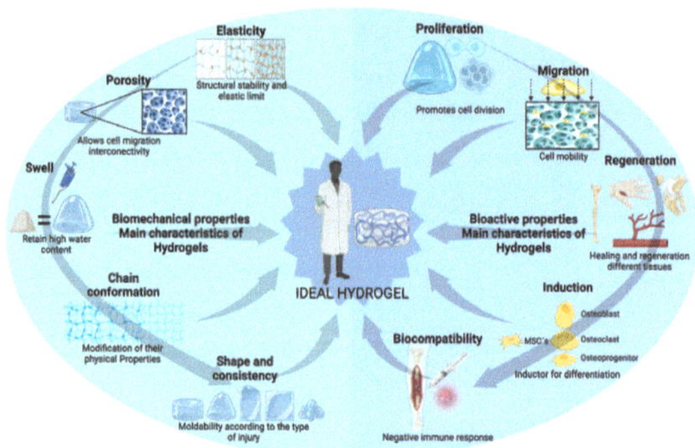

Figure 2. There are various properties and characteristics that make HGs ideal materials for the regeneration of different tissues, such as bone. These properties can be classified into biomechanical and bioactive categories. Biomechanical properties include attributes that facilitate the administration of HGs at the injury site, such as their shape and consistency relative to the conformation of their polymer chains, ensuring minimal procedural complexity. Additionally, their pore size and high water absorption capacity promote the retention and migration of exogenous agents incorporated into the HG, thereby supporting the regenerative process. Regarding bioactive properties, HGs exhibit high biocompatibility due to their natural or synthetic origins. Depending on the manufacturing process and their combination with cells and growth factors, these materials can enhance and improve their inductive and proliferative capabilities, ultimately leading to the successful regeneration of damaged tissues.

According to Calcei and Rodeo, osteoconductive, osteoinductive, and osteogenesis are the three major properties of orthobiologics used in bone healing [42]. In efforts to achieve bone regeneration using natural or synthetic materials, the principles of osteoconductivity, osteoinductivity, and osteogenesis, are fundamental. Osteoconductive refers to the use of scaffolds to passively support the growth of cells, tissues, and blood vessels. Osteoinductive involves the application of growth factors that induce the differentiation of MSCs, while osteogenic capacity denotes the presence of cellular elements within a material that enables it to generate new bone independently [42,43].

In this regard, HGs could achieve these three properties depending on the form of use: Osteoconductive HGs: hydrogels based on ECM have molecules such as proteoglycans, collagen, and/or gelatin, have molecules such as arginine–glycine–aspartic acid (RGD), and have a variety of low-molecular-weight peptides that give them properties that favor the regeneration of bone tissue without the need to be enriched with any other molecule, activating the migrations of resident cells of the tissue. Osteoinductive HGs: in addition to having their regenerative properties, they are enriched or functionalized with molecules that are not part of the HG itself, such as growth factors that amplify the therapeutic activation of resident cells by exogenous factors to generate new bone. Osteogenic HGs: these HGs are enriched with osteoprogenitor cells capable of generating new bones at the site of the injury and the capacity to elicit bone formation in ectopic sites (Figure 3).

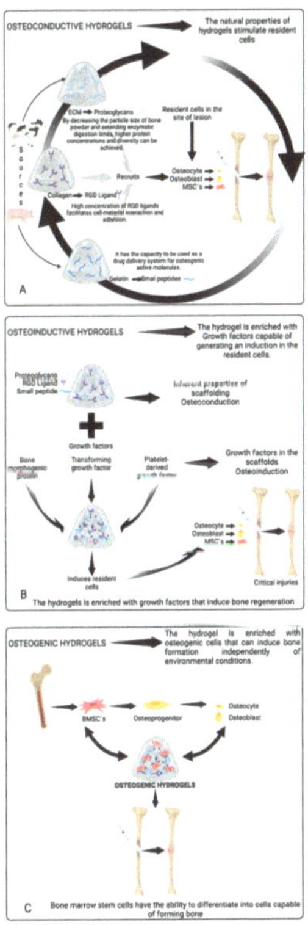

Figure 3. HGs could be classified according to their regenerative capacities. (**A**) Osteoconductive HGs: hydrogels based on ECM, collagen, and/or gelatin have their molecules such as proteoglycans, RGD, and a variety of low-molecular-weight peptides that give them properties that favor the regeneration of bone tissue without the need to be enriched with any other molecule, activating the resident cells of the tissue, and promoting cell migration to the injury site. (**B**) Osteoinductive HGs: in addition to having their regenerative properties, they are enriched or functionalized with molecules that are not part of the HG itself, such as growth factors that amplify the therapeutic action of the scaffold by inducing the resident cells to generate new bone by exogenous factors. (**C**) Osteogenic HGs: these HGs are characterized by being enriched by osteoprogenitor cells capable of generating new bone at the site of the injury or ectopic sites.

7.1. Natural Polymers

The most commonly used polymers in tissue engineering are derived from natural sources formed via photosynthesis or biochemical reactions in a biological environment [21]. These include materials such as collagen, alginate, starch, cellulose, gelatin, fibrin, silk, hyaluronic acid, and chitosan (Figure 4) [76].

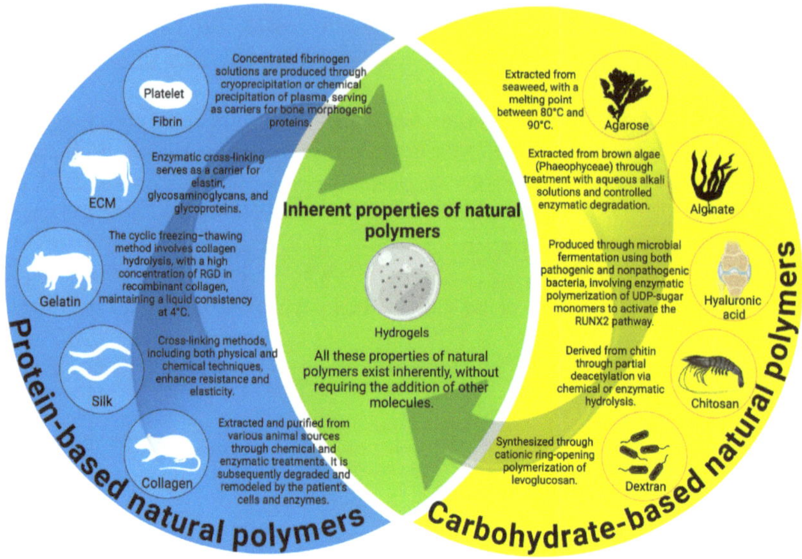

Figure 4. HGs derived from natural polymers, such as proteins or carbohydrates, exhibit distinct properties inherent to their molecular structure. These intrinsic characteristics confer these HGs (except alginate) osteoconductive capacities that support resident cell populations, eliminating the necessity for additional molecular modifications or supplements.

Natural polymers exhibit excellent biocompatibility with minimal immunogenicity. They facilitate cell adhesion and proliferation, support tissue regeneration, and contribute to the mechanical stability and structural integrity of bone. However, their primary limitations include low mechanical strength and susceptibility to rapid degradation in specific biological environments, influenced by enzymatic activity and implantation site (Table 2) [23].

Table 2. Natural polymer-based HGs.

Natural Polymer-Based HG	Properties	Applications	Cross-Linking	References
Collagen	Biocompatibility Biodegradability Cationic polymer Ready degradation High cross-linking potential Mechanical compression	Tissue engineering Bone regeneration	Enzymatic cross-linking	[3,16,17,21]
Fibrin	Biocompatibility Biodegradability Injectable application Elastic and plasticity Minimal inflammatory reaction	Drug carrier Bone regeneration Osteogenic capability Regeneration cartilage Skin tissues	Enzymatic cross-linking	[16,77,78]
Gelatin	Negative charge Chemical strength	Drug carrier Cartilage regeneration	Hydrophobic cross-linking	[17,36,79,80]

Table 2. *Cont.*

Natural Polymer-Based HG	Properties	Applications	Cross-Linking	References
Extracellular bone matrix (ECM)	Antibacterial activity support of essential molecules Maintain homeostasis Easily mixed with other materials	Bone regeneration Organoid formation Dental implants survival Drug carrier	Enzymatic cross-linking	[81–84]
Agarose	Biocompatibility Biodegradability Antibacterial activity conductive activity Porous structure Water insoluble No need for cross-linked agents	Tissue engineering	Self-gelling at low temperatures	[17,21,85,86]
Alginate	Biocompatibility Biodegradability Water solubility High fluid absorption Regulated homeostasis Adhesive properties Stable 3D structure Rigidity	Tissue engineering	Water cross-linking	[87–90]
Hyaluronic acid	Antibacterial activity anionic copolymer Injectable Regulated cell interaction Anti-inflammatory Antiedematous Low toxicity Hypoallergenic	Tissue engineering Wound healing Drug carrier Bone regeneration	Multifunctional derivative cross-linking	[17,21,91–93]
Chitosan	Biocompatibility Biodegradability Antibacterial Antifungal Non-toxicity Adhesive properties polycationic polymer Hydrophilic Injectable	Hemostasis Wound healing Transdermal patches Skin, cartilage, and bone regeneration Blood vessel embolization Drug delivery Gene expression modulation	Ionic cross-linking in acid solution	[17,21,27,44,94,95]
Dextran	Biocompatibility Biodegradability injectable Non-toxic Easy chemical modification	Tissue regeneration Drug carrier	Free radical polymerization cross-linking	[96–100]
Silk	Hypoallergenic Resistance and elasticity Cell adhesion	Tissue regeneration Bone regeneration Wound healing Cartilage and tendon regeneration Drug carrier	Hydrophobic interactions cross-linked	[16,28]
Silk fibroin	Biocompatibility Controllable degradation Mechanical properties	Tissue engineering Cell and drug carriers	Self-assembly Shear force application Ultrasonication Electric field exposure cross-linking	[101–103]

7.1.1. Protein-Based Natural Polymers

Collagen

Collagen, as a flexible, cationic polymer, is the most abundant protein in the extracellular matrix [104]. Its main functions are related to providing mechanical support to cells, tissues, and blood vessels, as well as offering resistance to compressive forces [105]. It is sourced from various animal species, with well-established protocols for its isolation and purification involving enzymatic or chemical treatments [106]. Collagen possesses a range of physicochemical properties, including biocompatibility, biodegradability, and cross-linking potential, due to the presence of various functional groups, such as hydroxyl, amino, carboxyl, guanidyl, and imidazole groups. These groups confer unique physical and chemical characteristics, making collagen highly suitable as a compound in tissue engineering applications [21].

COL-I is the most utilized form in tissue engineering and HG formulation. Additionally, its high concentration of RGD ligands facilitates material–cell interactions and adhesion [107]. Is primarily secreted by osteoblasts and undergoes mineralization in the later stages of bone development, providing an optimal microenvironment for tissue engineering applications [16]. In vitro studies have demonstrated that collagen micro- and nanostructured scaffolds possess mechanical compression properties that are approximately 50% comparable to human trabecular bone, without compromising porosity. Furthermore, viability assays—such as those assessing adhesion, proliferation, and metabolic activity, including alkaline phosphatase (ALP) activity and mineralized matrix synthesis—have yielded promising results when compared to lactic acid and PEG scaffolds using human osteoblasts (HOBs) [3]. These findings highlight the superior performance of collagen-based HGs over other types for bone regeneration [3]. It has also been demonstrated that the combination of implanting dense collagen HGs containing wild-type murine dental pulp stem cells (mDPSCs) with weekly systemic injections of a sclerostin antibody (Scl-Ab) significantly enhances bone regeneration in critical-sized calvarial defects created in mice [108]. There is also evidence of the development of collagen-based HG scaffolds infused with tacrolimus, which were surrounded by a PCL/gelatin membrane, the scaffolds were characterized for their potential use in bone tissue engineering applications, and the in vivo study provided preliminary evidence of the efficacy of these HGs in treating bone defects [109].

This biomaterial exhibits properties that promote the proliferation of urine-derived stem cells (USCs) in both 2D and 3D models. The HG consisted of a composite of chitosan microspheres loaded with BMP-2 embedded within a type I collagen HG that released growth factor for over 28 days, enhancing ALP and promoting calcium mineral deposition in 2D models. In vivo assays demonstrated a significant volume of newly formed bone eight weeks after HG administration in cranial defects in Sprague Dawley rats [110].

A significant drawback of collagen HGs as scaffolds for tissue engineering is their variability, influenced by factors like collagen source and gelation pH, which can result in difficulties during the synthesis of the material. Additionally, collagen's rapid degradation, weak mechanical strength, opacity, and high shrinkage limit its use in tissues that need more stiffness, such as bone or cartilage [107,111].

Fibrin

Fibrin is a commonly utilized scaffolding material in tissue engineering and regenerative medicine [112,113]; notably, fibrin exhibits minimal inflammatory responses and demonstrates excellent biocompatibility, making it a promising candidate for biomedical applications [78]. Is found within blood clots, and it has the unique capacity to carry bone morphogenetic proteins (BMPs); notably, fibrin can undergo enzymatic cross-linking, forming a gel with adhesive properties that facilitate injectable application [16]; however, its application is constrained by a lack of responsiveness to external stimuli in situ, inadequate mechanical properties, pronounced shrinkage, and rapid biodegradation, which limit its suitability for cell culture [112,113].

A straightforward method for fabricating fibrin HGs involves combining fibrinogen and thrombin at 37 °C, which also enables the modification of the gel's internal fiber structure and porosity [76]. In vitro and in vivo studies of fibrin HGs combined with BMP-2 (37.5 μg/mL) have demonstrated osteogenic properties. In vitro assays revealed that sustained BMP-2 release after 28 days, along with increased ALP expression compared to control groups and fibrin HGs without BMP-2. In vivo, statistically significant increases in total bone volume were observed at tibial defect sites in Wistar rats treated with the BMP-2-enriched fibrin HG, compared to those treated with fibrin HGs alone [77].

Another study reported the mechanical properties and biodegradability of a fibrin HG combined with konjac glucomannan. In vitro assays demonstrated biocompatibility and the ability to enhance osteogenesis of nasal mucosa-derived ectodermal mesenchymal stem cells (EMSCs) by upregulating mineralization and ALP. In vivo results further showed the improved reconstruction of alveolar bone defects, as evaluated by micro-computed tomography (μCT) in Sprague Dawley rat models [114].

Gelatin

Gelatin is a material that, due to its ability to be used as a drug delivery system for osteogenic active molecules and scaffolds, is widely used in tissue engineering. It is a natural polymer obtained by the hydrolysis of collagen obtained from pig skin and bone, and bovine skin and bone [115].

These materials offer physicochemical advantages, such as a naturally occurring negative charge that interacts with positively charged inductive molecules added to them, a property that can be adjusted during the manufacturing process [36]. In this context, these HGs have also been combined with molecules such as oxidized chondroitin sulfate and mesoporous bioactive glass nanoparticles, capable of cross-linking in situ under physiological conditions in the presence of borax. This combination improves compressive strength, enhances the differentiation of bone marrow-derived mesenchymal stem cells (BMSCs) into osteoblasts, and successfully facilitates the regeneration of cranial bone defects in rats [116].

The chemical coupling of polydopamine modified gelatin HGs with polyetheretherketone using glutaraldehyde solutions has demonstrated sustained release capabilities of bone BMP-2. Tests on MSC cultures also demonstrated an increase in run-related transcription factor 2 (RUNX2) expression associated with increased BMP-2. In addition, the production of OCN, ALP, and calcium deposition increased due to the incorporation of the growth factor, proving that this HG maintains a potential for bone regeneration [117].

This material also offers promising results in decreasing the risk of delayed bone consolidation when loaded with a black phosphorus nanosheet and deferoxamine. In vitro studies showed the overexpression of RUNX2, COL-I, and BMP-4 and the proliferation of BMSCs. In vivo results in Sprague Dawley rats demonstrated satisfactory results 4 weeks after the administration of the material on 3 mm diameter defects in tibial bone evaluated macroscopically and histologically [118].

Other combinations of this HG have similarly demonstrated preosteoblast differentiation and high cell viability when biphasic calcium phosphate is incorporated into gelatin HGs [119]. Studies are reporting the ability of gelatin-based HGs to be cross-linked using visible-wavelength light when combined with riboflavin as a photoinitiator. This approach achieves a stiff range that supports the proper differentiation of osteoblasts and the proliferation of KUSA-A1 cells [120].

Extracellular Matrix (ECM)

Natural scaffolds derived from the ECM are widely utilized in preclinical research and clinical trials [26]. For bone regeneration, they promote the development of new bone and maintain homeostasis within a complex environment of biochemical and biomechanical signals that collectively regulate pluripotency, proliferation, and differentiation of resident cells [84]. Additionally, they can be combined with other materials or polymers, such as collagen and chitosan [81]. It has been reported that the combination of bone ECM (bECM) with dental pulp stem cells enhances cell viability after three weeks in culture and promotes the expression of sialoproteins (BSPs) and RUNX-2, even in the absence of osteogenic inducers [121].

The ECM obtained from human bone tissue has emerged as a promising candidate in regenerative medicine. By decreasing the particle size of bone powder and extending enzymatic digestion times, it is possible to achieve higher protein concentrations and diversity, alongside improved gelation properties. Bone powder with particle sizes ranging from 45 to 250 μm, tested on MSCs for 7 days, significantly promoted osteogenic differentiation, evidenced by the upregulation of key genes such as ALP, RUNX2, COLA1, and OCN. This approach represents a valuable addition to the strategies for bone regeneration [122].

The ECM derived from porcine dermis loaded with biphasic calcium phosphate powder has also been utilized for bone regeneration. It has demonstrated the capability to be administered as an injectable material due to its thermosensitive properties and favorable mechanical characteristics. MC3T3-E1 pre-osteoblasts showed no evidence of cytotoxicity and exhibited enhanced osteogenic differentiation through the overexpression of genes such as bone BSPs, OPN, and ALP, as analyzed by quantitative real-time PCR. Additionally, in vivo assays on defects in the femoral head of rabbits demonstrated new bone formation at 4 and 8 weeks, with superior outcomes observed in HGs containing 15% biphasic calcium phosphate [123].

Silk and Silk Fibroin

Silks possess a distinctive combination of properties including biocompatibility, biodegradability, low immunogenicity, and ease of accessibility, and can be obtained from natural sources, such as silkworms [103]. Biomineralized materials from silk spider proteins are widely used in bone regeneration [124]. For instance, the combination of MaSp1 protein and bone BSP in the development of scaffolds promotes the adhesion of MSCs and contributes to their differentiation [125].

The gelation process can be achieved via various methods, including physical cross-linking techniques such as self-assembly, shear force application, ultrasonication, and electric field exposure. Chemical cross-linking methods, including photopolymerization, radiation exposure, and enzymatic cross-linking, are also employed for silk [78].

However, silk fibroin exhibits limited osteoconductive properties; therefore, researchers have incorporated various bioactive molecules to address this limitation [126]. When combined with Laponite (LAP), silk fibroin demonstrates the ability to regenerate bone in calvarial defects in rat models and elicits an osteogenic response in BMSC cultures [126].

The combination of silk fibroin and fibrin with amorphous calcium phosphate (ACP) and platelet-rich plasma (PRP) has been applied to the repair of 5 mm cranial bone defects in Sprague Dawley rats. New bone formation was observed 8 weeks later and evaluated through μCT. Additionally, the seedbed HGs demonstrated succinate-like activity, marked by the upregulation of solute carrier family 13 member 3 (SLC13A3), which subsequently activated the PI3K/Akt signaling pathway. These findings highlight its potential as a biomaterial with significant bone regeneration capabilities for addressing critical-sized defects [127].

In the same context, silk fibroin combined with calcium phosphate particles and methylcellulose enhance the biological properties of the material while also providing the opportunity for the HGs to be administered in an injectable form, gelating at a physiological temperature in approximately 40 min. An osteoblast precursor derived from *mus musculus* calvarial mouse cells (MCT3T3-E1) exhibits low cytotoxicity, with viability exceeding 98% under concentrations of 10% and 15% *w*/*v* of calcium phosphate particles. Furthermore, this material also demonstrates the ability to promote osteogenic differentiation of the cell line after 7 days [128].

7.1.2. Carbohydrate-Based Natural Polymer

Alginate

Alginate is the most widely used biomolecule of marine origin, just below cellulose [129], with good biocompatibility and biodegradability, and has been used for cell delivery or as a protein carrier in bone tissue engineering applications [87]. Extracted primarily from brown algae (*Phaeophyceae*) using alkaline solutions such as calcium chloride ($CaCl_2$) [21], alginate is a natural anionic biopolymer composed of alternating α-L-guluronic acid (G) and β-D-mannuronic acid (M) residues [44]. This polymer is capable of absorbing large amounts of biological fluids, making it effective in regulating homeostatic and wound-healing processes [21].

In vivo assays in rats have demonstrated the ability of these HGs, loaded with allogeneic BMSCs, to regenerate bone defects mimicking cleft lip and palate injuries. The regeneration was evaluated through μCT and histological sections at different time points [87].

In addition to combining these HGs with MSCs, the adhesive properties of alginate have been enhanced by conjugating it with other molecules, such as adhesion-promoting ligands like RGD [130].

This polymer can be combined with various materials, such as polyvinyl alcohol (PVA) and bioactive glass (BAG), demonstrating excellent biocompatibility in hemolysis and cytotoxicity assays using MC3T3-E1 cells. The fabrication of this HG also enhances ALP activity, and the expression of genes associated with osteogenic differentiation, indicating its potential for bone repair applications [131].

Finally, alginate methacrylate HGs have been developed as bioinks for bioprinting scaffolds, providing alternative approaches in manufacturing. These HGs have shown promising results in the expression of genes associated with osteogenesis [89].

Hyaluronic Acid

Hyaluronic acid is a naturally occurring glycosaminoglycan (polysaccharide) abundant in soft connective tissues, such as the skin, cartilage extracellular matrix, and synovial fluid [79]. With its low toxicity, hypoallergenic nature, biocompatibility, and biodegradability, hyaluronic acid is widely used in biomedicine for applications like scaffold manufacturing in tissue engineering, wound healing, and drug delivery systems [132]. As a non-sulfated glycosaminoglycan, hyaluronic acid significantly impacts properties within the ECM like cell motility mediation, and other capabilities of the material such as lubrication and water retention. Additionally, it plays a role in cell interaction with growth factors and osmotic pressure regulation [17].

In combination with niobium-doped BAG, it can promote new bone formation through the RUNX2 signaling pathway [92]. These HGs, in combination with hydroxyapatite and short peptides such as fluorenylmethoxycarbonyl-diphenylalanine, have demonstrated favorable outcomes in promoting the osteogenic differentiation of MC3T3-E1 preosteoblasts. Additionally, they facilitate calcium deposition, as evidenced by Alizarin Red staining [93].

Studies report the use of hyaluronic acid and PVA-Gel that can be rapidly administered via injection, forming through cross-linking of its multifunctional derivatives. When enriched with BMP-2, this material has demonstrated the capability to generate ectopic bone in muscle tissue, observed through computed axial tomography at 4 and 10 weeks post-injection. In vitro MTT assays on human dermal fibroblasts (HDFa) revealed no restrictions in cell proliferation, highlighting the material's promising osteogenic potential for bone regeneration [133]. They are also capable of supporting exosomes while preserving their biological activity and enabling sustained release at bone defect sites [134]. In addition, it can blend with synthetic polymers such as PEG and PCL to form HGs loaded with solid lipid nanoparticles, demonstrating satisfactory biocompatibility on BMSCs evaluated through live/dead staining. Furthermore, this HG combination exhibits a modulatory capacity on macrophage polarization between M1 and M2 phenotypes, reducing M1 polarization while enhancing M2 polarization, thereby contributing to bone regeneration. This effect has been confirmed through evaluations of critical cranial defects in rats using μCT analysis [135].

Chitosan

Chitosan ranks as the second most abundant biodegradable polymer after cellulose, commonly extracted from the exoskeletons of insects, fungal mycelia, and crustaceans [27,95]. Structurally, it consists of D-glucosamine and N-acetyl-D-glucosamine units linked by β-1,4-glycosidic bonds, with a molecular weight range from 10 to over 1000 kDa [27,44,79,136]. It is a natural polycationic polymer with hydrophilic properties, derived from the partial deacetylation (40% to 80%) of chitin [17,137]. Its unique attributes make it suitable for biomedical applications, including biocompatibility, biodegradability, non-toxicity, and strong mucoadhesive properties, as well as a wide spectrum of antibacterial activity.

Chitosan exhibits oxygen permeability and a structure closely resembling glycosaminoglycans, which also enhances its biological functionality. This biopolymer is widely applied in areas such as hemostasis, wound healing, transdermal patches, skin and bone regeneration, cartilage tissue engineering, blood vessel embolization, drug delivery systems, and gene expression modulation [17,27,138]. Chitosan HGs can be fabricated through physical or chemical methods. Chemical cross-linking involves agents, irradiation, and secondary polymerization, while physical cross-linking relies on electrostatic, hydrophobic, and hydrogen-bonding interactions [21]. Aqueous chitosan solutions at physiological pH values or above 6.2 precipitate due to ionic forces, making them challenging to handle. However, combining chitosan with glycerol phosphate enables gelation at physiological temperatures, which improves its usability in surgical applications [17].

Chitosan is also adaptable for injectable applications and can be combined with growth factors, such as BMPs, to enhance its regenerative potential. Luca et al. conducted in vivo and in vitro trials combining chitosan HGs with BMP-2 and β-tricalcium phosphate; this composite showed promising results: in murine models, it successfully induced ectopic bone formation, and in rabbit femur models, it contributed to lesion regeneration. Although the regeneration was incomplete, this may have been due to the limited duration of the study or the critical size of the bone defect, which is often challenging for natural healing processes [139].

Chitosan has also been combined with injectable rat-tail collagen and supplemented with extracellular vesicles derived from osteoblasts to promote bone regeneration. This approach yielded favorable results, particularly in enhancing the ECM mineralization of BMSCs compared to gels without vesicles [94]. Satisfactory results have also been observed in alveolar bone regeneration by incorporating basic fibroblast growth factor (bFGF) and transforming growth factor β3 (TGFβ3). Chitosan microspheres containing these growth factors have demonstrated the ability to regenerate alveolar bone defects in Sprague Dawley rats while ensuring sustained long-term release of TGFβ3 and 95% release of bFGF within the first 7 days. In vivo evaluations using μCT have shown bone regeneration, facilitated by the nutrients assimilated by stem cells due to microvascular formation promoted by bFGF [140]. Finally, it has been shown that chitosan upregulates OPN and OCN gene expression in human amnion-derived stem cells (HAMSCs) from amniotic tissue, as confirmed by RT-PCR assays after 21 days [81].

Dextran

Dextran is a biodegradable and biocompatible exopolysaccharide synthesized by lactic bacteria [141]. This biopolymer is composed of glucose monomers with an α-1,6 bond [142]. This neutral glucan, composed of glucose monomers with numerous active hydroxyl groups, allows for easy chemical modification, enhancing its utility in biomedical applications; it is biodegradable and exhibits exceptional solubility, biocompatibility, and non-immunogenic properties [21]. To enhance the bioactivity of the material, various strategies can be employed. One effective approach involves incorporating inorganic compounds, such as hydroxyapatite, β-tricalcium phosphate, or BAG, into the polymeric matrix [100]. When combined with PEG and gelatin, these HGs enhance the colonization of human BMSCs. Furthermore, the inclusion of osteogenic factors promotes mineralization after 28 days [98]. This combination of dextran with gelatin has also demonstrated the manipulation of pore size and distribution, a key feature in regenerative materials. When this combination is used with bioprinting, it promotes the proliferation, migration, and expansion of BMSCs while activating signaling pathways associated with Yes-Associated Protein (YAP) [143].

The capacity of dextran to generate apatite crystals is null; however, when combined with BAG, the formation of crystals has been demonstrated, and it shows favorable cell viability, as evaluated through MTT assays in human osteosarcoma cell cultures (SaOS-2) [99]. Other studies have demonstrated the potential of dextran microspheres incorporated within hyaluronic acid scaffolds. These experiments revealed promising cell viability when human adipose-derived stem cells (hASCs) were seeded onto the scaffold, with fluorescence microscopy indicating favorable live/dead cell ratios [96]. Furthermore, when photobiomodulation was applied, hASCs exhibited enhanced differentiation characteristics, with increased ALP and calcium deposition observed within 24 h [144].

In vivo experiments in C57BL/6J mice with femur defects treated with dextran–tyramine conjugate HGs showed satisfactory results, evaluating regeneration with μCT when the HG was loaded with basic fibroblastic growth factor (bFGF) [145]. Similar results have been observed in cranial defects in rats when dextran HGs are loaded with human umbilical vein endothelial cells (HUVECs). However, when this material is loaded with both HUVECs and human osteoblasts in co-culture, the outcomes are less favorable. Alternatively, the use of stromal-derived growth factor (SDF) presents a viable option for improving regenerative outcomes [146].

7.2. Synthetic Polymers

Synthetic HGs can control their physicochemical properties through adjustments during their synthesis. They are materials composed of synthetic polymer networks and can be mass-produced, which surpasses natural materials, making them specific for different scenarios (Table 3) [147].

They exhibit lower bioactivity and biocompatibility compared to natural polymers; however, they can be enhanced during manufacturing processes to enable drug and growth factor loading, thereby increasing their regenerative potential [148].

These materials play a crucial role in the field of tissue engineering, often being used in conjunction with growth factors and short peptides derived from proteins naturally found in tissues to support cell viability and the direct phenotype of proliferation to enhance their therapeutic potential [149]. While they exhibit little or no osteoconductive properties, their effectiveness in achieving regenerative outcomes is often limited when used alone; they typically require the incorporation of additional molecules or bioactive materials [76].

Synthetic polymers' versatility facilitates the modification of their physical properties, molecular weights, functional groups, configurations, and chain conformations. This adaptability enables the production of various shapes with high reproducibility. Furthermore, their exceptional tensile strength and resistance make them capable of providing mechanical support while simultaneously degrading over time [150,151].

Table 3. Synthetic polymer scaffolds; properties and function.

Synthetic Polymer Scaffolds	Properties	Applications	Scaffold Synthesis	References
Polyethylene glycol (PEG)	Injectable Cytocompatibility Increase collagen deposition	Bone regeneration Dental pulp regeneration Cell and drug carrier	Photopolymerized cross-linking	[35,152–154]
Polypropylene fumarate (PPF)	Biocompatibility Biodegradability Viscous (21 °C) Injectable Elasticity Hardness	Bone regeneration Substitute for trabecular bone Drug carrier	Covalent bond cross-linking	[155–158]
Polycaprolactone (PCL)	Biocompatibility Biodegradability Osteogenic properties	Bone regeneration Periodontal regeneration	Covalent bond cross-linking	[13,159–162]
Polyvinyl alcohol (PVA)	Biocompatibility Cytocompatibility Mechanical stability high water absorption capacity	Bone regeneration	Physically cross-linking	[57,163,164]
Poly (γ-glutamic acid) (γ-PGA)	Biocompatibility Non-toxicity Tailorable degradation Encapsulation characteristic	Bone regeneration	Glucose cross-linking	[165–167]

7.2.1. Polyethylene Glycol (PEG)

Poly(ethylene glycol) (PEG) is extensively utilized in the fabrication of well-defined HG due to its exceptional biocompatibility and flexibility. However, its application is hindered by limitations such as low mechanical strength and inadequate cell adhesion. These drawbacks can be effectively addressed through engineering strategies, such as the integration of cell adhesion motifs and matrix metalloproteinase (MMP)-sensitive cross-links, which facilitate cellular infiltration and matrix remodeling in both in vitro and in vivo environments [168].

When combined with polylactic acid (PLA), the surface hydrophilicity is enhanced, as evidenced by a reduction in the water contact angle. Additionally, its degradability increases upon exposure to phosphate-buffered saline (PBS) solutions over 8 weeks. Furthermore, the material demonstrates low cytotoxicity when tested in cells that has fibroblast morphology isolated from the bone patient with osteosarcoma (MG-63) cell cultures [152]. When combined with BAG, the material demonstrates enhanced osteogenic properties along with significant improvements in mechanical characteristics, including compressive strength, flexibility, and extensibility [153]. Furthermore, PEG HGs have been utilized for cell encapsulation; however, they may elicit foreign body responses. Studies involving murine macrophages (RAW 264.7) and MC3T3-E1 preosteoblasts encapsulated in PEG HGs exposed to lipopolysaccharide (LPS) have shown that macrophage stimulation leads to a

5.3-fold increase in apoptotic osteoblasts, a 4.2-fold reduction in ALP, and a 7-fold decrease in collagen deposition [154].

This compound exhibits excellent biocompatibility, as demonstrated by viability assays such as MTT and live/dead staining in human vascular endothelial cells (HUVECs) and BMSCs. When combined with poly(L-lysine) and dendrimers, it significantly enhances the expression of BMP-2, indicating its osteogenic potential [35]. In addition to enhancing BMP-2 expression, this polymer can encapsulate the growth factor to further support bone formation. The incorporation of RGD as a functional group promotes cell adhesion, particularly of osteoblasts, resulting in effective bone regeneration. This has been demonstrated in cranial defects in rats, with μCT evaluations at 4 weeks post-administration confirming significant bone regeneration [168].

7.2.2. Polypropylene Fumarate (PPF)

PPF is a material with medical applications due to its biological properties such as biodegradability. It has a fumaric acid-based structure with the main disadvantage of not having osteoinductive or osteoconductive properties [169]. As a derivative of fumaric acid, PPF exhibits a viscous consistency at room temperature (21 °C) and is characterized by its cross-linking mechanism, which occurs via covalent bonds. This property facilitates its manipulation in conjunction with other cross-linking agents, enabling in situ cross-linking upon injection [170]. Additionally, reports have highlighted its antibacterial capacity when antibiotics such as rifampicin and ciprofloxacin are incorporated. In vitro infection models demonstrate a reduction in bacterial load, while simultaneously supporting the formation of new bone tissue [156].

Further studies report that PPF effectively encapsulates PGA microspheres for the controlled release of BMP-2 growth factors. This system has been evaluated in bone stromal cell cultures (W20-17) by assessing ALP and drug release profiles. In in vivo bone defect in rats, PPF's regenerative capacity has been validated through μCT imaging, confirming its osteogenic effectiveness [158].

When PPF is combined with methoxy PEG and enriched with flavonoid nanoparticles, it improves the expression of osteogenesis-related genes, including COL I, OCN, and OPN. This effect is observed 14 days after the in vitro exposure of MG-63 cells to the HG. These in vitro results position the material as a promising candidate for applications in bone regeneration [171].

7.2.3. Polycaprolactone (PCL)

PCL is a polymer with high hydrophobicity as its main disadvantage; however, it is characterized by its high biocompatibility, design flexibility, low degradation rate, high stability, and cost-effectiveness, making it an ideal candidate for tissue engineering applications, particularly in periodontal and bone tissue regeneration. Its nanofibers have been widely employed as scaffolding materials for effective tissue repair and regeneration processes [159]. It has shown the ability to regenerate critical bone defects when combined with zinc after 8 weeks, as well as activating the Wnt/β-catenin and NF-κB signaling pathways, as verified in in vitro models using MC3T3-E1 and RAW264.7 cells [159].

Scaffolds based on PCL enriched with magnesium microparticles (3% weight) enhance osteogenic properties [13]. These scaffolds promote the expression of key genes involved in bone metabolism, such as OPN, OCN, Col1a1, and RUNX2, while also demonstrating biomineralization through Alizarin Red staining [162].

The functionality of PCL in controlled molecule release has been demonstrated, particularly with growth factors such as BMP-2. Studies have shown that PCL enables a complete and sustained release of BMP-2 over 21 days, as corroborated by in vitro assays utilizing MSCs, highlighting its potential in tissue regeneration applications [160].

7.2.4. Polyvinyl Alcohol (PVA)

Polyvinyl alcohol (PVA) is widely utilized due to its high hydrophilicity, biocompatibility, low toxicity, and chemical stability. It is a polymer derived from the hydrolysis of polyvinyl acetate [57]. When combined with natural molecules capable of self-assembling with PVA, supramolecular hydrogels can be formed, resulting in enhanced mechanical properties [163].

PVA-based hydrogels incorporating resveratrol have demonstrated favorable viscoelastic mechanical properties, as assessed through rheological tests. Additionally, in vitro studies report that hydrogels containing 0.4% resveratrol effectively promote alkaline phosphatase (ALP) activity and mineral deposition. Furthermore, this HG upregulates genes associated with osteogenic activity, including BMP-9, OCN, and ALP in MC3T3-E1 cells, positioning it as a promising material in the field of bone regeneration [57].

Other studies support the favorable mechanical properties of PVA-based hydrogels at various concentrations, particularly when combined with chitosan, which enhances mechanical performance. Notably, a 2% PVA concentration has been shown to provide high water absorption capacity.

Furthermore, the degradation behavior of these hydrogels in phosphate-buffered saline (PBS) containing 100 mg/L of lysozyme has yielded promising results. Finally, cytotoxicity assays using MG63 cells demonstrate a statistically significant increase in cell proliferation compared to a positive control (DMSO), further supporting the biocompatibility and potential biomedical applications of these materials [164].

7.2.5. Poly(γ-glutamic acid) (γ-PGA)

A water-soluble polyamino acid composed of glutamic acid units linked by peptide bonds through amino and carboxyl groups [165]. It is primarily produced by Gram-positive bacteria of the *Bacillus* species [166].

The combination of γ-PGA with whey protein isolate HGs has demonstrated a high water absorption capacity before degradation begins. Additionally, cytotoxicity assessments at days 3 and 14 indicated minimal toxicity. Regarding its osteogenic potential, evaluations using MC3T3-E1 cells showed significantly higher alkaline phosphatase (ALP) activity compared to the control group, suggesting its suitability for in vivo studies [166].

Collagen-based HGs containing γ-PGA loaded with BMP-2 have yielded promising results in bone regeneration, with evaluations conducted both in vitro and in vivo. These HGs exhibit low viscosity at room temperature, transitioning to high elastic and viscous moduli under physiological conditions. Furthermore, they support BMMSC proliferation, attributed to their high porosity and upregulation of ALP, BSP, and OCN at days 7 and 14. Finally, in vivo assays in calvarial defect models of mice demonstrated complete recovery within 4 weeks, as assessed by micro-CT (μCT) [167].

8. Biomodels, Type of Bone Defect and Tests for the Evaluation of Bone Defect Regeneration

For the in vivo evaluation of HGs in bone regeneration studies, the literature reports promising results across various materials. Regarding biomodels, the most commonly used are rodents, specifically Sprague Dawley rats, Wistar rats, C57BL/6J mice, and Fisher 344Jcl rats. Several factors contribute to the preference for these models, including their genetic similarity to humans, estimated at approximately 99%. Additionally, their handling in laboratory animal facilities is significantly easier compared to larger biomodels such as dogs or primates (Table 4).

Table 4. Biomodels used for the evaluation of bone regeneration with HG.

Biomodel	HG	Bone Defect	In Vivo Evaluation	In Vivo Results	References
Rodents biomodels					
Sprague Dawley rats	Composite of chitosan microspheres loaded with BMP-2 embedded within a type I collagen HG	Critical cranial defects	μCT Histological analysis	Significant volume of newly formed bone eight weeks after	[110]
	Gelatin with black phosphorus nanosheet and deferoxamine	Defects in tibial bone	Macroscopic and histological analysis	Satisfactory results 4 weeks after administration of the material on 3 mm diameter defects in tibial bone	[118]
	Silk fibroin and fibrin with amorphous calcium phosphate and platelet-rich plasma HG	Critical cranial defects	μCT	New bone formation was observed 8 weeks	[127]
	Alginate loaded with allogeneic BMSCs HG	Cleft lip and palate injuries	μCT Histological sections at different time points	Regenerate bone defects	[87]
	PEG and PCL loaded with solid lipid nanoparticles HG	Critical cranial defects	μCT	Regenerate bone defects	[135]
	Chitosan microspheres containing basic fibroblast growth factor (bFGF) and transforming growth factor β3 (TGFβ3) HG	Alveolar bone defects Spherical defect with a diameter of 2 mm and depth of 1 mm	μCT	Successfully bone regeneration	[140]
	PEG with RGD cell adhesion motifs and MBP2 HG	Critical cranial defects	μCT Histological analysis	Bone regeneration after 4 weeks	[168]
	Collagen hydroxyapatite with BMP-2 and microspheres with alendronate	Critical cranial defects	μCT Histological analysis	Accumulation and maturation after 2, 4, and 8 weeks	[172]
	Amorphous calcium phosphate nanoparticle–platelet injectable HG	Lower and middle parts of the femur with 2.4 mm diameter	μCT Histological analysis	Significant improvements in tissue density after 2 weeks	[19]
	Gelatin/chitosan cryogel with platelet derived growth factor and BMP-2	Cranial bone defects	μCT	After 4 weeks, notable degree of new bone formation After 8 weeks, more than 50% of the initial bone defect was cover with new bone	[5]

Table 4. *Cont.*

Biomodel	HG	Bone Defect	In Vivo Evaluation	In Vivo Results	References
Wistar rats	Fibrin HG combined with BMP-2 (37.5 μg/mL)	Tibial defect	μCT	Significant increases in total bone volume	[77]
	Chitosan derivative/collagen thermosensitive composite loads with fibroblastic growth factor	Infrabony periodontal defects	Hematoxylin–eosin staining Masson staining	Increase in alveolar bone height	[173]
	Gelatin with oxidized chondroitin sulfate and mesoporous bioactive glass nanoparticles	Cranial bone defects	μCT	Facilitates the regeneration	[116]
	Chitosan PEG with mineral-coated microparticle BMP-2	Cranial bone defects	μCT Histological analysis	Bone volume and tissue volume after 4 and 8 weeks	[174]
C57BL/6J mice	Dextran–tyramine conjugate with fibroblastic growth factor HG	Femur fracture	μCT	Satisfactory bone formation	[145]
	Photocross-linked dextran-based HGs loaded with human umbilical vein endothelial cells	Cranial bone defects	μCt Histology Immunohistology	Potential bone regeneration	[146]
BALB/c mice	Collagen with γ-PGA loaded with BMP-2 HG	Cranial bone defects	μCT Histological analysis	Potential bone regeneration	[167]
Fisher 344jcl rats	Resorbable collagen membrane with fibroblastic growth factor HG	4 mm mandibular bone defects	μCT Histological analysis	Volume of newly formed bone, bone mineral density and closure percentage of the defect after 6 weeks	[175]
Larger animals biomodels					
New Zealand rabbits	Extracellular matrix from porcine with biphasic calcium phosphate powder HG	Femoral head defect	μCT Immunohistochemistry Histological analysis	No immune responses, and new bone formation after 4 weeks After 8 weeks, no fibrous tissue bridge was evident	[123]
	Injectable rhBMP-2-loaded chitosan HG	Radius critical size defect	Radiographic evaluation Histological analysis	Discontinued bone was observed after 8 weeks	[139]
	Gelatin with BMP-9-coated nano-calcium-deficient hydroxyapatite/poly-amino acid	Femoral bone defects	X-ray observation μCT SEM	Satisfactory bone regeneration assessed at 1, 4, 12, and 24 weeks	[176]
Monkeys					
Cynomolgus monkeys	bFGF-incorporating gelatin HG	Four defects (each 6 mm in diameter) in each monkey's parietal bone	Dual-energy X-ray absorptiometry (DEXA)	Defects are visible radiographically until week 18 Finally, at week 21, defects are observed with radiopacity	[177]

Rodents also offer cost-effectiveness, as they require minimal maintenance, consume relatively little food, and are easy to manage. Furthermore, their rapid reproductive cycle facilitates the establishment of large populations in a short period, and they reach sexual maturity quickly, making them highly suitable for in vivo studies.

9. Functionalization of HGs Using Growth Factors and Peptides for Bone Regeneration

Bone regeneration is driven by a complex cascade of growth factors [178]. Scaffolds and HGs must incorporate chemical signals, including cell adhesion sites, growth factors, and other signaling molecules, to improve the bioactivity of the scaffold and support the maintenance of homeostasis [179]. Bioactive molecules like growth factors and peptides play a pivotal role in bone tissue engineering by enhancing the functionality of biomaterials to facilitate bone remodeling on cranial defects in rats [172]. Additionally, various components and factors significantly impact the effectiveness of HG, including the incorporation of MSCs, the bioactive molecules used for their functionalization or combination, and the selection of the most suitable cell lines for in vivo evaluations (Table 5).

Table 5. Key components and considerations for hydrogel-based bone regeneration: cell sources, biomaterials, bioactive factors, functional assessments, and potential challenges.

Mesenchymal stem cells used in bone regeneration		**Human derived**	
	Bone marrow-derived mesenchymal stem cells (BMMSCs)	Human mesenchymal stem cells (hMSCs)	Human adipose-derived stem cells (hASCs)
	Urine-derived stem cells (USCs)	Nasal mucosa-derived ectodermal mesenchymal stem cells (EMSCs)	Dental pulp stem cells (DPSCs)
	Human umbilical vein endothelial cells (HUVECs).	Human amnion-derived stem cells (HAMSCs)	
		Animal derived	
	Murine dental pulp stem cells (mDPSCs)	Rat-derived mesenchymal stem cells (rMSCs)	Bone marrow derived from rabbit mesenchymal stem cells (rBMMSCs)
	Murine dental pulp stem cells (mDPSCs)		
Cell lines used for in vitro evaluations in bone regeneration		**Human derived**	
	Endothelial cells	MC3T3-E1 pre-osteoblasts	Human osteoblast (hFOB)
	Macrophages isolated from peripheral blood	Human dermal fibroblasts (HDFa)	Human osteosarcoma cell cultures (SaOS-2)
	Human vascular endothelial cells (HUVECs)	Osteosarcoma (MG-63)	
		Animal derived	
	Bone stromal cell cultures (W20-17)	Murine macrophages (RAW 264.7)	KUSA-A1 cells
Assays to evaluate the osteogenic capacities of Hg	Alkaline phosphatase (ALP) activity	μCT density	Quantitative real-time PCR
	Calcium depositions with Alizarin Red staining	Critical bone defects	MTT
	Live/dead staining	M1 and M2 phenotypes	Activating signaling pathways associated with Yes-Associated Protein (YAP)
	Fluorescence microscopy	Cellular migration	
Biomaterials to combine with HGs in bone regeneration	Konjac glucomannan	Salts such as calcium and sodium	Platelet-rich plasma (PRP)
	Oxidized chondroitin sulfate	Mesoporous bioactive glass (BAG) nanoparticles	Biphasic calcium phosphate
	β-tricalcium phosphate	Laponite	Amorphous calcium phosphate (ACP
	Magnesium phosphate cement (MPC)	Calcium phosphate particles	Curcumin
	Zeolitic imidazolate	Flavonoid nanoparticles	Bone powder
	Poly(L-lysine)	Zn	Cu
	Mg	Solid lipid nanoparticles	Hydroxyapatite

Table 5. *Cont.*

Growth factors and peptides used in bone regeneration	Bone morphogenetic protein (BMP-2) Transforming growth factor β3 (TGFβ3) Platelet is a potent source of growth factors like (PDGF) LL37	Bone morphogenetic protein (BMP-9) Basic fibroblast growth factor (bFGF) Osteogenic growth peptide (OGP) Peptide MP196	Parathyroid hormone-related peptide (PTHrP Stromal-derived growth factor (SDF) Arginine–glycine–aspartic acid (RGD Fluorenylmethoxycarbonyl-diphenylalanine
Possible negative effects of HGs on bone regeneration	Ectopic bone in muscle tissue and joints High-production-cost peptides and growth factors	Epithelial tissue overgrowth	Fibrotic encapsulation
Genes related to osteogenic capacities	Runxs 2, Bone Sialoprotein (BSP) Osterix (Osx)	Collagen type I (CoL-I) Osteocalcin (OCN) STRO-1 gene	BMP-4, Osteopontin (OPN) ALP

9.1. Growth Factors

Bone defects can be effectively regenerated through the application of scaffolds enriched with appropriate growth factors (frequently BMPs) [16,36,176]. This is attributed to the electrostatic interactions between the positively charged growth factors and the negatively charged HG, which facilitate the physical immobilization of the factors within the scaffold [36]. Growth factors serve as crucial agents in regulating cellular behavior and enhancing the functionality of newly formed bone tissues by promoting angiogenesis, thereby ensuring an adequate supply of oxygen and nutrients [15,36]. These factors play pivotal roles in chondrogenesis, osteogenesis, and vascularization and have demonstrated promising outcomes when integrated into HGs [15]. Conversely, the use of growth factors administered without HGs, typically through direct injections at the target site, has been associated with limitations such as reduced bioavailability, a short half-life, and adverse effects, including the formation of ectopic spindles and significant inflammation in surrounding tissues [78].

9.1.1. Bone Morphogenic Protein (BMP)

BMPs function as stimulatory agents for the differentiation of MSCs into various cell types, including chondrocytes, osteoblasts, fibroblasts, myocytes, and adipocytes [15,34,43]. This makes them a viable therapeutic option for enhancing local bone regeneration [158,160,172].

BMP-2 has been shown to yield positive outcomes in osteogenesis and the healing of spinal injuries; however, its use is contraindicated in patients with a history of cancer, as it has been associated with the promotion of cancer cell proliferation [9]. Clinical trials involving BMP-2-enriched HGs have demonstrated favorable outcomes in bone formation within nasopalatine clefts at a concentration of 250 μg/mL, particularly when compared to iliac crest autografts, as assessed via μCT six months post-grafting [180].

When BMP-2 is encapsulated in chitosan and PEG HGs, it ensures a sustained release profile while promoting bone regeneration in cranial defects in rats. Additionally, in vitro assays using BMSCs have demonstrated that HGs containing this growth factor enhance the expression of osteogenic differentiation-related genes, such as RUNX2, OPN, and OCN. This activity facilitates osteoblast differentiation, bone formation, and development, while also contributing to the regulation of bone metabolism, highlighting the therapeutic potential of BMP-2 functionalized HGs in bone tissue engineering [174].

In vitro assays have demonstrated the potential of bone morphogenetic protein 9 (BMP-9) as a potent inducer of osteogenic differentiation in MG63 cell cultures. Notably, BMP-9 also enhances its expression, amplifying its osteogenic effects. Furthermore, in vivo studies conducted on femoral bone defects in rabbits corroborate these findings, showing satisfactory bone regeneration. This outcome was assessed at 1, 4, 12, and 24 weeks post-treatment using μCT, highlighting the consistent and robust regenerative capabilities of BMP-9 in bone tissue engineering applications [181].

9.1.2. Platelet-Derived Growth Factor (PDGF)

PDGF is an FDA-approved therapy specifically utilized to promote bone regeneration [178]. Platelet is a potent source of growth factors like PDGF that can be extracted through centrifugation methods from blood treated with anticoagulants [79]. Both in vitro and in vivo studies demonstrate the efficacy of platelets and PDGF in promoting tissue regeneration. In a study by Chen et al., an HG enriched with salts such as calcium and sodium, combined with platelets, was utilized for in vitro analysis with rat-derived MSCs and in vivo analysis in Sprague Dawley rats. Viability assays revealed significant increases in cell count and proliferation within the HG compared to controls at 48 h. Additionally, enhanced alkaline phosphatase activity and expression of osteogenic markers such as OCN, OPN, and RUNX2 were observed. The in vivo findings further indicated statistically significant improvements in tissue density, thickness, and bioavailability of the HG [19].

Studies suggest that the use of PDGF positively impacts the regeneration of gaps induced in primary osteoblast cell cultures and MC3T3-E1 cells, promoting junction formation through the upregulation of connexin via the activation of p-Akt signaling, thereby highlighting its regenerative potential [182].

PDGF has also been utilized in microspheres of nanohydroxyapatite incorporated into gelatin methacryloyl (GelMA) scaffolds. The results demonstrated that it not only induced the recruitment of MSCs derived from rabbit bone marrow but also promoted the osteogenic differentiation of these stem cells, as confirmed through cell migration assays, ALP, and Alizarin Red staining [183].

Finally, the potential of PDGF is further corroborated through the use of biomimetic periosteum-bone scaffolds, where electrospun poly-L-lactic acid (PLLA) fibers are deposited on the surface of a gelatin/chitosan cryogel, with platelet-derived growth factor-BB (PDGF-BB) encapsulated within the core of PLLA nanofibers in a core–shell structure. The PDGF-loaded scaffold exerts a synergistic effect on osteogenesis, significantly accelerating bone healing. In vitro experiments demonstrated that the biomimetic periosteum-bone scaffolds exhibit favorable biocompatibility and osteogenic capacity. Additionally, in vivo experiments showed that the composite scaffold effectively and rapidly repaired rat cranial defects [5].

9.1.3. Fibroblast Growth Factor (FGF)

FGF are a class of proteins fundamental to the regulation of various cellular processes and tissue development [79]. In vivo studies in monkeys have demonstrated promising outcomes for cranial defect repair within 21 days post-implantation [177]. A comparable effect is observed when bFGF is used in dental pulp stem cells and silver-loaded titanium implants, demonstrating osteogenic differentiation. Furthermore, bFGF exhibits anti-inflammatory properties by inhibiting M1 macrophage activity [184]. In the context of alveolar bone regeneration and height augmentation in Wistar rats, FGF has demonstrated promising results in addressing bone defects. Moreover, FGF effectively inhibits epithelial tissue overgrowth in the affected area, thereby creating a favorable environment for bone formation. These outcomes position FGF as an ideal growth factor for periodontal tissue regeneration, where bone plays a critical role in restoring structural and functional integrity [173].

Although FGF does not directly participate in signaling pathways that promote bone regeneration, its role is crucial in initiating the healing process of such defects. When combined with BMP-2 in gelatin methacrylate HG, FGF significantly enhances bone regeneration outcomes. Additionally, FGF effectively stimulates the proliferation of BMSCs, a finding supported by robust in vitro assays, further highlighting its potential in bone tissue engineering applications [185]. Additionally, in vivo studies on mandibular defects in rats, evaluated using μCT, have demonstrated that the use of resorbable collagen membranes loaded with 0.5 μg and 2 μg of FGF-2 effectively promotes bone defect regeneration. The release of the growth factor is sustained for up to six days before gradually declining, highlighting its controlled release profile and its potential to facilitate bone tissue repair [175].

9.2. Peptides

Functional peptides have demonstrated efficacy in promoting bone regeneration and macrophage polarization to M1 and M2 phenotypes, an important point since M1 macrophages are involved in the beginnings of bone healing, promoting the inflammation and infection control necessary for these processes, and finally, M2 is involved in regeneration itself. However, their rapid degradation, physicochemical instability, and limited in vivo half-life remain significant limitations [186].

9.2.1. Arginine–Glycine–Aspartic Acid (RGD)

The primary ligand studied, RGD, is a peptide that binds to integrins. The covalent binding of RGD to alginate scaffolds significantly enhances bone formation. However, high concentrations of adhesive ligands, while promoting adhesion, can impede cellular migration; thus, optimizing adhesive properties is essential to maintain a balance that supports both adhesion and migration [15]. RGD peptides exhibit an affinity for integrins expressed by osteoblasts. Studies on dental implants loaded with these peptides report higher levels of newly formed bone tissue, suggesting their contribution to improved implant stability [187]. RGD peptides are frequently incorporated into scaffolds to enhance cell adhesion. The results demonstrate that the addition of RGD peptides to titanium alloy, PCL, and phosphate complexes significantly promotes bone repair and healing [188].

9.2.2. LL37

Cathelicidin, also known as LL-37, is a 37-amino-acid peptide with a cationic charge, extensively involved in the innate immune response, exhibiting both pro-inflammatory and anti-inflammatory activities depending on the specific tissue context [189]. However, LL-37 is associated with several limitations, including high production costs, reduced activity under physiological conditions, susceptibility to proteolytic degradation, and significant cytotoxicity in human cells [190].

The sequential release of peptide W9 following LL-37 release has been shown to further enhance the osteogenic differentiation of BMSCs, demonstrating a synergistic effect in promoting bone tissue regeneration [191]. Studies on cranial defects in rats have provided evidence of LL-37's efficacy in bone regeneration. During the initial phases of tissue repair, fibroblast-like cells have been observed, indicative of early regenerative activity. Moreover, LL-37 is associated with the recruitment of stem cells, as demonstrated by the expression of the STRO-1 gene, further supporting its role in bone tissue engineering applications [192]. Other works have also shown that the peptide promotes the proliferation and migration of BMSCs. Regarding osteogenic differentiation, the presence of the peptide (10 μg/mL) promotes the expression of related genes such as ALP and OCN [193].

Studies utilizing 3D scaffolds composed of silk fibroin, chitosan, nanohydroxyapatite, and the peptide LL-37 have demonstrated a reduction in bacterial load and subsequent bone regeneration in murine models. These findings are supported by in vivo RNA sequencing analyses that reveal gene expression related to bone remodeling, as well as assessments of ALP and calcium mineral staining with Alizarin Red [194].

9.2.3. Osteogenic Growth Factor Peptide (OGP)

Osteogenic growth factor peptide (OGP) is a tetradecapeptide [195] with a highly conserved 14-amino-acid motif [196]. In vitro studies indicate that OGP-enriched gelatin HGs can upregulate the expression of BMP-2, OCN, and OPN genes, as well as facilitate calcium salt deposition by murine MCT3T-E1 pre-osteoblasts [197].

10. Difficulties and Adverse Effects of HGs and Their Components

A variety of products for tissue engineering are approved by the FDA and EMA, underscoring the importance of HGs in this field [198]. Numerous studies emphasize the limitations and complications of current clinical methods for bone repair, including autografts and allografts [73]. Adverse effects, such as ectopic bone formation in muscle tissue, have been reported with BMP-2-enriched HG. Studies show that incorporating bioceramic agents like hydroxyapatite into gelatin-based HGs reduces the required BMP-2 concentration from 2–15 μg to 600 ng, facilitating a more controlled release profile [198]. Challenges also include the need for HGs to maintain the stability of incorporated molecules, such as growth factors or peptides, which rely on cross-linking structures with precisely tuned degradation properties to ensure controlled, sustained drug release [138]. Synthetic or non-biological HGs often provoke a localized innate immune response, beginning with nonspecific protein adsorption, followed by macrophage persistence, chronic inflammation, and eventual fibrous encapsulation of the material [154].

Additionally, the degradation properties of HGs play a crucial role in achieving effective bone regeneration. Ideally, the degradation rate should be perfectly synchronized with new bone formation. Materials such as gelatin methacryloyl GelMA HGs combined with BAG have shown promising results in degradation assays using collagenase-containing media within the first 24 h, significantly enhancing the functional properties of HGs [199]. Other materials that contribute to the degradation rate of HGs, in addition to BAG, include chitin whiskers. After 14 days, the porosity of the HG increases, becoming more elongated and partially collapsed. When exposed to enzymatic solutions for 21 days, the degradation rate reaches 19.78 ± 1.49%, supporting cellular differentiation and in vitro osteogenesis [200].

Another challenge associated with HGs is their mechanical properties.

A clear example of the limitations of hydrogels, particularly injectable formulations, is their inability to regenerate bone tissue in load-bearing sites due to their lack of compressive strength [201].

For instance, while gelatin methacryloyl (GelMA) HGs offer advantages in tissue engineering—such as cell culturing, drug delivery, and tissue regeneration—due to their similarity to the extracellular matrix (ECM), they exhibit low mechanical strength. This limitation can be addressed by incorporating materials like magnesium phosphate cement (MPC), which has demonstrated promising results in enhancing the viscoelastic properties of HGs. Studies evaluating different MPC concentrations (2.5%, 5%, and 7.5%) have shown significant improvements in mechanical performance [202].

11. Conclusions

HG-based scaffolds represent a transformative advancement in the field of bone tissue engineering, offering a synergistic combination of biocompatibility, biodegradability, and tunable properties. By mimicking the ECM and providing a supportive environment for cellular adhesion, proliferation, and differentiation, these scaffolds facilitate the regeneration of bone tissue. The incorporation of natural polymers, such as collagen and chitosan, and synthetic polymers, like PEG and PCL, enhances the mechanical and bioactive characteristics of these materials, optimizing their performance in regenerative therapies. Furthermore, the functionalization of HGs with bioactive molecules, such as growth factors and peptides, amplifies their osteoconductive and osteoinductive capacities, driving superior outcomes in bone repair and regeneration. Despite their promise, ongoing research is essential to address challenges such as mechanical limitations and variability in degradation rates, ensuring the long-term efficacy of these innovative biomaterials. This approach holds significant potential for addressing the growing global demand for effective and minimally invasive treatments for bone defects.

12. Future Perspective

Tissue engineering has achieved remarkable progress in regenerative therapies, evolving from the use of natural materials capable of inducing bone growth to synthetic materials with tunable mechanical properties and reduced risks of rejection. Despite these advancements, significant challenges remain in bridging the gap between experimental outcomes and clinical trials.

While the mechanical properties of synthetic materials have improved considerably, critical fractures in long bones often necessitate rigid fixation devices to ensure stability. Although minimally invasive surgical techniques have advanced significantly in accelerating bone regeneration, certain cases still require pre-formed scaffolds specifically designed to match the topographical and anatomical characteristics of the defect.

The integration of bioprinting and bio-ink technologies, incorporating cells and growth factors [203], represents a key area of focus in tissue regeneration research within the biomedical sciences. These innovative approaches offer the potential to provide the structural rigidity necessary to emulate the properties of bone [204] while simultaneously reducing the dependence on expensive fixation materials. As a result, these strategies hold significant promise, warranting further development to enhance treatments for bone degeneration and improve patient outcomes. When used with 3D bioprinting technology in combination with stimuli-responsive bioinks, this approach is referred to as 4D bioprinting [205].

Finally, it is important to highlight smart HGs, which are designed to respond to external stimuli such as temperature, pH, electric and magnetic fields, ionic strength, and enzymatic environments. Their key characteristics include controlled drug delivery, biomedical applications, biosensors, and their use in tissue engineering and regenerative medicine [50].

Author Contributions: J.L.C.Q., E.M.A.-M., R.R.-P. and M.B. conceived and designed the content of this review and wrote the paper. J.G.R.-Q. and A.A.-H. contributed to the final version of the manuscript. All authors have read and agreed to the published version of the manuscript.

Funding: This research received no external funding.

Data Availability Statement: No new data were created or analyzed in this study. Data sharing is not applicable to this article.

Conflicts of Interest: The authors declare no conflicts of interest.

References

1. El-Husseiny, H.M.; Mady, E.A.; Hamabe, L.; Abugomaa, A.; Shimada, K.; Yoshida, T.; Tanaka, T.; Yokoi, A.; Elbadawy, M.; Tanaka, R. Smart/stimuli-responsive hydrogels: Cutting-edge platforms for tissue engineering and other biomedical applications. *Mater. Today Bio* **2022**, *13*, 100186. [CrossRef]

2. Kamarehei, F.; Saleh, G.N. Tissue regeneration therapy by Nano composite scaffolds based on PLGA hydrogel embedded with human dental pulp stem cells: A systematic review. *Am. J. Transl. Res.* **2024**, *16*, 1521–1530. [CrossRef] [PubMed]
3. Aravamudhan, A.; Ramos, D.M.; Nip, J.; Harmon, M.D.; James, R.; Deng, M.; Laurencin, C.T.; Yu, X.; Kumbar, S.G. Cellulose and collagen derived micro-nano structured scaffolds for bone tissue engineering. *J. Biomed. Nanotechnol.* **2013**, *9*, 719–731. [CrossRef] [PubMed]
4. Zöller, K.; To, D.; Bernkop-Schnürch, A. Biomedical applications of functional hydrogels: Innovative developments, relevant clinical trials and advanced products. *Biomaterials* **2025**, *312*, 122718. [CrossRef]
5. Zhan, Z.; Li, R.; Wu, Y.; Shen, X.; Fu, D.; Han, H.; Jing, P.; Li, B.; Han, F.; Meng, B. Biomimetic periosteum-bone scaffolds with codelivery of BMP-2 and PDGF-BB for skull repair. *Bone* **2025**, *190*, 117315. [CrossRef]
6. Stowers, R.S. Advances in Extracellular Matrix-Mimetic Hydrogels to Guide Stem Cell Fate. *Cells Tissues Organs* **2022**, *211*, 703–720. [CrossRef] [PubMed]
7. Capitelli-McMahon, H.; Kahlar, N.; Rahman, S. Titanium Versus Autologous Bone-Based Cranioplasty: A Systematic Review and Meta-Analysis. *Cureus* **2023**, *15*, e39516. [CrossRef] [PubMed]
8. Ferraz, M.P. Bone Grafts in Dental Medicine: An Overview of Autografts, Allografts and Synthetic Materials. *Materials* **2023**, *16*, 4117. [CrossRef] [PubMed]
9. White, K.A.; Olabisi, R.M. Spatiotemporal Control Strategies for Bone Formation through Tissue Engineering and Regenerative Medicine Approaches. *Adv. Healthc. Mater.* **2019**, *8*, e1801044. [CrossRef] [PubMed]
10. Matichescu, A.; Ardelean, L.C.; Rusu, L.C.; Craciun, D.; Bratu, E.A.; Babucea, M.; Leretter, M. Advanced Biomaterials and Techniques for Oral Tissue Engineering and Regeneration—A Review. *Materials* **2020**, *13*, 5303. [CrossRef] [PubMed]
11. Zumbo, B.; Guagnini, B.; Medagli, B.; Porrelli, D.; Turco, G. Fibronectin Functionalization: A Way to Enhance Dynamic Cell Culture on Alginate/Hydroxyapatite Scaffolds. *J. Funct. Biomater.* **2024**, *15*, 222. [CrossRef] [PubMed]
12. VK, A.D.; Udduttula, A.; Jaiswal, A.K. Unveiling the secrets of marine-derived fucoidan for bone tissue engineering—A review. *Front. Bioeng. Biotechnol.* **2022**, *10*, 1100164.
13. Dong, Q.; Zhang, M.; Zhou, X.; Shao, Y.; Li, J.; Wang, L.; Chu, C.; Xue, F.; Yao, Q.; Bai, J. 3D-printed Mg-incorporated PCL-based scaffolds: A promising approach for bone healing. *Mater. Sci. Eng. C Mater. Biol. Appl.* **2021**, *129*, 112372. [CrossRef] [PubMed]
14. Mikos, A.G.; Herring, S.W.; Ochareon, P.; Elisseeff, J.; Lu, H.H.; Kandel, R.; Schoen, F.J.; Toner, M.; Mooney, D.; Atala, A.; et al. Engineering complex tissues. *Tissue Eng.* **2006**, *12*, 3307–3339. [CrossRef] [PubMed]
15. Fedorovich, N.E.; Alblas, J.; de Wijn, J.R.; Hennink, W.E.; Verbout, A.J.; Dhert, W.J. Hydrogels as extracellular matrices for skeletal tissue engineering: State-of-the-art and novel application in organ printing. *Tissue Eng.* **2007**, *13*, 1905–1925. [CrossRef] [PubMed]
16. Lee, S.H.; Shin, H. Matrices and scaffolds for delivery of bioactive molecules in bone and cartilage tissue engineering. *Adv. Drug Deliv. Rev.* **2007**, *59*, 339–359. [CrossRef]
17. Fatimi, A.; Okoro, O.V.; Podstawczyk, D.; Siminska-Stanny, J.; Shavandi, A. Natural Hydrogel-Based Bio-Inks for 3D Bioprinting in Tissue Engineering: A Review. *Gels* **2022**, *8*, 179. [CrossRef] [PubMed]
18. Ding, Q.; Zhang, S.; Liu, X.; Zhao, Y.; Yang, J.; Chai, G.; Wang, N.; Ma, S.; Liu, W.; Ding, C. Hydrogel Tissue Bioengineered Scaffolds in Bone Repair: A Review. *Molecules* **2023**, *28*, 7039. [CrossRef]
19. Chen, X.; Yan, J.; Jiang, Y.; Fan, Y.; Ying, Z.; Tan, S.; Zhou, Z.; Liu, J.; Chen, F.; He, S. Platelet-Activating Biominerals Enhanced Injectable Hydrogels With Superior Bioactivity for Bone Regeneration. *Front. Bioeng. Biotechnol.* **2022**, *10*, 826855. [CrossRef] [PubMed]
20. Munarin, F.; Petrini, P.; Bozzini, S.; Tanzi, M.C. New perspectives in cell delivery systems for tissue regeneration: Natural-derived injectable hydrogels. *J. Appl. Biomater. Funct. Mater.* **2012**, *10*, 67–81. [CrossRef]
21. Zhao, L.; Zhou, Y.; Zhang, J.; Liang, H.; Chen, X.; Tan, H. Natural Polymer-Based Hydrogels: From Polymer to Biomedical Applications. *Pharmaceutics* **2023**, *15*, 2514. [CrossRef] [PubMed]
22. Samiei, M.; Dalir Abdollahinia, E.; Amiryaghoubi, N.; Fathi, M.; Barar, J.; Omidi, Y. Injectable thermosensitive chitosan/gelatin hydrogel for dental pulp stem cells proliferation and differentiation. *Bioimpacts* **2023**, *13*, 63–72. [CrossRef] [PubMed]
23. Bashir, M.H.; Korany, N.S.; Farag, D.B.E.; Abbass, M.M.S.; Ezzat, B.A.; Hegazy, R.H.; Dörfer, C.E.; Fawzy El-Sayed, K.M. Polymeric Nanocomposite Hydrogel Scaffolds in Craniofacial Bone Regeneration: A Comprehensive Review. *Biomolecules* **2023**, *13*, 205. [CrossRef] [PubMed]
24. Whulanza, Y.; Azadi, A.; Supriadi, S.; Rahman, S.F.; Chalid, M.; Irsyad, M.; Nadhif, M.H.; Kreshanti, P. Tailoring mechanical properties and degradation rate of maxillofacial implant based on sago starch/polylactid acid blend. *Heliyon* **2022**, *8*, e08600. [CrossRef] [PubMed]
25. Åkerlund, E.; Diez-Escudero, A.; Grzeszczak, A.; Persson, C. The Effect of PCL Addition on 3D-Printable PLA/HA Composite Filaments for the Treatment of Bone Defects. *Polymers* **2022**, *14*, 3305. [CrossRef] [PubMed]
26. Amirazad, H.; Dadashpour, M.; Zarghami, N. Application of decellularized bone matrix as a bioscaffold in bone tissue engineering. *J. Biol. Eng.* **2022**, *16*, 1. [CrossRef] [PubMed]

27. Hu, D.; Ren, Q.; Li, Z.; Zhang, L. Chitosan-Based Biomimetically Mineralized Composite Materials in Human Hard Tissue Repair. *Molecules* **2020**, *25*, 4785. [CrossRef] [PubMed]
28. Hadjidakis, D.J.; Androulakis, I.I. Bone remodeling. *Ann. N. Y. Acad. Sci.* **2006**, *1092*, 385–396. [CrossRef] [PubMed]
29. Duda, G.N.; Geissler, S.; Checa, S.; Tsitsilonis, S.; Petersen, A.; Schmidt-Bleek, K. The decisive early phase of bone regeneration. *Nat. Rev. Rheumatol.* **2023**, *19*, 78–95. [CrossRef]
30. Zhu, M.; Fan, Z. The role of the Wnt signalling pathway in the energy metabolism of bone remodelling. *Cell Prolif.* **2022**, *55*, e13309. [CrossRef] [PubMed]
31. Kenkre, J.S.; Bassett, J. The bone remodelling cycle. *Ann. Clin. Biochem.* **2018**, *55*, 308–327. [CrossRef] [PubMed]
32. Qiu, P.; Li, M.; Chen, K.; Fang, B.; Chen, P.; Tang, Z.; Lin, X.; Fan, S. Periosteal matrix-derived hydrogel promotes bone repair through an early immune regulation coupled with enhanced angio- and osteogenesis. *Biomaterials* **2020**, *227*, 119552. [CrossRef]
33. Ledda, M.; Merco, M.; Sciortino, A.; Scatena, E.; Convertino, A.; Lisi, A.; Del Gaudio, C. Biological Response to Bioinspired Microporous 3D-Printed Scaffolds for Bone Tissue Engineering. *Int. J. Mol. Sci.* **2022**, *23*, 5383. [CrossRef] [PubMed]
34. Bahney, C.S.; Zondervan, R.L.; Allison, P.; Theologis, A.; Ashley, J.W.; Ahn, J.; Miclau, T.; Marcucio, R.S.; Hankenson, K.D. Cellular biology of fracture healing. *J. Orthop. Res.* **2019**, *37*, 35–50. [CrossRef] [PubMed]
35. Roumani, S.; Jeanneau, C.; Giraud, T.; Cotten, A.; Laucournet, M.; Sohier, J.; Pithioux, M.; About, I. Osteogenic Potential of a Polyethylene Glycol Hydrogel Functionalized with Poly-Lysine Dendrigrafts (DGL) for Bone Regeneration. *Materials* **2023**, *16*, 862. [CrossRef]
36. Tabata, Y. Tissue regeneration based on growth factor release. *Tissue Eng.* **2003**, *9* (Suppl. 1), S5–S15. [CrossRef]
37. Wang, J.Z.; You, M.L.; Ding, Z.Q.; Ye, W.B. A review of emerging bone tissue engineering via PEG conjugated biodegradable amphiphilic copolymers. *Mater. Sci. Eng. C Mater. Biol. Appl.* **2019**, *97*, 1021–1035. [CrossRef] [PubMed]
38. Klosterhoff, B.S.; Vantucci, C.E.; Kaiser, J.; Ong, K.G.; Wood, L.B.; Weiss, J.A.; Guldberg, R.E.; Willett, N.J. Effects of osteogenic ambulatory mechanical stimulation on early stages of BMP-2 mediated bone repair. *Connect. Tissue Res.* **2022**, *63*, 16–27. [CrossRef]
39. Global, regional, and national burden of bone fractures in 204 countries and territories, 1990–2019: A systematic analysis from the Global Burden of Disease Study 2019. *Lancet Healthy Longev.* **2021**, *2*, e580–e592. [CrossRef]
40. Bernardo, M.P.; da Silva, B.C.R.; Hamouda, A.E.I.; de Toledo, M.A.S.; Schalla, C.; Rütten, S.; Goetzke, R.; Mattoso, L.H.C.; Zenke, M.; Sechi, A. PLA/Hydroxyapatite scaffolds exhibit in vitro immunological inertness and promote robust osteogenic differentiation of human mesenchymal stem cells without osteogenic stimuli. *Sci. Rep.* **2022**, *12*, 2333. [CrossRef]
41. Townsend, J.M.; Kiyotake, E.A.; Easley, J.; Seim, H.B., III; Stewart, H.L.; Fung, K.M.; Detamore, M.S. Comparison of a Thiolated Demineralized Bone Matrix Hydrogel to a Clinical Product Control for Regeneration of Large Sheep Cranial Defects. *Materialia* **2023**, *27*, 101690. [CrossRef] [PubMed]
42. Calcei, J.G.; Rodeo, S.A. Orthobiologics for Bone Healing. *Clin. Sports Med.* **2019**, *38*, 79–95. [CrossRef] [PubMed]
43. Rodríguez-Merchán, E.C. Bone Healing Materials in the Treatment of Recalcitrant Nonunions and Bone Defects. *Int. J. Mol. Sci.* **2022**, *23*, 3352. [CrossRef]
44. Joshi, J.; Homburg, S.V.; Ehrmann, A. Atomic Force Microscopy (AFM) on Biopolymers and Hydrogels for Biotechnological Applications-Possibilities and Limits. *Polymers* **2022**, *14*, 1267. [CrossRef] [PubMed]
45. DiStefano, T.J.; Shmukler, J.O.; Danias, G.; Iatridis, J.C. The Functional Role of Interface Tissue Engineering in Annulus Fibrosus Repair: Bridging Mechanisms of Hydrogel Integration with Regenerative Outcomes. *ACS Biomater. Sci. Eng.* **2020**, *6*, 6556–6586. [CrossRef]
46. Madhusudanan, P.; Raju, G.; Shankarappa, S. Hydrogel systems and their role in neural tissue engineering. *J. R. Soc. Interface* **2020**, *17*, 20190505. [CrossRef]
47. Parimala Chelvi Ratnamani, M.; Zhang, X.; Wang, H. A Comprehensive Assessment on the Pivotal Role of Hydrogels in Scaffold-Based Bioprinting. *Gels* **2022**, *8*, 239. [CrossRef]
48. Zhou, B.; Jiang, X.; Zhou, X.; Tan, W.; Luo, H.; Lei, S.; Yang, Y. GelMA-based bioactive hydrogel scaffolds with multiple bone defect repair functions: Therapeutic strategies and recent advances. *Biomater. Res.* **2023**, *27*, 86. [CrossRef]
49. Qi, J.; Wu, H.; Liu, G. Novel Strategies for Spatiotemporal and Controlled BMP-2 Delivery in Bone Tissue Engineering. *Cell Transplant.* **2024**, *33*, 9636897241276733. [CrossRef] [PubMed]
50. Mantha, S.; Pillai, S.; Khayambashi, P.; Upadhyay, A.; Zhang, Y.; Tao, O.; Pham, H.M.; Tran, S.D. Smart Hydrogels in Tissue Engineering and Regenerative Medicine. *Materials* **2019**, *12*, 3323. [CrossRef] [PubMed]
51. Zhang, Y.; Xu, C.; Huang, Y.; Tan, D.; Luo, W.; Zhang, Y.; Tan, Y. Establishment of immortalized rabbit bone marrow mesenchymal stem cells and a preliminary study of their osteogenic differentiation capability. *Anim. Model. Exp. Med.* **2024**, *7*, 824–834. [CrossRef] [PubMed]
52. Yang, J.; Zhang, X.; Lu, B.; Mei, J.; Xu, L.; Zhang, X.; Su, Z.; Xu, W.; Fang, S.; Zhu, C.; et al. Inflammation-Responsive Hydrogel Spray for Synergistic Prevention of Traumatic Heterotopic Ossification via Dual-Homeostatic Modulation Strategy. *Adv. Sci.* **2023**, *10*, e2302905. [CrossRef] [PubMed]

53. Zhuo, H.; Zhang, S.; Wang, H.; Deng, J.; Zhang, X. Gelatin methacryloyl @MP196/exos hydrogel induced neutrophil apoptosis and macrophage M2 polarization to inhibit periodontal bone loss. *Colloids Surf. B Biointerfaces* **2025**, *248*, 114466. [CrossRef] [PubMed]
54. Zhao, T.; Chen, L.; Yu, C.; He, G.; Lin, H.; Sang, H.; Chen, Z.; Hong, Y.; Sui, W.; Zhao, J. Effect of injectable calcium alginate-amelogenin hydrogel on macrophage polarization and promotion of jawbone osteogenesis. *RSC Adv.* **2024**, *14*, 2016–2026. [CrossRef]
55. Zhou, X.; Cao, W.; Chen, Y.; Zhu, Z.; Chen, Y.; Ni, Y.; Liu, Z.; Jia, F.; Lu, Z.; Ye, Y.; et al. Poly(Glutamic Acid-Lysine) Hydrogels with Alternating Sequence Resist the Foreign Body Response in Rodents and Non-Human Primates. *Adv. Sci.* **2024**, *11*, e2308077. [CrossRef]
56. Zuev, Y.F.; Derkach, S.R.; Lunev, I.V.; Nikiforova, A.A.; Klimovitskaya, M.A.; Bogdanova, L.R.; Skvortsova, P.V.; Kurbanov, R.K.; Kazantseva, M.A.; Zueva, O.S. Water as a Structural Marker in Gelatin Hydrogels with Different Cross-Linking Nature. *Int. J. Mol. Sci.* **2024**, *25*, 11738. [CrossRef]
57. Li, P.; Chen, S.; Meng, Y.; Wang, C.; Ni, X. Simple Preparation and Bone Regeneration Effects of Poly(vinyl alcohol)-Resveratrol Self-Cross-Linked Hydrogels. *ACS Omega* **2024**, *9*, 49043–49053. [CrossRef]
58. Zhu, M.; Wang, Q.; Gu, T.; Han, Y.; Zeng, X.; Li, J.; Dong, J.; Huang, H.; Qian, P. Hydrogel-based microenvironment engineering of haematopoietic stem cells. *Cell Mol. Life Sci.* **2023**, *80*, 49. [CrossRef]
59. Xu, D.; Hernandez Miranda, M.L.; Evans, N.D.; Sengers, B.G.; Browne, M.; Cook, R.B. Depth profiling via nanoindentation for characterisation of the elastic modulus and hydraulic properties of thin hydrogel layers. *J. Mech. Behav. Biomed. Mater.* **2023**, *148*, 106195. [CrossRef]
60. Wu, Y.; Li, X.; Sun, Y.; Tan, X.; Wang, C.; Wang, Z.; Ye, L. Multiscale design of stiffening and ROS scavenging hydrogels for the augmentation of mandibular bone regeneration. *Bioact. Mater.* **2023**, *20*, 111–125. [CrossRef]
61. Wang, Z.; Zhao, Y.; Bai, H.; Chang, F.; Yang, X.; Wang, X.; Liu, J.; Wu, M.; Lin, Q.; Wang, J.; et al. Bioactive prosthesis interface compositing variable-stiffness hydrogels regulates stem cells fates to facilitate osseointegration through mechanotransduction. *Int. J. Biol. Macromol.* **2024**, *259*, 129073. [CrossRef]
62. Xue, J.; Gurav, N.; Elsharkawy, S.; Deb, S. Hydrogel Composite Magnetic Scaffolds: Toward Cell-Free In Situ Bone Tissue Engineering. *ACS Appl. Bio Mater.* **2024**, *7*, 168–181. [CrossRef]
63. Arpornmaeklong, P.; Jaiman, N.; Apinyauppatham, K.; Fuongfuchat, A.; Boonyuen, S. Effects of Calcium Carbonate Microcapsules and Nanohydroxyapatite on Properties of Thermosensitive Chitosan/Collagen Hydrogels. *Polymers* **2023**, *15*, 416. [CrossRef]
64. Abudukelimu, K.; Aierken, A.; Tuerxuntayi, A.; Yilihamu, Y.; Abulizi, S.; Wufuer, D.; Dong, H. Preliminary study on the preparation of antler powder/chitosan/β-glycerophosphate sodium/polyvinyl alcohol porous hydrogel scaffolds and their osteogenic effects. *Front. Bioeng. Biotechnol.* **2024**, *12*, 1421718. [CrossRef]
65. Zhu, Y.; Liu, H.; Wu, P.; Chen, Y.; Deng, Z.; Cai, L.; Wu, M. Multifunctional injectable hydrogel system as a mild photothermal-assisted therapeutic platform for programmed regulation of inflammation and osteo-microenvironment for enhanced healing of diabetic bone defects in situ. *Theranostics* **2024**, *14*, 7140–7198. [CrossRef]
66. Wang, X.; Zeng, J.; Gan, D.; Ling, K.; He, M.; Li, J.; Lu, Y. Recent Strategies and Advances in Hydrogel-Based Delivery Platforms for Bone Regeneration. *Nanomicro Lett.* **2024**, *17*, 73. [CrossRef]
67. Zhu, W.; Chen, R.; Wang, W.; Liu, Y.; Shi, C.; Tang, S.; Tang, G. Fabrication of Naturally Derived Double-Network Hydrogels With a Sustained Aspirin Release System for Facilitating Bone Regeneration. *Front. Chem.* **2022**, *10*, 874985. [CrossRef]
68. Zhu, L.; Hou, Q.; Yan, M.; Gao, W.; Tang, G.; Liu, Z. Flexible Fabrication and Hybridization of Bioactive Hydrogels with Robust Osteogenic Potency. *Pharmaceutics* **2023**, *15*, 2384. [CrossRef]
69. Zheng, L.; Zhou, D.; Ju, F.; Liu, Z.; Yan, C.; Dong, Z.; Chen, S.; Deng, L.; Chan, S.; Deng, J.; et al. Oscillating Fluid Flow Activated Osteocyte Lysate-Based Hydrogel for Regulating Osteoblast/Osteoclast Homeostasis to Enhance Bone Repair. *Adv. Sci.* **2023**, *10*, e2204592. [CrossRef]
70. Zhang, W.; Sun, T.; Zhang, J.; Hu, X.; Yang, M.; Han, L.; Xu, G.; Zhao, Y.; Li, Z. Construction of artificial periosteum with methacrylamide gelatin hydrogel-wharton's jelly based on stem cell recruitment and its application in bone tissue engineering. *Mater. Today Bio* **2023**, *18*, 100528. [CrossRef]
71. Soleymani, H.; Moghaddam, M.M.; Naderi-Manesh, H.; Taheri, R.A. Single-layer graphene oxide nanosheets induce proliferation and Osteogenesis of single-cell hBMSCs encapsulated in Alginate Microgels. *Sci. Rep.* **2024**, *14*, 25272. [CrossRef]
72. Bu, W.; Wu, Y.; Ghaemmaghami, A.M.; Sun, H.; Mata, A. Rational design of hydrogels for immunomodulation. *Regen. Biomater.* **2022**, *9*, rbac009. [CrossRef]
73. Bektas, C.; Mao, Y. Hydrogel Microparticles for Bone Regeneration. *Gels* **2023**, *10*, 28. [CrossRef]
74. Sood, A.; Ji, S.M.; Kumar, A.; Han, S.S. Enzyme-Triggered Crosslinked Hybrid Hydrogels for Bone Tissue Engineering. *Materials* **2022**, *15*, 6383. [CrossRef]

75. Song, J.; Michas, C.; Chen, C.S.; White, A.E.; Grinstaff, M.W. From Simple to Architecturally Complex Hydrogel Scaffolds for Cell and Tissue Engineering Applications: Opportunities Presented by Two-Photon Polymerization. *Adv. Healthc. Mater.* **2020**, *9*, e1901217. [CrossRef]
76. Wu, D.T.; Munguia-Lopez, J.G.; Cho, Y.W.; Ma, X.; Song, V.; Zhu, Z.; Tran, S.D. Polymeric Scaffolds for Dental, Oral, and Craniofacial Regenerative Medicine. *Molecules* **2021**, *26*, 7043. [CrossRef]
77. Koolen, M.; Longoni, A.; van der Stok, J.; Van der Jagt, O.; Gawlitta, D.; Weinans, H. Complete regeneration of large bone defects in rats with commercially available fibrin loaded with BMP-2. *Eur. Cell Mater.* **2019**, *38*, 94–105. [CrossRef]
78. Lyu, Y.; Liu, Y.; He, H.; Wang, H. Application of Silk-Fibroin-Based Hydrogels in Tissue Engineering. *Gels* **2023**, *9*, 431. [CrossRef]
79. Alturki, A.M. Rationally design of electrospun polysaccharides polymeric nanofiber webs by various tools for biomedical applications: A review. *Int. J. Biol. Macromol.* **2021**, *184*, 648–665. [CrossRef]
80. Gegg, C.; Yang, F. Spatially patterned microribbon-based hydrogels induce zonally-organized cartilage regeneration by stem cells in 3D. *Acta Biomater.* **2020**, *101*, 196–205. [CrossRef]
81. Datta, S.; Rameshbabu, A.P.; Bankoti, K.; Roy, M.; Gupta, C.; Jana, S.; Das, A.K.; Sen, R.; Dhara, S. Decellularized bone matrix/oleoyl chitosan derived supramolecular injectable hydrogel promotes efficient bone integration. *Mater. Sci. Eng. C Mater. Biol. Appl.* **2021**, *119*, 111604. [CrossRef]
82. Prinelli, A.; Silva-Almeida, C.; Parks, S.; Pasotti, A.; Telopoulou, A.; Dunlop, S.; Sutherland, E.; Lynch, M.; Ewart, M.A.; Wilde, C.J.; et al. In-Plate Cryopreservation of 2D and 3D Cell Models: Innovative Tools for Biomedical Research and Preclinical Drug Discovery. *SLAS Discov.* **2021**, *26*, 32–43. [CrossRef]
83. Mendes, B.B.; Daly, A.C.; Reis, R.L.; Domingues, R.M.A.; Gomes, M.E.; Burdick, J.A. Injectable hyaluronic acid and platelet lysate-derived granular hydrogels for biomedical applications. *Acta Biomater.* **2021**, *119*, 101–113. [CrossRef]
84. Benmassaoud, M.M.; Gultian, K.A.; DiCerbo, M.; Vega, S.L. Hydrogel screening approaches for bone and cartilage tissue regeneration. *Ann. N. Y. Acad. Sci.* **2020**, *1460*, 25–42. [CrossRef]
85. Yu, X.; Deng, Z.; Li, H.; Ma, Y.; Zheng, Q. In situ fabrication of an anisotropic double-layer hydrogel as a bio-scaffold for repairing articular cartilage and subchondral bone injuries. *RSC Adv.* **2023**, *13*, 34958–34971. [CrossRef]
86. Ge, Y.; Li, Y.; Wang, Z.; Li, L.; Teng, H.; Jiang, Q. Effects of Mechanical Compression on Chondrogenesis of Human Synovium-Derived Mesenchymal Stem Cells in Agarose Hydrogel. *Front. Bioeng. Biotechnol.* **2021**, *9*, 697281. [CrossRef]
87. Naudot, M.; Davrou, J.; Djebara, A.E.; Barre, A.; Lavagen, N.; Lardière, S.; Azdad, S.Z.; Zabijak, L.; Lack, S.; Devauchelle, B.; et al. Functional Validation of a New Alginate-based Hydrogel Scaffold Combined with Mesenchymal Stem Cells in a Rat Hard Palate Cleft Model. *Plast. Reconstr. Surg. Glob. Open* **2020**, *8*, e2743. [CrossRef]
88. Barre, A.; Naudot, M.; Colin, F.; Sevestre, H.; Collet, L.; Devauchelle, B.; Lack, S.; Marolleau, J.P.; Le Ricousse, S. An Alginate-Based Hydrogel with a High Angiogenic Capacity and a High Osteogenic Potential. *Biores Open Access* **2020**, *9*, 174–182. [CrossRef]
89. Kim, J.; Choi, Y.J.; Gal, C.W.; Sung, A.; Park, H.; Yun, H.S. Development of an alginate-gelatin bioink enhancing osteogenic differentiation by gelatin release. *Int. J. Bioprint* **2023**, *9*, 660. [CrossRef]
90. Shi, P.; Laude, A.; Yeong, W.Y. Investigation of cell viability and morphology in 3D bio-printed alginate constructs with tunable stiffness. *J. Biomed. Mater. Res. A* **2017**, *105*, 1009–1018. [CrossRef]
91. Gibbs, D.M.; Black, C.R.; Dawson, J.I.; Oreffo, R.O. A review of hydrogel use in fracture healing and bone regeneration. *J. Tissue Eng. Regen. Med.* **2016**, *10*, 187–198. [CrossRef] [PubMed]
92. Zhao, F.; Yang, Z.; Xiong, H.; Yan, Y.; Chen, X.; Shao, L. A bioactive glass functional hydrogel enhances bone augmentation via synergistic angiogenesis, self-swelling and osteogenesis. *Bioact. Mater.* **2023**, *22*, 201–210. [CrossRef] [PubMed]
93. Halperin-Sternfeld, M.; Pokhojaev, A.; Ghosh, M.; Rachmiel, D.; Kannan, R.; Grinberg, I.; Asher, M.; Aviv, M.; Ma, P.X.; Binderman, I.; et al. Immunomodulatory fibrous hyaluronic acid-Fmoc-diphenylalanine-based hydrogel induces bone regeneration. *J. Clin. Periodontol.* **2023**, *50*, 200–219. [CrossRef] [PubMed]
94. Man, K.; Brunet, M.Y.; Federici, A.S.; Hoey, D.A.; Cox, S.C. An ECM-Mimetic Hydrogel to Promote the Therapeutic Efficacy of Osteoblast-Derived Extracellular Vesicles for Bone Regeneration. *Front. Bioeng. Biotechnol.* **2022**, *10*, 829969. [CrossRef] [PubMed]
95. Zhang, L.; He, G.; Yu, Y.; Zhang, Y.; Li, X.; Wang, S. Design of Biocompatible Chitosan/Polyaniline/Laponite Hydrogel with Photothermal Conversion Capability. *Biomolecules* **2022**, *12*, 1089. [CrossRef]
96. Kim, B.S.; Choi, J.S.; Kim, J.D.; Yeo, T.Y.; Cho, Y.W. Improvement of stem cell viability in hyaluronic acid hydrogels using dextran microspheres. *J. Biomater. Sci. Polym. Ed.* **2010**, *21*, 1701–1711. [CrossRef] [PubMed]
97. Sartore, L.; Manferdini, C.; Saleh, Y.; Dey, K.; Gabusi, E.; Ramorino, G.; Zini, N.; Almici, C.; Re, F.; Russo, D.; et al. Polysaccharides on gelatin-based hydrogels differently affect chondrogenic differentiation of human mesenchymal stromal cells. *Mater. Sci. Eng. C Mater. Biol. Appl.* **2021**, *126*, 112175. [CrossRef] [PubMed]
98. Re, F.; Sartore, L.; Borsani, E.; Ferroni, M.; Baratto, C.; Mahajneh, A.; Smith, A.; Dey, K.; Almici, C.; Guizzi, P.; et al. Mineralization of 3D Osteogenic Model Based on Gelatin-Dextran Hybrid Hydrogel Scaffold Bioengineered with Mesenchymal Stromal Cells: A Multiparametric Evaluation. *Materials* **2021**, *14*, 3852. [CrossRef] [PubMed]

99. Nikpour, P.; Salimi-Kenari, H.; Rabiee, S.M. Biological and bioactivity assessment of dextran nanocomposite hydrogel for bone regeneration. *Prog. Biomater.* **2021**, *10*, 271–280. [CrossRef]
100. Alves, P.; Simão, A.F.; Graça, M.F.P.; Mariz, M.J.; Correia, I.J.; Ferreira, P. Dextran-Based Injectable Hydrogel Composites for Bone Regeneration. *Polymers* **2023**, *15*, 4501. [CrossRef] [PubMed]
101. Yucel, T.; Cebe, P.; Kaplan, D.L. Vortex-induced injectable silk fibroin hydrogels. *Biophys. J.* **2009**, *97*, 2044–2050. [CrossRef] [PubMed]
102. Sun, W.; Gregory, D.A.; Tomeh, M.A.; Zhao, X. Silk Fibroin as a Functional Biomaterial for Tissue Engineering. *Int. J. Mol. Sci.* **2021**, *22*, 1499. [CrossRef] [PubMed]
103. Li, G.; Sun, S. Silk Fibroin-Based Biomaterials for Tissue Engineering Applications. *Molecules* **2022**, *27*, 2757. [CrossRef] [PubMed]
104. Wosicka-Frąckowiak, H.; Poniedziałek, K.; Woźny, S.; Kuprianowicz, M.; Nyga, M.; Jadach, B.; Milanowski, B. Collagen and Its Derivatives Serving Biomedical Purposes: A Review. *Polymers* **2024**, *16*, 2668. [CrossRef] [PubMed]
105. Zhao, P.; Yang, F.; Jia, X.; Xiao, Y.; Hua, C.; Xing, M.; Lyu, G. Extracellular Matrices as Bioactive Materials for In Situ Tissue Regeneration. *Pharmaceutics* **2023**, *15*, 2771. [CrossRef] [PubMed]
106. Miranda-Nieves, D.; Chaikof, E.L. Collagen and Elastin Biomaterials for the Fabrication of Engineered Living Tissues. *ACS Biomater. Sci. Eng.* **2017**, *3*, 694–711. [CrossRef]
107. Antoine, E.E.; Vlachos, P.P.; Rylander, M.N. Review of collagen I hydrogels for bioengineered tissue microenvironments: Characterization of mechanics, structure, and transport. *Tissue Eng. Part B Rev.* **2014**, *20*, 683–696. [CrossRef] [PubMed]
108. Maillard, S.; Sicard, L.; Andrique, C.; Torrens, C.; Lesieur, J.; Baroukh, B.; Coradin, T.; Poliard, A.; Slimani, L.; Chaussain, C. Combining sclerostin neutralization with tissue engineering: An improved strategy for craniofacial bone repair. *Acta Biomater.* **2022**, *140*, 178–189. [CrossRef] [PubMed]
109. Nabavi, M.H.; Salehi, M.; Ehterami, A.; Bastami, F.; Semyari, H.; Tehranchi, M.; Nabavi, M.A.; Semyari, H. A collagen-based hydrogel containing tacrolimus for bone tissue engineering. *Drug Deliv. Transl. Res.* **2020**, *10*, 108–121. [CrossRef] [PubMed]
110. Wu, S.; Chen, Z.; Yu, X.; Duan, X.; Chen, J.; Liu, G.; Gong, M.; Xing, F.; Sun, J.; Huang, S.; et al. A sustained release of BMP2 in urine-derived stem cells enhances the osteogenic differentiation and the potential of bone regeneration. *Regen. Biomater.* **2022**, *9*, rbac015. [CrossRef] [PubMed]
111. Sarrigiannidis, S.O.; Rey, J.M.; Dobre, O.; González-García, C.; Dalby, M.J.; Salmeron-Sanchez, M. A tough act to follow: Collagen hydrogel modifications to improve mechanical and growth factor loading capabilities. *Mater. Today Bio* **2021**, *10*, 100098. [CrossRef]
112. Sanz-Horta, R.; Matesanz, A.; Jorcano, J.L.; Velasco, D.; Acedo, P.; Gallardo, A.; Reinecke, H.; Elvira, C. Preparation and Characterization of Plasma-Derived Fibrin Hydrogels Modified by Alginate di-Aldehyde. *Int. J. Mol. Sci.* **2022**, *23*, 4296. [CrossRef] [PubMed]
113. Linsley, C.S.; Sung, K.; White, C.; Abecunas, C.A.; Tawil, B.J.; Wu, B.M. Functionalizing Fibrin Hydrogels with Thermally Responsive Oligonucleotide Tethers for On-Demand Delivery. *Bioengineering* **2022**, *9*, 25. [CrossRef]
114. Zou, Y.; Mei, X.; Wang, X.; Zhang, X.; Wang, X.; Xiang, W.; Lu, N. Fibrin-konjac glucomannan-black phosphorus hydrogel scaffolds loaded with nasal ectodermal mesenchymal stem cells accelerated alveolar bone regeneration. *BMC Oral Health* **2024**, *24*, 878. [CrossRef]
115. Wu, E.; Huang, L.; Shen, Y.; Wei, Z.; Li, Y.; Wang, J.; Chen, Z. Application of gelatin-based composites in bone tissue engineering. *Heliyon* **2024**, *10*, e36258. [CrossRef] [PubMed]
116. Zhou, L.; Fan, L.; Zhang, F.M.; Jiang, Y.; Cai, M.; Dai, C.; Luo, Y.A.; Tu, L.J.; Zhou, Z.N.; Li, X.J.; et al. Hybrid gelatin/oxidized chondroitin sulfate hydrogels incorporating bioactive glass nanoparticles with enhanced mechanical properties, mineralization, and osteogenic differentiation. *Bioact. Mater.* **2021**, *6*, 890–904. [CrossRef]
117. Zhang, R.; Jo, J.I.; Kanda, R.; Nishiura, A.; Hashimoto, Y.; Matsumoto, N. Bioactive Polyetheretherketone with Gelatin Hydrogel Leads to Sustained Release of Bone Morphogenetic Protein-2 and Promotes Osteogenic Differentiation. *Int. J. Mol. Sci.* **2023**, *24*, 12741. [CrossRef] [PubMed]
118. Xu, D.; Gan, K.; Wang, Y.; Wu, Z.; Wang, Y.; Zhang, S.; Peng, Y.; Fang, X.; Wei, H.; Zhang, Y.; et al. A Composite Deferoxamine/Black Phosphorus Nanosheet/Gelatin Hydrogel Scaffold for Ischemic Tibial Bone Repair. *Int. J. Nanomed.* **2022**, *17*, 1015–1030. [CrossRef] [PubMed]
119. Choi, J.B.; Kim, Y.K.; Byeon, S.M.; Park, J.E.; Bae, T.S.; Jang, Y.S.; Lee, M.H. Fabrication and Characterization of Biodegradable Gelatin Methacrylate/Biphasic Calcium Phosphate Composite Hydrogel for Bone Tissue Engineering. *Nanomaterials* **2021**, *11*, 617. [CrossRef] [PubMed]
120. Goto, R.; Nishida, E.; Kobayashi, S.; Aino, M.; Ohno, T.; Iwamura, Y.; Kikuchi, T.; Hayashi, J.I.; Yamamoto, G.; Asakura, M.; et al. Gelatin Methacryloyl-Riboflavin (GelMA-RF) Hydrogels for Bone Regeneration. *Int. J. Mol. Sci.* **2021**, *22*, 1635. [CrossRef] [PubMed]
121. Paduano, F.; Marrelli, M.; Alom, N.; Amer, M.; White, L.J.; Shakesheff, K.M.; Tatullo, M. Decellularized bone extracellular matrix and human dental pulp stem cells as a construct for bone regeneration. *J. Biomater. Sci. Polym. Ed.* **2017**, *28*, 730–748. [CrossRef]

122. Kim, Y.H.; Cidonio, G.; Kanczler, J.M.; Oreffo, R.O.; Dawson, J.I. Human bone tissue-derived ECM hydrogels: Controlling physicochemical, biochemical, and biological properties through processing parameters. *Bioact. Mater.* **2025**, *43*, 114–128. [CrossRef] [PubMed]
123. Ventura, R.D.; Padalhin, A.R.; Kim, B.; Park, M.; Lee, B.T. Evaluation of bone regeneration potential of injectable extracellular matrix (ECM) from porcine dermis loaded with biphasic calcium phosphate (BCP) powder. *Mater. Sci. Eng. C Mater. Biol. Appl.* **2020**, *110*, 110663. [CrossRef]
124. Branković, M.; Zivic, F.; Grujovic, N.; Stojadinovic, I.; Milenkovic, S.; Kotorcevic, N. Review of Spider Silk Applications in Biomedical and Tissue Engineering. *Biomimetics* **2024**, *9*, 169. [CrossRef] [PubMed]
125. Gomes, S.; Leonor, I.B.; Mano, J.F.; Reis, R.L.; Kaplan, D.L. Spider silk-bone sialoprotein fusion proteins for bone tissue engineering. *Soft Matter* **2011**, *7*, 4964–4973. [CrossRef]
126. Sun, L.; Lu, M.; Chen, L.; Zhao, B.; Yao, J.; Shao, Z.; Chen, X.; Liu, Y. Silk-Inorganic Nanoparticle Hybrid Hydrogel as an Injectable Bone Repairing Biomaterial. *J. Funct. Biomater.* **2023**, *14*, 86. [CrossRef] [PubMed]
127. Zhu, Y.; Gu, H.; Yang, J.; Li, A.; Hou, L.; Zhou, M.; Jiang, X. An Injectable silk-based hydrogel as a novel biomineralization seedbed for critical-sized bone defect regeneration. *Bioact. Mater.* **2024**, *35*, 274–290. [CrossRef]
128. Phewchan, P.; Laoruengthana, A.; Tiyaboonchai, W. Calcium phosphate incorporated in silk fibroin/methylcellulose based injectable hydrogel: Preparation, characterization, and in vitro biological evaluation for bone defect treatment. *J. Biomed. Mater. Res. B Appl. Biomater.* **2023**, *111*, 1640–1652. [CrossRef]
129. Chen, X.; Wu, T.; Bu, Y.; Yan, H.; Lin, Q. Fabrication and Biomedical Application of Alginate Composite Hydrogels in Bone Tissue Engineering: A Review. *Int. J. Mol. Sci.* **2024**, *25*, 7810. [CrossRef] [PubMed]
130. Rowley, J.A.; Madlambayan, G.; Mooney, D.J. Alginate hydrogels as synthetic extracellular matrix materials. *Biomaterials* **1999**, *20*, 45–53. [CrossRef]
131. Wu, H.; Zhang, X.; Wang, Z.; Chen, X.; Li, Y.; Fang, J.; Zheng, S.; Zhang, L.; Li, C.; Hao, L. Preparation, properties and in vitro osteogensis of self-reinforcing injectable hydrogel. *Eur. J. Pharm. Sci.* **2024**, *192*, 106617. [CrossRef] [PubMed]
132. Hwang, H.S.; Lee, C.S. Recent Progress in Hyaluronic-Acid-Based Hydrogels for Bone Tissue Engineering. *Gels* **2023**, *9*, 588. [CrossRef] [PubMed]
133. Bergman, K.; Engstrand, T.; Hilborn, J.; Ossipov, D.; Piskounova, S.; Bowden, T. Injectable cell-free template for bone-tissue formation. *J. Biomed. Mater. Res. A* **2009**, *91*, 1111–1118. [CrossRef] [PubMed]
134. Deng, H.; Wang, J.; An, R. Hyaluronic acid-based hydrogels: As an exosome delivery system in bone regeneration. *Front. Pharmacol.* **2023**, *14*, 1131001. [CrossRef]
135. Zhou, P.; Yan, B.; Wei, B.; Fu, L.; Wang, Y.; Wang, W.; Zhang, L.; Mao, Y. Quercetin-solid lipid nanoparticle-embedded hyaluronic acid functionalized hydrogel for immunomodulation to promote bone reconstruction. *Regen. Biomater.* **2023**, *10*, rbad025. [CrossRef]
136. Fakhri, E.; Eslami, H.; Maroufi, P.; Pakdel, F.; Taghizadeh, S.; Ganbarov, K.; Yousefi, M.; Tanomand, A.; Yousefi, B.; Mahmoudi, S.; et al. Chitosan biomaterials application in dentistry. *Int. J. Biol. Macromol.* **2020**, *162*, 956–974. [CrossRef] [PubMed]
137. Ye, R.; Liu, S.; Zhu, W.; Li, Y.; Huang, L.; Zhang, G.; Zhang, Y. Synthesis, Characterization, Properties, and Biomedical Application of Chitosan-Based Hydrogels. *Polymers* **2023**, *15*, 2482. [CrossRef]
138. Yang, X.; Guo, J.L.; Han, J.; Si, R.J.; Liu, P.P.; Zhang, Z.R.; Wang, A.M.; Zhang, J. Chitosan hydrogel encapsulated with LL-37 peptide promotes deep tissue injury healing in a mouse model. *Mil. Med. Res.* **2020**, *7*, 20. [CrossRef] [PubMed]
139. Luca, L.; Rougemont, A.L.; Walpoth, B.H.; Boure, L.; Tami, A.; Anderson, J.M.; Jordan, O.; Gurny, R. Injectable rhBMP-2-loaded chitosan hydrogel composite: Osteoinduction at ectopic site and in segmental long bone defect. *J. Biomed. Mater. Res. A* **2011**, *96*, 66–74. [CrossRef] [PubMed]
140. Yu, F.; Geng, D.; Kuang, Z.; Huang, S.; Cheng, Y.; Chen, Y.; Leng, F.; Bei, Y.; Zhao, Y.; Tang, Q.; et al. Sequentially releasing self-healing hydrogel fabricated with TGFβ3-microspheres and bFGF to facilitate rat alveolar bone defect repair. *Asian J. Pharm. Sci.* **2022**, *17*, 425–434. [CrossRef]
141. Díaz-Montes, E. Dextran: Sources, Structures, and Properties. *Polysaccharides* **2021**, *2*, 554–565. [CrossRef]
142. Kaur, H.; Gogoi, B.; Sharma, I.; Das, D.K.; Azad, M.A.; Pramanik, D.D.; Pramanik, A. Hydrogels as a Potential Biomaterial for Multimodal Therapeutic Applications. *Mol. Pharm.* **2024**, *21*, 4827–4848. [CrossRef] [PubMed]
143. Tao, J.; Zhu, S.; Liao, X.; Wang, Y.; Zhou, N.; Li, Z.; Wan, H.; Tang, Y.; Yang, S.; Du, T.; et al. DLP-based bioprinting of void-forming hydrogels for enhanced stem-cell-mediated bone regeneration. *Mater. Today Bio* **2022**, *17*, 100487. [CrossRef] [PubMed]
144. Silva, D.D.; Crous, A.; Abrahamse, H. Photobiomodulation Dose-Response on Adipose-Derived Stem Cell Osteogenesis in 3D Cultures. *Int. J. Mol. Sci.* **2024**, *25*, 9176. [CrossRef] [PubMed]
145. Shoji, S.; Uchida, K.; Saito, W.; Sekiguchi, H.; Inoue, G.; Miyagi, M.; Kuroda, A.; Takaso, M. Acceleration of Bone Healing by In Situ-Forming Dextran-Tyramine Conjugates Containing Basic Fibroblast Growth Factor in Mice. *Cureus* **2020**, *12*, e10085. [CrossRef] [PubMed]

146. Ritz, U.; Eberhardt, M.; Klein, A.; Frank, P.; Götz, H.; Hofmann, A.; Rommens, P.M.; Jonas, U. Photocrosslinked Dextran-Based Hydrogels as Carrier System for the Cells and Cytokines Induce Bone Regeneration in Critical Size Defects in Mice. *Gels* **2018**, *4*, 63. [CrossRef] [PubMed]
147. Zhang, P.; Qin, Q.; Cao, X.; Xiang, H.; Feng, D.; Wusiman, D.; Li, Y. Hydrogel microspheres for bone regeneration through regulation of the regenerative microenvironment. *Biomater. Transl.* **2024**, *5*, 205–235. [PubMed]
148. Ye, B.; Wu, B.; Su, Y.; Sun, T.; Guo, X. Recent Advances in the Application of Natural and Synthetic Polymer-Based Scaffolds in Musculoskeletal Regeneration. *Polymers* **2022**, *14*, 4566. [CrossRef]
149. Jansen, L.E.; Kim, H.; Hall, C.L.; McCarthy, T.P.; Lee, M.J.; Peyton, S.R. A poly(ethylene glycol) three-dimensional bone marrow hydrogel. *Biomaterials* **2022**, *280*, 121270. [CrossRef]
150. Wang, S.; Li, R.; Xia, D.; Zhao, X.; Zhu, Y.; Gu, R.; Yoon, J.; Liu, Y. The impact of Zn-doped synthetic polymer materials on bone regeneration: A systematic review. *Stem Cell Res. Ther.* **2021**, *12*, 123. [CrossRef]
151. Carles-Carner, M.; Saleh, L.S.; Bryant, S.J. The effects of hydroxyapatite nanoparticles embedded in a MMP-sensitive photoclickable PEG hydrogel on encapsulated MC3T3-E1 pre-osteoblasts. *Biomed. Mater.* **2018**, *13*, 045009. [CrossRef]
152. Salehi, S.; Ghomi, H.; Hassanzadeh-Tabrizi, S.A.; Koupaei, N.; Khodaei, M. The effect of polyethylene glycol on printability, physical and mechanical properties and osteogenic potential of 3D-printed poly (l-lactic acid)/polyethylene glycol scaffold for bone tissue engineering. *Int. J. Biol. Macromol.* **2022**, *221*, 1325–1334. [CrossRef] [PubMed]
153. Huang, P.; Yang, P.; Liu, K.; Tao, W.; Tao, J.; Ai, F. Evaluation of 'surgery-friendly' bone scaffold characteristics: 3D printed ductile BG/PCL scaffold with high inorganic content to repair critical bone defects. *Biomed. Mater.* **2022**, *18*, 015021. [CrossRef] [PubMed]
154. Saleh, L.S.; Carles-Carner, M.; Bryant, S.J. The in vitro effects of macrophages on the osteogenic capabilities of MC3T3-E1 cells encapsulated in a biomimetic poly(ethylene glycol) hydrogel. *Acta Biomater.* **2018**, *71*, 37–48. [CrossRef]
155. Wilson, J.A.; Luong, D.; Kleinfehn, A.P.; Sallam, S.; Wesdemiotis, C.; Becker, M.L. Magnesium Catalyzed Polymerization of End Functionalized Poly(propylene maleate) and Poly(propylene fumarate) for 3D Printing of Bioactive Scaffolds. *J. Am. Chem. Soc.* **2018**, *140*, 277–284. [CrossRef]
156. Qayoom, I.; Prasad, A.; Srivastava, E.; Fazili, K.M.; Nussler, A.K.; Kumar, A. Organic-inorganic composite of polypropylene fumarate and nanohydroxyapatite as carrier of antibiotics for the treatment of bone infections. *Biomater. Adv.* **2024**, *157*, 213714. [CrossRef]
157. Wojda, S.J.; Marozas, I.A.; Anseth, K.S.; Yaszemski, M.J.; Donahue, S.W. Impact of Release Kinetics on Efficacy of Locally Delivered Parathyroid Hormone for Bone Regeneration Applications. *Tissue Eng. Part A* **2021**, *27*, 246–255. [CrossRef] [PubMed]
158. Olthof, M.G.L.; Kempen, D.H.R.; Herrick, J.L.; Yaszemski, M.J.; Dhert, W.J.A.; Lu, L. Effect of different sustained bone morphogenetic protein-2 release kinetics on bone formation in poly(propylene fumarate) scaffolds. *J. Biomed. Mater. Res. B Appl. Biomater.* **2018**, *106*, 477–487. [CrossRef] [PubMed]
159. Wang, S.; Gu, R.; Wang, F.; Zhao, X.; Yang, F.; Xu, Y.; Yan, F.; Zhu, Y.; Xia, D.; Liu, Y. 3D-Printed PCL/Zn scaffolds for bone regeneration with a dose-dependent effect on osteogenesis and osteoclastogenesis. *Mater. Today Bio* **2022**, *13*, 100202. [CrossRef] [PubMed]
160. Qin, X.; Wu, Y.; Liu, S.; Yang, L.; Yuan, H.; Cai, S.; Flesch, J.; Li, Z.; Tang, Y.; Li, X.; et al. Surface Modification of Polycaprolactone Scaffold With Improved Biocompatibility and Controlled Growth Factor Release for Enhanced Stem Cell Differentiation. *Front. Bioeng. Biotechnol.* **2021**, *9*, 802311. [CrossRef]
161. Kim, H.Y.; Kim, B.H.; Kim, M.S. Amine Plasma-Polymerization of 3D Polycaprolactone/β-Tricalcium Phosphate Scaffold to Improving Osteogenic Differentiation In Vitro. *Materials* **2022**, *15*, 366. [CrossRef]
162. Niknam, Z.; Golchin, A.; Rezaei-Tavirani, M.; Ranjbarvan, P.; Zali, H.; Omidi, M.; Mansouri, V. Osteogenic Differentiation Potential of Adipose-Derived Mesenchymal Stem Cells Cultured on Magnesium Oxide/Polycaprolactone Nanofibrous Scaffolds for Improving Bone Tissue Reconstruction. *Adv. Pharm. Bull.* **2022**, *12*, 142–154. [CrossRef] [PubMed]
163. Zheng, J.; Fan, R.; Wu, H.; Yao, H.; Yan, Y.; Liu, J.; Ran, L.; Sun, Z.; Yi, L.; Dang, L.; et al. Directed self-assembly of herbal small molecules into sustained release hydrogels for treating neural inflammation. *Nat. Commun.* **2019**, *10*, 1604. [CrossRef]
164. Huang, C.L.; Huang, H.Y.; Lu, Y.C.; Cheng, C.J.; Lee, T.M. Development of a flexible film made of polyvinyl alcohol with chitosan based thermosensitive hydrogel. *J. Dent. Sci.* **2023**, *18*, 822–832. [CrossRef]
165. Xue, C.; Chen, L.; Wang, N.; Chen, H.; Xu, W.; Xi, Z.; Sun, Q.; Kang, R.; Xie, L.; Liu, X. Stimuli-responsive hydrogels for bone tissue engineering. *Biomater. Transl.* **2024**, *5*, 257–273. [PubMed]
166. Baines, D.K.; Platania, V.; Tavernaraki, N.N.; Parati, M.; Wright, K.; Radecka, I.; Chatzinikolaidou, M.; Douglas, T.E.L. The Enrichment of Whey Protein Isolate Hydrogels with Poly-γ-Glutamic Acid Promotes the Proliferation and Osteogenic Differentiation of Preosteoblasts. *Gels* **2023**, *10*, 18. [CrossRef] [PubMed]
167. Cho, S.H.; Shin, K.K.; Kim, S.Y.; Cho, M.Y.; Oh, D.B.; Lim, Y.T. In Situ-Forming Collagen/poly-γ-glutamic Acid Hydrogel System with Mesenchymal Stem Cells and Bone Morphogenetic Protein-2 for Bone Tissue Regeneration in a Mouse Calvarial Bone Defect Model. *Tissue Eng. Regen. Med.* **2022**, *19*, 1099–1111. [CrossRef]

168. Vallmajo-Martin, Q.; Millan, C.; Müller, R.; Weber, F.E.; Ehrbar, M.; Ghayor, C. Enhanced bone regeneration in rat calvarial defects through BMP2 release from engineered poly(ethylene glycol) hydrogels. *Sci. Rep.* **2024**, *14*, 4916. [CrossRef]
169. Periferakis, A.; Periferakis, A.T.; Troumpata, L.; Dragosloveanu, S.; Timofticiuc, I.A.; Georgatos-Garcia, S.; Scheau, A.E.; Periferakis, K.; Caruntu, A.; Badarau, I.A.; et al. Use of Biomaterials in 3D Printing as a Solution to Microbial Infections in Arthroplasty and Osseous Reconstruction. *Biomimetics* **2024**, *9*, 154. [CrossRef]
170. Kasper, F.K.; Tanahashi, K.; Fisher, J.P.; Mikos, A.G. Synthesis of poly(propylene fumarate). *Nat. Protoc.* **2009**, *4*, 518–525. [CrossRef]
171. Zaghian, M.; Varshosaz, J.; Rostami, M.; Mirian, M. A forsterite-reinforced polypropylene fumarate/methoxy polyethylene glycol-hydrogel enriched with flavonoid nanoparticles enhances osteoconductivity. *Mater. Adv.* **2024**, *5*, 4324–4344. [CrossRef]
172. Lee, D.; Wufuer, M.; Kim, I.; Choi, T.H.; Kim, B.J.; Jung, H.G.; Jeon, B.; Lee, G.; Jeon, O.H.; Chang, H.; et al. Sequential dual-drug delivery of BMP-2 and alendronate from hydroxyapatite-collagen scaffolds for enhanced bone regeneration. *Sci. Rep.* **2021**, *11*, 746. [CrossRef]
173. Tan, Z.; Luo, Y.; Yang, L. Basic fibroblast growth factor/chitosan derivatives/collagen composite thermosensitive hydrogel promotes perio-dontal tissue regeneration in rats. *West China J. Stomatol.* **2023**, *41*, 21–28.
174. Xu, H.; Luo, H.; Chen, J.; Chen, G.; Yu, X.; Ye, Z. BMP-2 releasing mineral-coated microparticle-integrated hydrogel system for enhanced bone regeneration. *Front. Bioeng. Biotechnol.* **2023**, *11*, 1217335. [CrossRef]
175. Furuhata, M.; Takayama, T.; Yamamoto, T.; Ozawa, Y.; Senoo, M.; Ozaki, M.; Yamano, S.; Sato, S. Real-time assessment of guided bone regeneration in critical size mandibular bone defects in rats using collagen membranes with adjunct fibroblast growth factor-2. *J. Dent. Sci.* **2021**, *16*, 1170–1181. [CrossRef]
176. Yazdanian, M.; Alam, M.; Abbasi, K.; Rahbar, M.; Farjood, A.; Tahmasebi, E.; Tebyaniyan, H.; Ranjbar, R.; Hesam Arefi, A. Synthetic materials in craniofacial regenerative medicine: A comprehensive overview. *Front. Bioeng. Biotechnol.* **2022**, *10*, 987195. [CrossRef]
177. Tabata, Y.; Yamada, K.; Hong, L.; Miyamoto, S.; Hashimoto, N.; Ikada, Y. Skull bone regeneration in primates in response to basic fibroblast growth factor. *J. Neurosurg.* **1999**, *91*, 851–856. [CrossRef]
178. Gillman, C.E.; Jayasuriya, A.C. FDA-approved bone grafts and bone graft substitute devices in bone regeneration. *Mater. Sci. Eng. C Mater. Biol. Appl.* **2021**, *130*, 112466. [CrossRef]
179. He, W.; Reaume, M.; Hennenfent, M.; Lee, B.P.; Rajachar, R. Biomimetic hydrogels with spatial- and temporal-controlled chemical cues for tissue engineering. *Biomater. Sci.* **2020**, *8*, 3248–3269. [CrossRef]
180. Neovius, E.; Lemberger, M.; Docherty Skogh, A.C.; Hilborn, J.; Engstrand, T. Alveolar bone healing accompanied by severe swelling in cleft children treated with bone morphogenetic protein-2 delivered by hydrogel. *J. Plast. Reconstr. Aesthet. Surg.* **2013**, *66*, 37–42. [CrossRef]
181. Yang, Q.; Li, Y.; Wan, R.; Dong, L.; He, A.; Zuo, D.; Dai, Z. Multilayer Gelatin-Supported BMP-9 Coating Promotes Osteointegration and Neo-Bone Formation at the n-CDHA/PAA Composite Biomaterial-Bone Interface. *Front. Biosci.* **2024**, *29*, 326. [CrossRef]
182. Zuo, T.; Liu, Y.; Duan, M.; Pu, X.; Huang, M.; Zhang, D.; Xie, J. Platelet-derived growth factor PDGF-AA upregulates connexin 43 expression and promotes gap junction formations in osteoblast cells through p-Akt signaling. *Biochem. Biophys. Rep.* **2023**, *34*, 101462. [CrossRef]
183. Zhou, J.; Zhao, Y.; Ling, Y.; Zhao, P.; Gao, H.; Yang, Y.; Chen, J. Microsphere-Composite Hydrogel for Recruiting Stem Cells and Promoting Osteogenic Differentiation. *ACS Appl. Bio Mater.* **2024**, *7*, 6520–6528. [CrossRef]
184. Albashari, A.A.; He, Y.; Albaadani, M.A.; Xiang, Y.; Ali, J.; Hu, F.; Zhang, Y.; Zhang, K.; Luo, L.; Wang, J.; et al. Titanium Nanotube Modified With Silver Cross-Linked Basic Fibroblast Growth Factor Improves Osteoblastic Activities of Dental Pulp Stem Cells and Antibacterial Effect. *Front. Cell Dev. Biol.* **2021**, *9*, 654654. [CrossRef]
185. Zhou, X.; Chen, J.; Sun, H.; Wang, F.; Wang, Y.; Zhang, Z.; Teng, W.; Ye, Y.; Huang, D.; Zhang, W.; et al. Spatiotemporal regulation of angiogenesis/osteogenesis emulating natural bone healing cascade for vascularized bone formation. *J. Nanobiotechnol.* **2021**, *19*, 420. [CrossRef]
186. Zhang, F.; Lv, M.; Wang, S.; Li, M.; Wang, Y.; Hu, C.; Hu, W.; Wang, X.; Wang, X.; Liu, Z.; et al. Ultrasound-triggered biomimetic ultrashort peptide nanofiber hydrogels promote bone regeneration by modulating macrophage and the osteogenic immune microenvironment. *Bioact. Mater.* **2024**, *31*, 231–246. [CrossRef]
187. Rappe, K.S.; Ortiz-Hernandez, M.; Punset, M.; Molmeneu, M.; Barba, A.; Mas-Moruno, C.; Guillem-Marti, J.; Caparrós, C.; Rupérez, E.; Calero, J.; et al. On-Growth and In-Growth Osseointegration Enhancement in PM Porous Ti-Scaffolds by Two Different Bioactivation Strategies: Alkali Thermochemical Treatment and RGD Peptide Coating. *Int. J. Mol. Sci.* **2022**, *23*, 1750. [CrossRef]
188. Lin, Y.; Huang, S.; Zou, R.; Gao, X.; Ruan, J.; Weir, M.D.; Reynolds, M.A.; Qin, W.; Chang, X.; Fu, H.; et al. Calcium phosphate cement scaffold with stem cell co-culture and prevascularization for dental and craniofacial bone tissue engineering. *Dent. Mater.* **2019**, *35*, 1031–1041. [CrossRef]

189. Vaivads, M.; Pilmane, M. Distribution of Immunomodulation, Protection and Regeneration Factors in Cleft-Affected Bone and Cartilage. *Diagnostics* **2024**, *14*, 2217. [CrossRef]
190. Ridyard, K.E.; Overhage, J. The Potential of Human Peptide LL-37 as an Antimicrobial and Anti-Biofilm Agent. *Antibiotics* **2021**, *10*, 650. [CrossRef]
191. Ma, S.; Wang, C.; Dong, Y.; Jing, W.; Wei, P.; Peng, C.; Liu, Z.; Zhao, B.; Wang, Y. Microsphere-Gel Composite System with Mesenchymal Stem Cell Recruitment, Antibacterial, and Immunomodulatory Properties Promote Bone Regeneration via Sequential Release of LL37 and W9 Peptides. *ACS Appl. Mater. Interfaces* **2022**, *14*, 38525–38540. [CrossRef]
192. Chinipardaz, Z.; Zhong, J.M.; Yang, S. Regulation of LL-37 in Bone and Periodontium Regeneration. *Life* **2022**, *12*, 1533. [CrossRef]
193. Yu, X.; Quan, J.; Long, W.; Chen, H.; Wang, R.; Guo, J.; Lin, X.; Mai, S. LL-37 inhibits LPS-induced inflammation and stimulates the osteogenic differentiation of BMSCs via P2X7 receptor and MAPK signaling pathway. *Exp. Cell Res.* **2018**, *372*, 178–187. [CrossRef] [PubMed]
194. Ye, P.; Yang, Y.; Qu, Y.; Yang, W.; Tan, J.; Zhang, C.; Sun, D.; Zhang, J.; Zhao, W.; Guo, S.; et al. LL-37 and bisphosphonate co-delivery 3D-scaffold with antimicrobial and antiresorptive activities for bone regeneration. *Int. J. Biol. Macromol.* **2024**, *277*, 134091. [CrossRef]
195. Chen, Z.X.; Chang, M.; Peng, Y.L.; Zhao, L.; Zhan, Y.R.; Wang, L.J.; Wang, R. Osteogenic growth peptide C-terminal pentapeptide [OGP(10-14)] acts on rat bone marrow mesenchymal stem cells to promote differentiation to osteoblasts and to inhibit differentiation to adipocytes. *Regul. Pept.* **2007**, *142*, 16–23. [CrossRef]
196. Bermúdez, M.; Hoz, L.; Montoya, G.; Nidome, M.; Pérez-Soria, A.; Romo, E.; Soto-Barreras, U.; Garnica-Palazuelos, J.; Aguilar-Medina, M.; Ramos-Payán, R.; et al. Bioactive Synthetic Peptides for Oral Tissues Regeneration. *Front. Mater.* **2021**, *8*, 655495. [CrossRef]
197. Qiao, Y.; Liu, X.; Zhou, X.; Zhang, H.; Zhang, W.; Xiao, W.; Pan, G.; Cui, W.; Santos, H.A.; Shi, Q. Gelatin Templated Polypeptide Co-Cross-Linked Hydrogel for Bone Regeneration. *Adv. Healthc. Mater.* **2020**, *9*, e1901239. [CrossRef] [PubMed]
198. Echave, M.C.; Erezuma, I.; Golafshan, N.; Castilho, M.; Kadumudi, F.B.; Pimenta-Lopes, C.; Ventura, F.; Pujol, A.; Jimenez, J.J.; Camara, J.A.; et al. Bioinspired gelatin/bioceramic composites loaded with bone morphogenetic protein-2 (BMP-2) promote osteoporotic bone repair. *Biomater. Adv.* **2022**, *134*, 112539. [CrossRef]
199. Yu, C.; Chen, J.; Wang, T.; Wang, Y.; Zhang, X.; Zhang, Z.; Wang, Y.; Yu, T.; Wu, T. GelMA hydrogels reinforced by PCL@GelMA nanofibers and bioactive glass induce bone regeneration in critical size cranial defects. *J. Nanobiotechnol.* **2024**, *22*, 696. [CrossRef]
200. Yuan, S.; Zheng, B.; Zheng, K.; Lai, Z.; Chen, Z.; Zhao, J.; Li, S.; Zheng, X.; Wu, P.; Wang, H. Immunoregulation in Skull Defect Repair with a Smart Hydrogel Loaded with Mesoporous Bioactive Glasses. *Biomater. Res.* **2024**, *28*, 0074. [CrossRef]
201. Yu, T.; Hu, Y.; He, W.; Xu, Y.; Zhan, A.; Chen, K.; Liu, M.; Xiao, X.; Xu, X.; Feng, Q.; et al. An injectable and self-healing hydrogel with dual physical crosslinking for in-situ bone formation. *Mater. Today Bio* **2023**, *19*, 100558. [CrossRef] [PubMed]
202. Zhang, X.; Gong, C.; Wang, X.; Wei, Z.; Guo, W. A Bioactive Gelatin Methacrylate Incorporating Magnesium Phosphate Cement for Bone Regeneration. *Biomedicines* **2024**, *12*, 228. [CrossRef] [PubMed]
203. Yazdanpanah, Z.; Johnston, J.D.; Cooper, D.M.L.; Chen, X. 3D Bioprinted Scaffolds for Bone Tissue Engineering: State-Of-The-Art and Emerging Technologies. *Front. Bioeng. Biotechnol.* **2022**, *10*, 824156. [CrossRef] [PubMed]
204. Lewns, F.K.; Tsigkou, O.; Cox, L.R.; Wildman, R.D.; Grover, L.M.; Poologasundarampillai, G. Hydrogels and Bioprinting in Bone Tissue Engineering: Creating Artificial Stem-Cell Niches for In Vitro Models. *Adv. Mater.* **2023**, *35*, e2301670. [CrossRef] [PubMed]
205. Neumann, M.; di Marco, G.; Iudin, D.; Viola, M.; van Nostrum, C.F.; van Ravensteijn, B.G.P.; Vermonden, T. Stimuli-Responsive Hydrogels: The Dynamic Smart Biomaterials of Tomorrow. *Macromolecules* **2023**, *56*, 8377–8392. [CrossRef]

gels

MDPI

Article

Functional Properties of Gelatin–Alginate Hydrogels for Use in Chronic Wound Healing Applications

Olha Maikovych [1,*], Pamela Pasetto [2], Nataliia Nosova [1], Olena Kudina [3], Dmytro Ostapiv [4], Volodymyr Samaryk [1] and Serhii Varvarenko [1]

[1] Department of Organic Chemistry, Lviv Polytechnic National University, 12 Bandera Str., 79013 Lviv, Ukraine; nataliia.h.nosova@lpnu.ua (N.N.); volodymyr.y.samaryk@lpnu.ua (V.S.); serhii.m.varvarenko@lpnu.ua (S.V.)
[2] Institut des Molécules et Matériaux du Mans, Faculté des Sciences & Techniques, Le Mans Université, Avnue Olivier Messiaen, 72085 Le Mans, Cedex 09, France; pamela.pasetto@univ-lemans.fr
[3] CÚRAM Research Ireland Centre for Medical Devices, University of Galway, H92 W2TY Galway, Ireland; olena.kudina@universityofgalway.ie
[4] Institute of Animal Biology NAAS, 38 V. Stusa Str., 79034 Lviv, Ukraine; oddost@ukr.net
* Correspondence: olha.v.maikovych@lpnu.ua

Abstract: In this study, a hydrogel material based on porcine gelatin and sodium alginate was synthesized for use as a dressing for chronic wound treatment. The hydrogels were covalently cross-linked using polyethylene glycol diglycidyl ether (PEGDE 500), and the interaction between the components was confirmed via FTIR. The properties of the resulting hydrogels were examined, including gel-fraction volume, swelling degree in different media, mechanical properties, pore size, cytotoxicity, and the ability to absorb and release analgesics (lidocaine, novocaine, sodium diclofenac). The hydrogel's resistance to enzymatic action by protease was enhanced both through chemical cross-linking and physical interactions between gelatin and alginate. The absorption capacity of the hydrogels, reaching 90 g per dm^2 of the hydrogel dressing, indicates their potential for absorbing wound exudates. It was demonstrated that the antiseptic (chlorhexidine) contained in the structured gelatin–alginate hydrogels can be released into an infected substrate, resulting in a significant inhibition of pathogenic microorganisms (*Escherichia coli, Staphylococcus aureus, Pseudomonas aeruginosa*, and *Aspergillus niger*). These results clearly demonstrate that the obtained hydrogel materials can serve as non-traumatic dressings for the treatment of chronic and/or infected wounds.

Keywords: hydrogel; gelatin; sodium alginate; crosslinking; wound dressing; drug delivery

Academic Editors: Daihua Fu and Jieyu Zhang

Received: 19 November 2024
Revised: 29 December 2024
Accepted: 24 February 2025
Published: 27 February 2025

Citation: Maikovych, O.; Pasetto, P.; Nosova, N.; Kudina, O.; Ostapiv, D.; Samaryk, V.; Varvarenko, S. Functional Properties of Gelatin–Alginate Hydrogels for Use in Chronic Wound Healing Applications. *Gels* **2025**, *11*, 174. https://doi.org/10.3390/gels11030174

1. Introduction

Chronic wounds, including diabetic ulcers, pressure ulcers, and burns, pose a serious health problem worldwide [1,2]. Effective wound care requires modern materials that can create a favorable environment for wound healing while protecting the wound from external irritants and contaminants.

All types of wounds (acute and chronic) require quality clinical care to prevent delayed healing, which can be caused by microbial infections and other adverse factors. Over 300 types of wound dressings are available on the market; however, the same dressing cannot be used for treating all wound types [3]. The global market for chronic wound treatment was valued at USD 11.61 billion in 2021. It is expected to grow from USD 12.36 billion in 2022 to USD 19.52 billion by 2029, demonstrating a CAGR of 6.7% during the forecast period [4].

The development of new drug forms with an enhanced bioavailability and prolonged, controlled drug release is highly relevant, especially when these forms are convenient to synthesize and use. The development of new drug forms with immobilized drugs is expanding. Hydrogels have become promising candidates for wound dressings due to their ability to maintain a moist environment, facilitate gas exchange, and support tissue regeneration [5–7]. They are used to make contact lenses, membranes, biosensors, materials for artificial tissues (such as skin and cartilage), and drug delivery systems [8,9]. Polymers are typically used as carriers, and biopolymers are the most desirable in this regard [10]. For effective wound treatment, biopolymer-based hydrogels are considered modern materials that can create a favorable healing environment while simultaneously protecting the wound from external irritants and contaminants.

Since the structure of hydrogels is similar to the extracellular matrix of many tissues, they can fill wounds, absorb exudates, and deliver drugs without causing additional damage, thus promoting an accelerated tissue regeneration [10,11].

Another advantage is the possibility of designing hydrogels with a high bioavailability and minimal immune response. The biomimetic nature of hydrogels, and their ability to maintain a moist environment for wound healing and skin regeneration are highly beneficial for achieving positive treatment outcomes, especially in cases of burn wounds [12,13]. The incorporation of antimicrobial agents and various biological molecules into hydrogels enables their use as dressings for the effective treatment of chronic/infected wounds [14,15]. The development of ideal wound dressing materials with desirable antibacterial and good wound healing properties remains a major challenge affecting the regeneration of bacterially infected wound tissues [16].

Hydrogel materials often exhibit insufficient strength and a limited release of hydrophobic molecules. Their ability to be modified can expand the potential applications of such materials. As noted in [17], incorporating a hydrophobic component imparts unique amphiphilic properties to hydrogels. The design of an amphiphilic inter-penetrating network can be extended to other polymeric hydrogel systems, opening new opportunities for drug delivery and the development of smart materials.

In [18], the possibility of modifying cyclodextrin-based hydrogels for the hydrophobic delivery of baicalein was demonstrated, showing potential for biomedical applications, particularly as a controlled delivery system for hydrophobic drugs.

Previously, in [19], the development of resorbable and biodegradable enzyme-crosslinked gelatin–alginate semi-IPN hydrogel dressings with curcumin was proposed. Moreover, in [20], a method for creating gelatin–alginate hydrogel with adipose-derived stem cells for skin regeneration was described, utilizing crosslinking in a 5% $CaCl_2$ solution. However, these methods have certain limitations and are rather time-consuming due to the multistep process required for obtaining hydrogel materials.

In recent years, hydrogels based on natural polymers such as gelatin, pectin, chitosan, and sodium alginate have attracted considerable attention in wound care. Gelatin, derived from collagen, has excellent biocompatibility and biodegradability, making it suitable for medical use [21,22]. Sodium alginate, derived from seaweed, has unique gel-forming properties and has been widely studied for its wound-healing potential [23,24].

The purpose of this article is to describe the synthesis and characterization of gelatin and sodium alginate hydrogels crosslinked with polyethylene glycol diepoxide as a basis for wound dressings. We will discuss the physicochemical properties of these hydrogels, including their swelling behavior, mechanical strength, and degradation kinetics, which are crucial factors affecting their effectiveness as wound dressings.

In addition, we will study the conditions of the saturation and release from gelatin–alginate hydrogels of a number of drugs, such as novocaine, lidocaine, diclofenac sodium, and chlorhexidine, which are used to increase the therapeutic effectiveness of hydrogels, due to their anesthetic effect and antiseptic properties [25,26].

To summarize, hydrogels based on gelatin and sodium alginate hold great promise as versatile wound coverage platforms [27–29]. With a full understanding of their properties and efficacy in wound healing applications, we can pave the way for the development of advanced biomaterials that meet the needs of treating hard-to-heal and chronic wounds and improve patient care [30–33].

2. Results and Discussion

2.1. The Synthesis Characteristics of Gelatin–Alginate Hydrogels

A recurring challenge for hydrogels is their low mechanical strength, which has significantly limited their applications, especially in the medical field. Researchers have been working to improve this since the development of hydrogels [34–36]. There are two main strategies for improving the mechanical properties of hydrogel materials. The first approach involves producing multimodified materials, such as double networks, double-crosslinked networks, and interpenetrating networks. The second strategy is to incorporate reinforcing elements, including fibers, meshes, and nonwoven polypropylene materials [37–39].

This paper focuses on the preparation of hydrogels based on gelatin and sodium alginate using a diepoxy crosslinker derived from polyethylene glycol 400 as a structuring agent. Mechanical properties were enhanced by maximizing the number of crosslinking points within the hydrogel structure. The synthesis scheme is shown in Figure 1. Gelatin–alginate hydrogel samples were prepared at varying crosslinker-to-(gelatin–alginate) ratios, ranging from 1:1 to 1:50, as described in the experimental section, and their properties are summarized in Table 1.

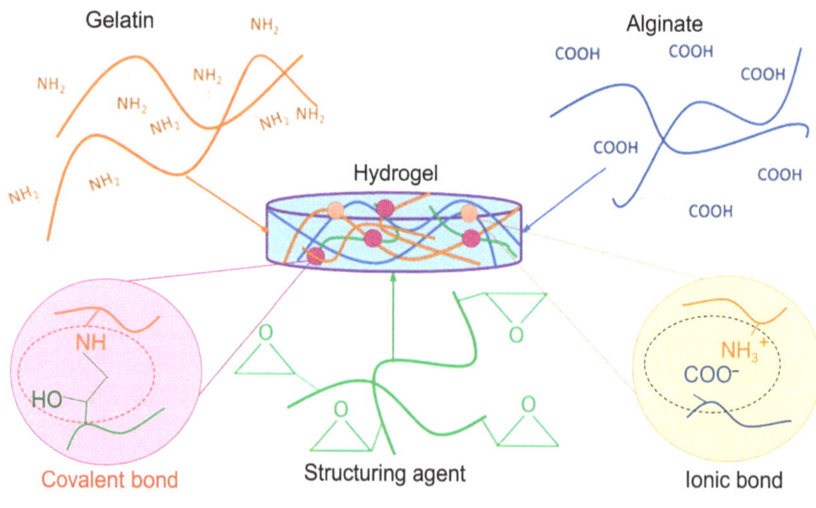

Figure 1. A scheme of the combined gelatin–alginate hydrogel synthesis.

The structure of hydrogels is formed due to several factors of its spatial network formation [40,41]. In particular, gelatin complexes with ionic polysaccharides are primarily formed by intermolecular electrostatic interactions and hydrogen bonds [42,43]. Ionic interactions of the complementary functional groups allow fragments of gelatin macromolecules to bind to sodium alginate macromolecules. The formation and stability of such polyelectrolyte complexes depend on various factors, including the ionization degree of polyelectrolytes, the charge distribution along the polymer chains, the nature and position of ionic groups, the flexibility of the polymer chain, and the molecular weight and ratio of polyelectrolytes, as well as the temperature, ionic strength, and pH of the reaction medium [44]. Under specific conditions—such as certain ratios of biopolymers, their concentration, pH, and ionic strength of the medium, these complexes can form hydrogels independently [45–47].

Table 1. The hydrogel composition and components ratio during PEGDE 500 crosslinking.

Sample No.	Weight Ratios		PEGDE 500 Content, %	Gelatin Content, %	Sodium Alginate Content, %	Water Content, %
	PEGDE 500	Gelatin–Sodium Alginate Blend				
1	1	1	8.571	8.571	0.857	
2	1	2	5.625	11.250	1.125	
3	1	3	4.186	12.558	1.256	
4	1	5	2.769	13.846	1.385	
5	1	8	2.00	14.694	1.470	82
6	1	10	1.500	15.000	1.500	
7	1	15	1.029	15.429	1.543	
8	1	25	0.632	15.789	1.579	
9	1	50	0.321	16.071	1.607	
10 *	0	1	0.00	16.364	1.636	
11 *	1	1	5.714	5.714	0.571	
12	1	2	3.750	7.500	0.750	
13	1	3	2.791	8.372	0.837	
14	1	5	1.846	9.231	0.923	88
15	1	8	1.224	9.796	0.980	
16	1	10	1.000	10.000	1.000	
17	1	15	0.686	10.286	1.023	
18	1	25	0.421	10.526	1.053	

* Under these conditions, the formation of a hydrogel structure was not observed at 37 °C.

We propose to increase the number of crosslinks by structuring a mixture of biopolymers that have already formed polyelectrolyte complexes (PECs) using a diepoxy crosslinker based on polyethylene glycol 400 (PEGDE 500). Previous studies have demonstrated PEGDE 500's ability to crosslink gelatin macromolecules to form hydrogels [25,26]. To implement this approach of creating intermolecular bonds in the mixture, it was crucial to identify the optimal range of biopolymer ratios at which a water-soluble PEC is formed. Such a PEC can be considered as a high molecular weight polymer, of whose further structuring can lead to hydrogels with improved properties. In this case, the change in the viscosity of the biopolymer solution serve as a qualitative and quantitative indicator of the adduct formation. The formation of the optimal number of intermolecular bonds, at which the PEC macromolecules remain in the most possible unfolded state in the solution, depends specifically on the ratio of the two polymers. Data shown in Figure 2, which illustrate the results of the rheological studies of 1% aqueous solutions of the polymer mixture at different ratios, indicate a significant increase in viscosity within the gelatin–alginate ratio range of 8:1 to 12:1. The analysis of this section reveals that the maximum viscosity, nearly 10–12 times higher, occurs at a gelatin–alginate ratio of 10:1. The established ratio was used in further hydrogel syntheses to investigate the effect of the covalent structuring of this polymer complex using the diepoxy crosslinker PEGDE 500.

An essential factor in hydrogel synthesis is the total polymer content, as it directly affects their properties. In a previous study of gelatin hydrogels obtained using PEGDE 500, the optimal total polymer concentration of 18% was found [25]. At lower concentrations, the spatial structure of the hydrogel was not formed, and samples of an unsatisfactory quality were obtained. At higher concentrations, no significant advantages were achieved. For gelatin–alginate hydrogels, a polymer concentration above 18% led to a rapid increase in solution viscosity and gelation occurs, hindering the removal of air bubbles and mixing homogeneity. The studies showed that the lowest limit of polymer concentration, at which gelation occurred, is 12%, and this concentration was selected for comparison in further studies. This fact alone favorably distinguishes the new hydrogels from the previously studied gelatin hydrogels, where gelation occurred at a total content of at least 18% of the polymer.

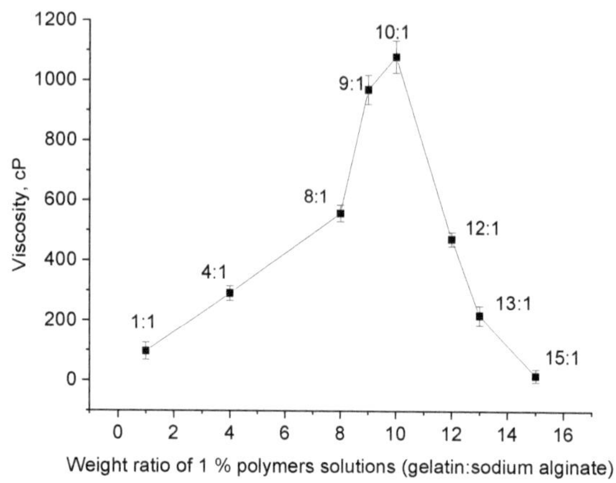

Figure 2. The viscosity of a mixture of 1% aqueous polymer solutions vs. their ratio.

Hydrogel samples were synthesized at the reaction mixture pH of 5.5–6.0, which was achieved spontaneously without additional pH regulators. Under these conditions, hydroxyl and carboxyl groups from both gelatin and sodium alginate most likely do not react with epoxy groups. Structuring occurs mainly through the interaction of the structuring agent with the lysine amino group in the gelatin macromolecules [25].

The gel-fraction value for hydrogels obtained in different contents of the structuring agent indicates that under certain synthesis conditions and ratios in the range of 1:2–1:15, the maximum number of crosslinks is achieved resulting in a hydrogel that maintains stable forms when heated to 50 °C. Unlike the sample of the unstructured gelatin–sodium alginate mixture, the obtained hydrogel swells in water but does not dissolve (Figure 3).

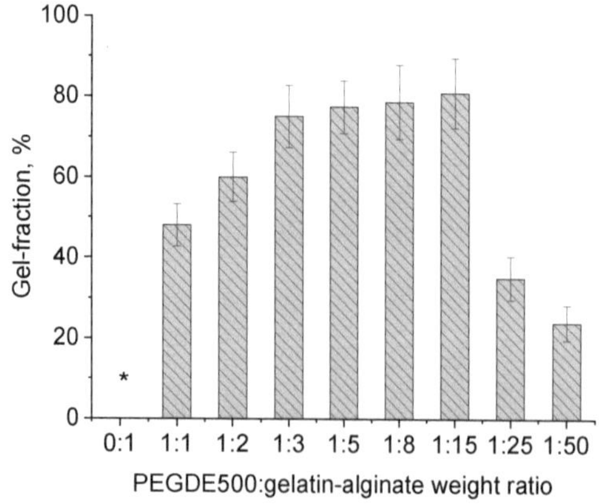

Figure 3. Gel-fraction values of the hydrogel obtained at different PEGDE 500/gelatin–alginate weight ratios. (*—unstructured gelatin-sodium alginate mixture).

Based on the above-described data, synthesis conditions were selected to maximize the involvement of reagents in the creation of a three-dimensional polymeric hydrogel network.

2.2. The Chemical Characterization of Gelatin–Alginate Hydrogel

Figure 4 shows the infrared spectra of the dried samples of gelatin, sodium alginate, PEGDE 500 structuring agent, and the polymer of the hydrogel gel fraction. The absorption peaks at 1650 cm^{-1} and 1540 cm^{-1} in the infrared spectrum of pure gelatin corresponding to the stretching vibrations of C-O and C-N (amide I band) and the bending vibrations of the -NH group (amide II band), respectively. For sodium alginate, the characteristic absorption bands at 1616 and 1419 cm^{-1} correspond to the peaks of the asymmetric and symmetric stretching of -COO-, respectively. In addition, the bands at 1300 cm^{-1} (C-O stretching), 1086 cm^{-1} (mannuronic units), 1033 cm^{-1} (guluronic units), and 817 cm^{-1} (α-configuration of the guluronic units) also relate to the structure of the saccharide. In the infrared spectrum of the structured polymer network (a washed hydrogel, devoid of ungrafted fragments), the absorption peaks at 1616 cm^{-1} and 1419 cm^{-1} of the sodium alginate became less prominent and overlapped with the peaks of pure gelatin. Since the formation of strong hydrogen bonds leads to a shift in the peaks in the IR spectra due to a change in the electron density at the hydrogen bond site, we believed that hydrogen bonds were formed between the chains of gelatin and sodium alginate.

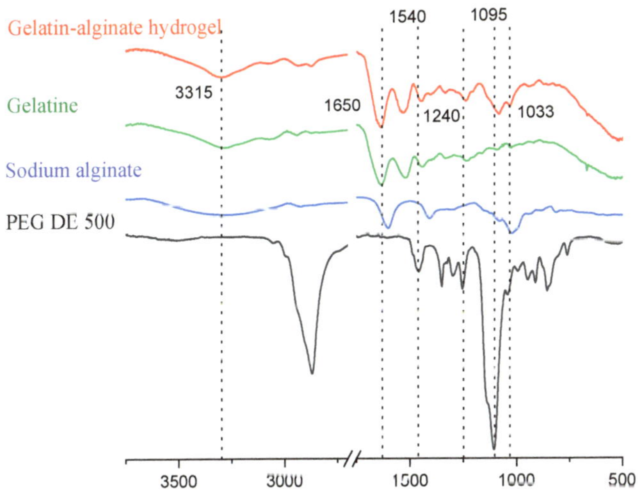

Figure 4. The FTIR spectra of the PEGDE 500, alginate, gelatin, and gel fraction of the reaction product isolated after structuring gelatin–alginate.

The absorption band at 1095 cm^{-1} corresponds to the vibrations of the C-O-C groups in PEGDE 500, which is absent in the spectrum of both gelatin and sodium alginate. The absorption band of guluronic residues at 1033 cm^{-1} in the spectrum of sodium alginate (not present in gelatin and the structuring agent) also appears. A peak at 3315 cm^{-1} assigned to -N-H stretching and hydrogen bonding (amide A), stretching vibrations of the gelatin amide group carbonyl (C=O) at 1650 cm^{-1} (amide I), and deformational vibrations of the -N-H group at 1540 cm^{-1} (amide II) remained present in the spectrum, with only a slight change in the ratio of functional groups compared to the intact gelatin involved in the reaction. The spectrum of the obtained hydrogel shows no asymmetric vibrations of the C-O bond stretching in the oxirane ring at 1240 cm^{-1}, associated with its opening. This indicates a chemical modification of the oxirane groups due to their reaction with gelatin amino groups and a change in their chemical environment in the structured hydrogel matrix. This serves as evidence of the effectiveness of covalent crosslinking between the components. This suggests that the three-dimensional structure of the polymer consists of gelatin macromolecules crosslinked by a diepoxide crosslinker and is connected with alginate macromolecules through hydrogen and ionic interactions.

2.3. The Mechanical Properties of Gelatin–Alginate Hydrogels

The study on mechanical properties was conducted by measuring the force in cylindrical samples of hydrogels with a diameter of 10 mm under uniaxial compression. Since the lowest values were obtained for the gelatin hydrogel, which was destructed at a 4.5 mm strain, the values for other hydrogel compositions are given at 4.5 mm of strain.

The analysis of the force developed under the uniaxial compression of hydrogel samples (Figure 5) revealed that, as follows: (1) replacing 1/10 of gelatin with sodium alginate significantly enhances the strength of the hydrogel, increasing it from 5.1 kPa to 17 kPa, and (2) the force depends on the amount of the PEGDE 500 structuring agent in the hydrogel. A maximum force of 34 kPa is achieved at a ratio of 1:8. At this ratio, hydrogel contains a total of 12% of the gelling agent polymer with 16.66% being the structuring agent. This composition results in the maximum number of effective cross-linking nodes within the polymer network, forming a robust hydrogel framework.

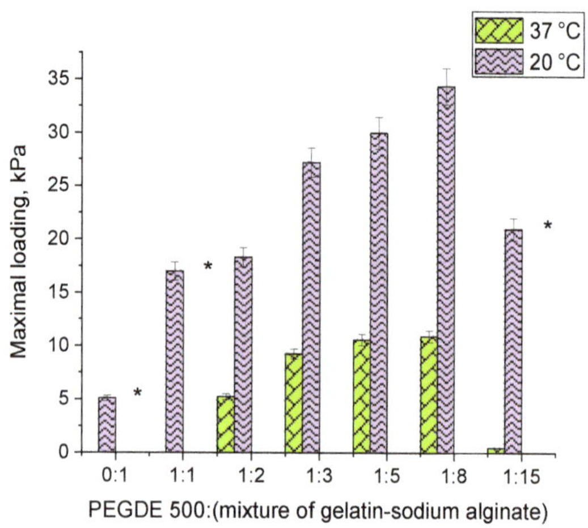

Figure 5. The force transmitted by the hydrogel under uniaxial compression at a strain of 4.5 mm. (For hydrogels obtained at 20 °C *—hydrogels are destroyed under compression, and the rest of the samples remained intact).

These findings correlate well with the gel-fraction data, which also have a maximum value with the composition (Figure 3, ratio 1:8). It is worth noting that the elasticity of the samples varies with the PEGDE 500 content. At ratios ranging from 1:2 to 1:8, the hydrogels withstand a maximum deformation without destruction at both 20 °C and 37 °C. In contrast, at a ratio of 1:15 (corresponding to a 6.25% structuring agent content in the gelling polymer), the hydrogel exhibits a similar mechanical behavior to unstructured gelatin, becoming destructed upon reaching a specific deformation threshold (slightly higher than that for intact gelatin), even at 20 °C. Thus, the optimal concentration range for the structuring agent in the hydrogel lies between 9.11% and 25%. This range produces hydrogels with improved mechanical properties. In terms of the total concentration of components in the hydrogel, this value ranges from 1.09 to 3.0%.

2.4. Rheological Characteristics of Gelatin–Alginate Hydrogels

The mechanical behavior of the hydrogels obtained from the gelatin–alginate mixture structured with PEGDE 500 was evaluated using the storage modulus (G′) and loss modulus (G″) to quantify the materials' viscoelastic properties. Initially, an amplitude sweep was performed at a fixed frequency ω = 10 rad/s to determine the linear viscoelastic region (LVR) of each hydrogel sample. The obtained data reveal that the linear region of both the dynamic moduli (G′, G″) for all of the samples (except for the sample without PEGDE 500) at temperatures of 20 °C and 37 °C is observed in up to approximately 7% of the strain (Figure 6a,b).

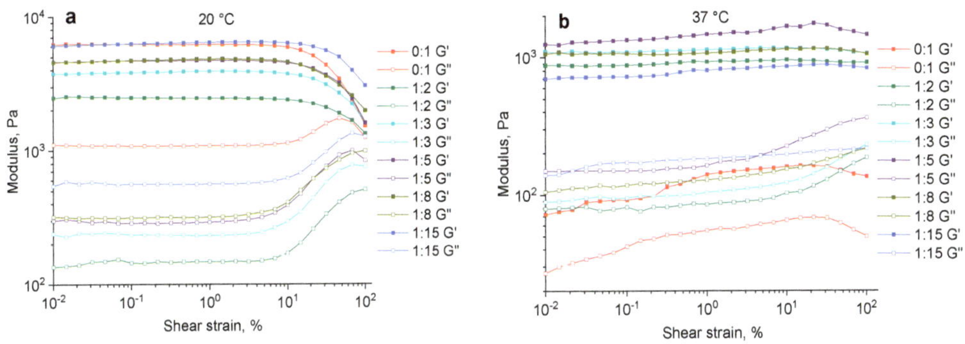

Figure 6. Strain (amplitude sweep) for gelatin–alginate hydrogels crosslinked with different amounts of PEGDE 500 at a temperature of 20 °C (**a**) and 37 °C (**b**).

The frequency sweep was conducted by varying the angular frequency from 1 to 150 rad/s at a strain of 1%, which corresponds to the previously determined LVR for all of the samples and temperatures (Figure 6). The dependence of the dynamic moduli on the angular frequency shows a characteristic response of a well-developed elastic polymer network, with the storage modulus (G′) being significantly higher than the loss modulus (G″) (Figure 7).

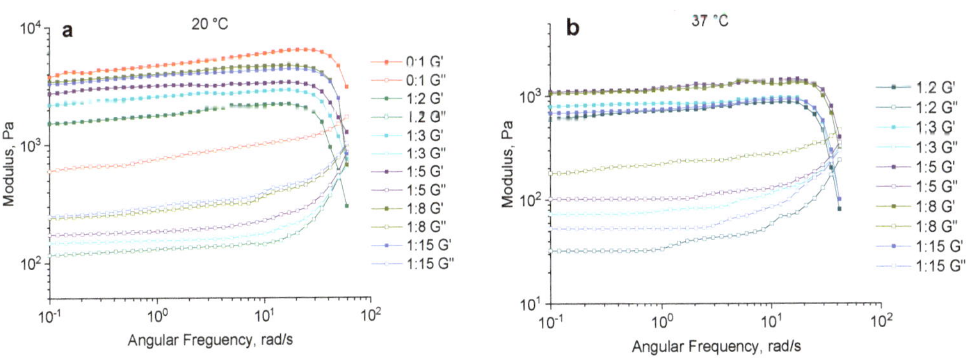

Figure 7. A frequency sweep for gelatin–alginate hydrogels crosslinked with different amounts of PEGDE 500 at a temperature of 20 °C (**a**) and 37 °C (**b**).

The independence of G′ and G″ values from the angular frequency (ω) in the range of 0.1–10 rad/s, both at 20 °C and 37 °C, indicates the presence of a well-structured polymer network that forms a hydrogel. However, the effect of the structuring agent content on these moduli shows an unclear dependence at different temperatures. At 20 °C, the storage modulus has maximum values of up to 3000 Pa at structuring agent ratios from 1:5 to 1:15.

At higher contents of the structuring agent, the value decreases to 1500–2000 Pa (Figure 7a). At 37 °C, the dependence of G′ is distinct with a maximum at the ratio of 1:5. The highest loss modulus G″ is observed at a ratio of 1:8 (Figure 7b).

Multicomponent composites containing gelatin as the main component typically have thermoreversible properties, resulting in a significant decrease in their mechanical properties with increasing temperature. Figure 8 illustrates the dependence of the dynamic modulus of hydrogels on the content of the structuring agent. The data show that at 20 °C, the storage modulus increases as the crosslinker content decreases, demonstrating maximum values at ratios of 1:15 and 1:8. At a temperature of 37 °C, while this general trend is maintained, the optimal range is shifted towards a higher PEGDE 500 concentration, with the peak values observed at ratios of 1:5 and 1:8. Thus, the observed behavior with increasing temperatures from 20 °C to 37 °C aligns with the previously reported gelatin hydrogels, leading to a decrease in the storage modulus G′, which reflects the material's elastic properties. This reduction in G′ ranges from a 2.6-fold decrease at a structuring agent ratio of 1:2 to a 4.8-fold decrease at a ratio of 1:15.

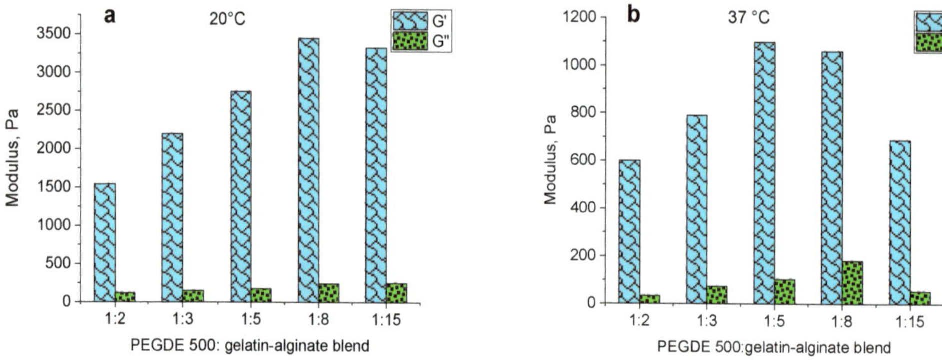

Figure 8. The viscoelastic characteristics of gelatin–alginate hydrogels crosslinked with different amounts of PEGDE 500 at a temperature of 20 °C (**a**) and 37 °C (**b**).

The mechanical loss coefficient calculated at a strain of 1% (Figure 9a) at 20 °C remains stable within the ratios from 1:2 to 1:8. However, at a ratio of 1:15, the coefficient starts to increase. At this ratio, corresponding to 6.25% of the PEGDE 500 in an anhydrous mixture of gelling agents, the mechanical loss coefficient is two times lower than that of an unstructured gelatin–alginate hydrogel at the same total polymer concentration (12%). At 37 °C, a slight increase in the tg δ value is observed, starting at a ratio of 1:5. At a 1:15 ratio, the coefficient increases approximately 2.5 times, which indicates the significant effect of the increased gelatin content and the decrease in elastic properties inherent for this biopolymer at higher temperatures. The temperature sweep reveals that, up to 30 °C, the mechanical loss coefficient for all hydrogel ratios is similar, but significantly lower than that of the unstructured gelatin–alginate hydrogel (Figure 9b). As was reported previously, at temperatures above 30 °C, specifically at 37.5 °C, a pronounced maximum is observed, indicating the transition of an unstructured material from a viscoelastic to a viscous flow state. This transition does not occur in the structured gelatin–alginate hydrogels, even at the lowest content of the structuring agent. Additionally, there is a decrease in the value of the mechanical loss coefficient with an increasing content of PEGDE 500 in the range of ratios from 1:15 to 1:8. At higher PEGDE 500 contents (1:2 to 1:5 ratio), the hydrogel retains its elastic properties over the entire temperature range with the mechanical loss coefficient remaining almost independent of both the ratio and the temperature.

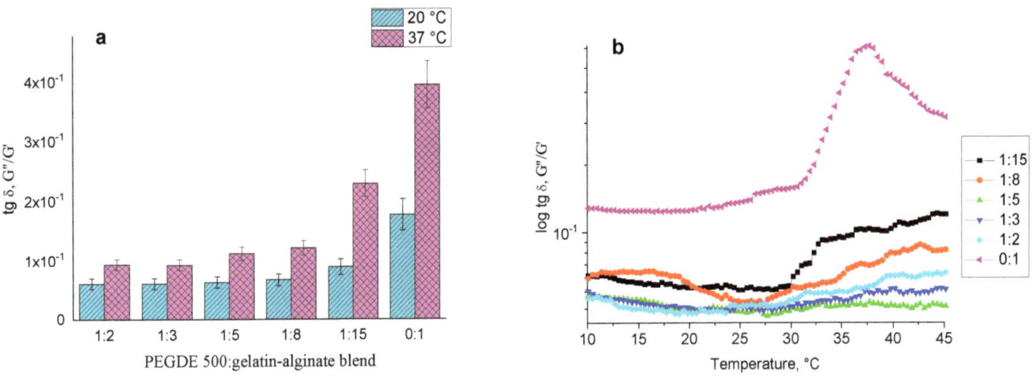

Figure 9. The loss tangent (mechanical loss coefficient tg δ = G″/G′) of hydrogels at different ratios of the structuring agent PEGDE 500 to the gelatin–alginate blend at 20 °C and 37 °C (**a**). The dependence of the mechanical loss coefficient on temperature for hydrogels with varying ratios of the structuring agent PEGDE 500 to the gelatin–alginate blend (**b**).

The developed material can be promising for the creation of hydrogel dressings for wound care, so it is important to assess its performance within temperature changes from storage to use.

For the hydrogel samples, as well as for the unstructured gelatin–alginate blend (a gelatin–alginate ratio of 9:1), a decrease in both of the dynamic moduli (G′ and G″) with increasing temperature is observed, which is typical for a hydrogel of unstructured gelatin. At a higher temperature, the number of binding sites responsible for the formation of a gelatin-based gel (a gel of the 2nd kind) decreases. This reduction leads to a mechanically weaker gel, i.e., with decreased elastic moduli.

The hydrogel sample without a structuring agent demonstrates the strongest temperature-dependent property change, as at 37 °C it becomes a viscous liquid (Figure 10). In contrast, for chemically structured samples, this decrease is slower, and even at 45 °C they still retain the properties of a solid (unlike liquids for which the moduli are identical). At the minimum ratio of 1:15, a strong temperature dependence is observed, but it differs significantly from the unstructured hydrogel sample. For all of the other ratios, the behavior is similar, supporting the previously obtained results of how within the studied temperature range, all of the samples of structured hydrogels retain the properties of elastic solids.

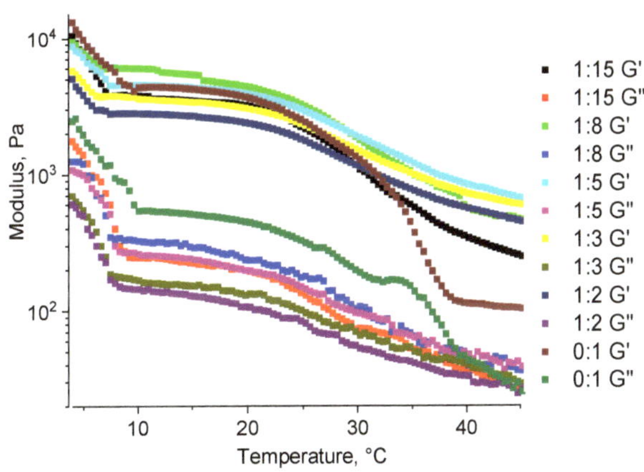

Figure 10. A temperature sweep for gelatin–alginate hydrogels at different ratios of the structuring agent PEGDE 500 to the gelatin–alginate blend.

In general, the mechanical properties of the hydrogels obtained according to the developed approach are well suited for use as wound dressings [25,28,48,49].

2.5. The Morphological Characterization of Gelatin–Alginate Hydrogels

Understanding pore size is crucial for evaluating the transport properties of hydrogels and the drug release kinetics in wound healing applications [50]. Scanning electron microscopy (SEM) was used to visualize the hydrogel structure, investigating hydrogels partially swollen in water for comparison. This approach was necessary because the morphology of the hydrogel polymer network in its non-swollen state cannot be effectively evaluated by SEM. The relative differences between the samples (obtained by the same methodology at a constant magnification value of the micrograph) are quite well visualized (Figure 11). The pore size of gelatin–alginate hydrogels in the swollen state ranges from 2 to 8 μm (Figures S1 and S2) for all of the samples, except for the one obtained with a crosslinker-to-gelatin–alginate base ratio of 1:1.

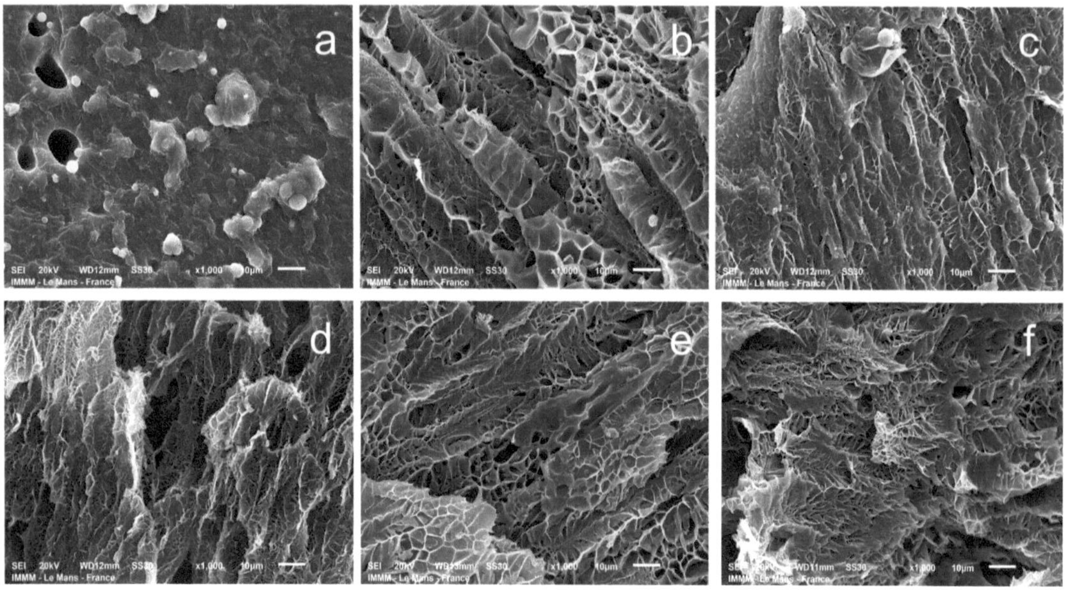

Figure 11. SEM images of gelatin–alginate hydrogels at different ratios of the structuring agent PEGDE 500 to the gelatin–alginate blend: 1:1 (**a**); 1:2 (**b**); 1:3 (**c**); 1:5 (**d**); 1:5 (**e**); and 1:15 (**f**), swollen twice their original size in water.

The hydrogels exhibit a complex spatially oriented morphology. At a ratio of 1:1, the lamellar-layered structure is observed. At other ratios, a microcellular morphology develops, with a noticeable thinning of the structure up to a ratio of 1:5. At higher ratios, it is difficult to observe much difference. This is quite natural since, according to the chosen scheme of hydrogel syntheses, when the ratios are above 1:5, there is no significant change in the concentration of the structuring agent in the system (Table 1).

2.6. Swelling Capacity in Model Media

Table 2 shows the swelling degree of the obtained hydrogels as a function of the structuring agent content and temperature of the medium. These values are commonly used in the literature to characterize and predict the behavior of hydrogels as wound dressings. The general trend observed is an increase in the swelling degree with increasing temperature and with a decreasing amount of structuring agent.

Table 2. Swelling values after 24 h of incubation.

Sample No.	Water, g/g		Saline, g/g		PBS, g/g		Ringer Solution, g/g		Exudate, g/g	
	20 °C	37 °C	20 °C	37 °C	20 °C	37 °C	20 °C	37 °C	20 °C	37 °C *
1 (1:1)	12.8 ± 1.4	14.0 ± 1.7	13.1 ± 1.8	16.0 ± 1.6	12.2 ± 2.1	14.9 ± 0.9	12.6 ± 1.3	15.8 ± 1.5	9.8 ± 1.2	10.3 ± 1.3
2 (1:2)	12.5 ± 1.1	16.7 ± 1.5	12.6 ± 1.9	14.4 ± 1.4	11.8 ± 1.5	14.7 ± 0.8	11.4 ± 0.9	14.8 ± 1.2	9.4 ± 0.7	9.5 ± 0.9
3 (1:3)	11.2 ± 1.2	17.1 ± 1.4	11.7 ± 1.6	14.8 ± 0.9	11.5 ± 0.8	14.6 ± 1.2	11.3 ± 0.8	14.9 ± 1.3	9.3 ± 0.8	10.7 ± 1.1
4 (1:5)	9.7 ± 0.8	19.1 ± 2.2	11.2 ± 0.8	15.3 ± 1.3	11.0 ± 1.0	16.4 ± 1.4	10.5 ± 1.2	15.0 ± 1.6	8.6 ± 0.5	10.4 ± 0.7
5 (1:8)	8.7 ± 0.8	26.8 ± 2.0	10.9 ± 0.7	15.6 ± 0.8	10.1 ± 0.6	15.3 ± 1.3	10.1 ± 1.1	14.3 ± 1.2	8.4 ± 0.6	11.2 ± 0.8
0:1	13.6 ± 1.6	-	15.0 ± 2.2	-	14.4 ± 1.2	-	14.8 ± 1.5	-	10.2 ± 1.1	-

* At 37 °C, a slight dissolution of the samples in the exudate was observed after 24 h of incubation, therefore the data are given after 12 h.

For most media at 20 °C and 37 °C, hydrogel samples synthesized with the maximum amount of structuring agent, swelling similarly within the margin of error. The decreasing of the structuring agent content leads to a change in the swelling capacity but these changes vary across the different media. The most notable behavior is observed in distilled water and exudate at 20 °C and 37 °C. At 20 °C, the swelling degree in water decreases symbiotically with the decreasing content of the structuring agent, which correlates well with an increase in the gel fraction of hydrogel samples. However, at 37 °C, the swelling degree increases. This can be attributed to the increased influence of gelatin, whose content is highest in sample 5 (Table 2). As was mentioned above, gelatin forms thermally reversible physical gels which are partially destroyed with increasing temperatures. In this case, the collagen-like clusters in the gelatin partially lose their intermolecular bonds, which leads to an increased swelling degree.

In saline solution, changes in the ionic strength of the medium cause minor differences in the swelling degree of hydrogel samples. These slight differences are observed with both changes in the content of the structuring agent and temperature, while in general, the swelling degree in saline is higher than in water. A similar pattern is observed in the phosphate buffer and Ringer's solution, where there is little dependence on the hydrogel composition. The lowest swelling degree is observed in an exudate, likely due to a significant amount of calcium ions, which interact with the functional groups of both gelatin and, more significantly, sodium alginate. These interactions lead to additional ionic structuring, and, consequently, a reduced swelling degree. This effect counters the temperature-dependent changes, so, unlike swelling in water, the swelling degree in exudate at 20 °C and 37 °C remains nearly the same.

In conclusion, despite certain features, the swelling degree of the synthesized hydrogels in all of the investigated media (60–90 g of per 1 dm^2 of hydrogel material with a thickness of 4 mm) is sufficient to absorb wound exudates and function effectively as a dressing [51].

2.7. The Enzymatic Degradation of Gelatin–Alginate Hydrogels

Proteolytic enzymes (proteases) are present in various types of wounds at different stages of healing. Their primary function is to degrade necrotic tissue and cleanse the wound, which is a crucial step in the healing process [52,53]. Additionally, the presence of exogenous proteins in the wound site can promote healing and significantly affect the release of therapeutic agents from drug delivery systems.

An investigation of the hydrogels degradation under enzymatic action is essential for assessing and predicting their behavior in vivo and in specific clinical cases [54]. The enzymatic degradation of the synthesized hydrogels was studied in vitro under model conditions using the protease enzyme (2.5 mg/mL in PBS). The percentage of residual polymers in the hydrogel was determined at specified intervals (Figure 12).

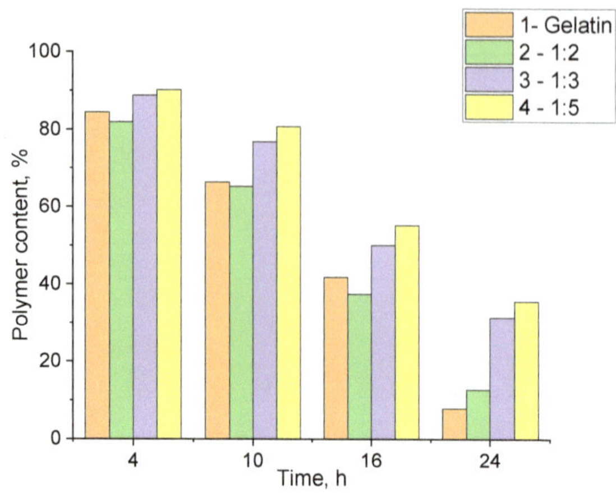

Figure 12. The enzymatic degradation of hydrogels at 20 °C: 1—unstructured gelatin; 2, 3, 4—gelatin–alginate hydrogels with a ratio of structuring PEGDE 500 to a gelatin–alginate base of 1:2, 1:3, and 1:5, respectively.

The data indicate that different structuring conditions when using PEGDE 500 have a mixed effect on biodegradation. As expected, unstructured gelatin degrades the most rapidly when exposed to proteolytic enzymes. However, the sample containing the highest amount of structuring agent also exhibited a similar degradation rate (Figure 12, samples 1,2). In contrast, hydrogels synthesized with lower ratios of PEGDE 500 demonstrated a significantly reduced degradation rate over the same time (Figure 12, samples 3,4). It can be assumed that this behavior is due to the incomplete utilization of PEGDE 500's potential as a crosslinker in systems with its high concentrations. It appears that a significant part of PEGDE 500 interacts with the amino groups in the gelatin via only one epoxy group to form comb-shaped macromolecules with PEG chains as side substituents. This results in a reduced crosslinking degree, leading to the lower gel-fraction values and accelerated enzymatic degradation, as proteases can more effectively target the gelatin fragments, breaking down the three-dimensional network of the hydrogel.

The ability to control the degradation rate of hydrogels within specific limits could be advantageous for the development of hydrogel dressings with controlled degradation properties. Hydrogels that degrade more slowly can also be used for a prolonged, degradation-mediated release of therapeutic agents over an extended period [55]. In addition, the release of protein macromolecule fragments from gelatin into the wound can stimulate wound cleansing, thereby promoting tissue regeneration and accelerating the healing process [56,57].

2.8. A Study of Drug Release

The release of drugs from hydrogels was evaluated using a vertical Franz cell and hydrogel sample 4 (preparation conditions are specified in Table 1) saturated with the corresponding drug. The resulting release profiles exhibit patterns (Figure 13a,b). For lidocaine and novocaine, the typical initial rapid release for such systems was observed [25]. During this phase, 25–30% of the drugs were released, then the process slows down and 35–50% were released over the next 24 h. The presence of the proteolytic enzyme appears to accelerate the release of these drugs after the initial rapid phase, likely due to the gradual degradation of the hydrogel.

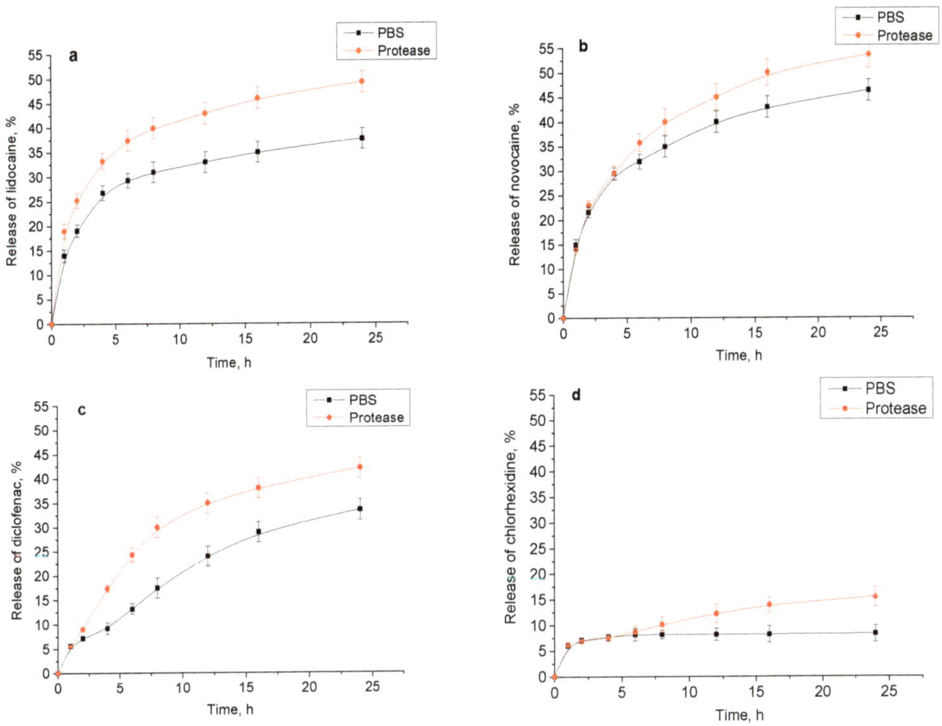

Figure 13. Kinetics of the drug release into phosphate-buffered saline (PBS) and into a 0.001% protease solution in PBS at 37 °C: lidocaine hydrochloride (**a**); novocaine (**b**); diclofenac sodium salt (**c**); and chlorhexidine digluconate (**d**).

The release profile of diclofenac sodium without protease in the medium demonstrates an almost perfectly linear direct proportional dependence. However, the presence of protease accelerates the release, slightly disrupting this "ideal" release pattern. Given that the release experiments were performed using a Franz cell, the enzyme had the access to only one side of the sample. Therefore, while the enzyme's impact was significant, it was not as pronounced as expected based on the previous section data.

For the antiseptic drug chlorhexidine, the hydrogel exhibited a limited release, achieving only 7.5% within the first 3 h, with no further increase observed within the experimental error margin. However, in the presence of the enzyme, the release continued, indicating that the release of chlorhexidine is controlled exclusively by the enzymatic degradation of the hydrogel polymer matrix (Figure 13d).

2.9. Antibacterial Properties

The low release capacity of the hydrogel for chlorhexidine prompted further investigation into its direct antimicrobial efficacy against pathogenic microorganisms via direct contact with the contaminated substrate. For this purpose, chlorhexidine was loaded into hydrogel disks with a diameter of 5 mm and a thickness of 2 mm. The samples were placed on the medium contaminated with *Escherichia coli* (*E. coli*), *Staphylococcus aureus* (*S. aureus*), *Pseudomonas aeruginosa* (*P. aeruginosa*), and *Aspergillus niger* (*A. niger*). The zone of microbial growth inhibition was measured to evaluate the hydrogel's antimicrobial activity. The choice of these strains in antimicrobial studies is due to their importance as model organisms for different categories of infections: bacterial (gram-positive and gram-negative) and fungal. This allows for a more comprehensive assessment of the effectiveness of antimicrobial agents. These types of strains are very common in hospitals, in particular, *S. aureus* and *P. aeruginosa*, and cause hospital-acquired infections, which is a serious healthcare problem.

The experimental results showed that chlorhexidine loaded in the structured gelatin–alginate hydrogels can be effectively released into the contaminated substrate. This is confirmed by the significant zones of the growth inhibition of *E. coli* and *S. aureus* (Table 3). Depending on the type of bacteria, the radius of the growth inhibition varies. For *E. coli* the radius of the growth inhibition is from 12.9 to 16.5 mm, for *S. aureus* from 7.3 to 9.8 mm, for *P. aeruginosa* from 7.5 to 8.5 mm, and for *A. niger* from 1.2 to 3.6 mm.

Table 3. The types of microorganisms and growth inhibition zones due to the chlorhexidine bigluconate released from the samples * of gelatin–alginate hydrogels, cm^2.

Microorganism Strain	Day 1	Day 2	Day 3
Aspergillus niger	 3.45 ± 0.15	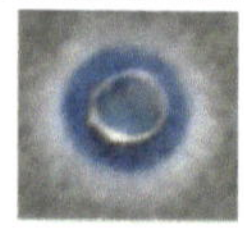 1.64 ± 0.11	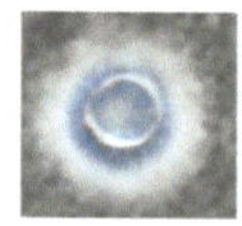 1.38 ± 0.12
Staphylococcus aureus	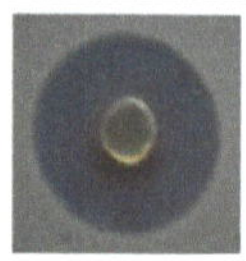 7.65 ± 0.36	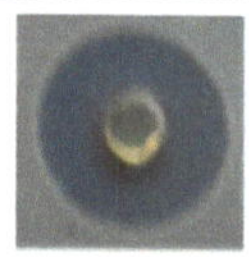 9.20 ± 0.54	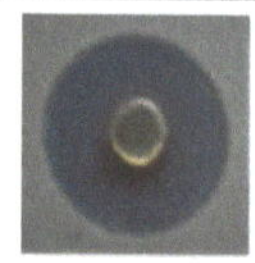 9.28 ± 0.48
Escherichia coli	 13.60 ± 0.74	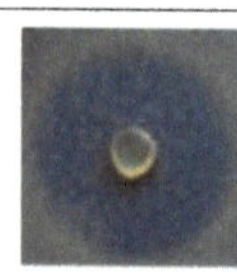 14.7 ± 0.75	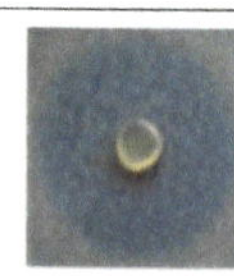 15.5 ± 0.87
Pseudomonas aeruginosa	 8.29 ± 0.22	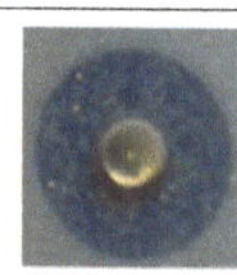 8.05 ± 0.19	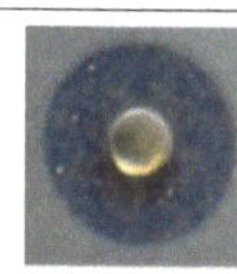 7.75 ± 0.20

* hydrogel sample 3 (preparation conditions are specified in Table 1).

3. Conclusions

The results of studies on obtaining gelatin–alginate hydrogels structured with polyethylene glycol diglycidyl ether (PEGDE 500) are presented. The conditions for synthesis were established, and the optimal ratio of the biopolymers and structuring agent in the hydrogel matrix was determined to produce materials with high strength values while maintaining the swelling capacity in water, saline, phosphate-buffered saline, Ringer's solution, and exudate. The introduction of sodium alginate into the hydrogels, compared to the unstructured gelatin hydrogel, had a positive effect on the material properties, notably allowing for the retention of physico-mechanical properties while reducing the prepolymer content in the initial composition from 18% to 12%, and simultaneously increasing the resistance of the samples to enzymatic degradation.

The possibility of loading the samples with various medications (lidocaine, novocaine, and sodium diclofenac) was demonstrated, and the release kinetics of these drugs were studied. The results of the experiments indicate that drugs are released from the gelatin–

alginate hydrogel matrix, which can provide an anesthetic effect for patients requiring pain relief.

The antibacterial properties of the hydrogels with incorporated chlorhexidine were confirmed in cultures of *E. coli*, *S. aureus*, *P. aeruginosa*, and *A. niger*, which may provide a prolonged antiseptic effect to prevent infectious complications in non-healing wounds. The obtained hydrogel material has the potential to serve as a basis for dressings in the care of chronic and/or infected wounds.

Future studies, especially in vivo experiments, are necessary to validate the findings and ensure their practical applicability.

4. Materials and Methods

4.1. Materials

Gelatin (type A), isolated from porcine skin (CAS Number: 9000-70-8, Sigma-Aldrich, St. Louis, MO, USA) was used without additional purification, with a moisture content of 9.2%, isoelectric point~8, and Bloom's power~175. Sodium alginate (CAS Number: 9005-38-3, Sigma-Aldrich) was used without further purification. Poly(ethylene glycol) diglycidyl ether (PEGDE 500, CAS Number: 26403-72-5, molecular weight: 500, Sigma-Aldrich), was used without further purification. Lidocaine hydrochloride, purity $\geq$ 98% (CAS Number: 6108-05-0, molecular weight: 288.81, Sigma-Aldrich), was used without further purification. Novocaine (synonym—procaine (hydrochloride)), purity $\geq$ 98% (CAS Number: 51-05-8, molecular weight: 272.8, Cayman Chemical Co., Ann Arbor, MI, USA) was used without further purification. Diclofenac sodium salt (CAS Number: 15307-79-6, molecular weight: 318.13, Sigma-Aldrich) was used without further purification. Chlorhexidine digluconate solution, 20–25%, in water (CAS Number: 18472-51-0, Sigma-Aldrich) was used without additional purification, and the dry matter content was checked before use. Model exudate—a solution with pH = 7.4–7.5, was prepared according to the following procedure [58]: 0.222 g of $CaCl_2$, 2.338 g of NaCl, 0.968 g of TRIS (2-Amino-2-hydroxymethyl-propane-1,3-diol), and 2 g of 5% aqueous bovine albumin solution (BSA) were added to a 100 mL volumetric flask; the salts were dissolved with stirring and made up to the mark with distilled water; and the resulting solution was stored in the refrigerator at 2–4 °C and was suitable for use within 7 days. The saline–sodium chloride solution, 0.9%, in water (CAS Number: 7647-14-5, Sigma-Aldrich) was used without additional purification. Phosphate buffered saline (PBS, Sigma-Aldrich) powder was used to prepare the aqueous solution with pH = 7.4. Ringer tablets (Sigma-Aldrich) were used for the preparation of Ringer's solution. Distilled water, with a specific electrical conductivity of 5–7 μS/cm and pH = 6.5 ± 0.1, was used.

4.2. The Synthesis of Gelatin–Alginate Hydrogels

For the preparation of the hydrogel, gelatin type A, sodium alginate, the structuring agent PEGDE 500, and distilled water were used. An aqueous gelatin solution with a concentration of 20.0% and an aqueous sodium alginate solution at 4% were prepared, with each solution stirred in a water bath at 40 °C until fully dissolved. The gelatin and alginate solutions were mixed in a 10:1 ratio, respectively, based on dry matter, and the required amount of PEGDE 500 and distilled water was added. The mixture was homogenized, loaded into prepared containers, sealed, and heated at 80 °C for 4 h. This method yielded a series of hydrogel samples that varied in polymer content and component ratios.

4.3. Gel Fraction

The gel fraction in hydrogel samples was determined using a water extraction method. Samples weighing approximately 2.5 g were weighed to an accuracy of 0.0001 g to determine the initial mass of polymers (W_1) in the hydrogel: $W_1 = Ws \cdot C_p/100$, where Ws is the initial sample mass and C_p is the polymer content in the hydrogel (%). Extraction was conducted with a large excess (100 g of water per 1 g of hydrogel sample) of distilled water at 50 °C for 24 h, with the water changed three times. The resulting insoluble fraction was dried to a constant weight in a drying oven at 100 °C. The gel fraction (G_f, %) was determined using the following equation: $G_f = W_2/W_1 \cdot 100\%$, where W_1 is the mass of polymers in the hydrogel before extraction, and W_2 is the mass of polymers remaining after extraction.

4.4. Fourier-Transform Infrared Spectroscopy (FTIR)

Fourier-Transform Infrared spectra were recorded using a FTIR Vertex 70V (Bruker, Billerica, MA, USA) spectrometer with the use of the Attenuated Total Reflectance (ATR) adapter (diamond crystal). The absorption bands were recorded in the range of 400–4000 cm^{-1} over 256 scans and a resolution of 2 cm^{-1}.

4.5. Mechanical Properties

Mechanical properties of the hydrogels were assessed by applying a stepwise uniaxial compressive load in 500 μm increments over a contact area of 0.95 cm^2, with continuous force measurement. The tests were conducted using standardized hydrogel samples with a diameter of 11 mm and a height of approximately 5.5 mm [28,59].

4.6. Rheological Characteristics of Polymer Solutions

The viscosity parameters of biopolymer solutions and their mixtures were studied using a Rheotest 2 rotational viscometer, which can measure viscosity in the range of 10^{-2} to 10^{-4} Pa·s over a shear rate range from 0.2 to 1.3×10^3 s^{-1}. A series of solutions was prepared at various gelatin-to-sodium alginate ratios (from 1:1 to 15:1, respectively) with a total biopolymer concentration of 1% in distilled water. The dynamic viscosity was calculated using the Equation (1):

$$\eta = (\tau)/(\dot{\gamma}) \quad (1)$$

where η—dynamic viscosity, [cP]; τ—shear stress, [Pa]; and $\dot{\gamma}$—shear rate, [s^{-1}].

4.7. Rheological Characteristics of Hydrogels

4.7.1. Viscoelastic Properties of Hydrogels

The testing of viscoelastic properties was performed using a Discovery HR-3 Hybrid rheometer (TA Instruments, New Castle, DE, USA) with a 20 mm parallel plate geometry. Hydrogels, prepared with a height of 2000 μm, were placed onto the Peltier plate, and the probe was lowered to create a controlled gap by applying a constant axial force of 0.2 ± 0.1 N, ensuring a consistent contact between the probe and the hydrogel. Samples were allowed to equilibrate for 120 s before testing, and all experiments were conducted at 20 °C in oscillatory mode. An amplitude sweep was first conducted from a 0.01% to 100% strain at an oscillation frequency of 1 Hz to identify the linear viscoelastic region (LVR), where stress and strain remain proportional. A strain of 0.25% was selected for evaluating the frequency dependence of the dynamic moduli, G' (storage modulus) and G'' (loss modulus). Subsequently, a frequency sweep was performed from 0.01 to 10 Hz at a 0.25% strain, assessing the frequency dependence of G' and G'' to characterize the viscoelastic behavior of the gelatin-derived hydrogels.

4.7.2. Strain (Amplitude) Sweep

Hydrogel samples were incubated for 2 h at 20 °C and 37 °C before the measurement. The rheological characterization of the hydrogels samples was performed using an Anton

Paar modular compact rotational rheometer MCR302 using a parallel plate probe (plate diameter 9.975 mm) at the corresponding temperatures (20 °C and 37 °C). A strain sweep was performed from 0.1 to 100%, and a shear strain (ramped logarithmic) was then conducted at a constant angular frequency of ω = 10 rad/s. Each sample was measured in triplicate and the results were averaged for each measured point. Th resulting curves were used to determine the linear viscoelastic region (LVR) for a further frequency sweep.

4.7.3. Frequency Sweep

The hydrogel samples were incubated for 2 h at 20 °C and 37 °C before the measurement. A frequency sweep was performed using Anton Paar modular compact rotational rheometer MCR302 (Anton Paar, Graz, Austria) using a parallel plate probe (plate diameter 9.975 mm) at the corresponding temperatures (20 °C and 37 °C). The frequency sweep was done by varying the angular frequency from a 1 to 150 rad/s at 1% strain, which fits to the previously determined LVR. The frequency-dependence of the storage (G′) and loss (G″) modulus was evaluated as the indicator of the viscoelastic response of the hydrogels.

4.7.4. Temperature Sweep

The hydrogel samples were incubated for 2 h at 6 °C before the measurement. The temperature sweep was performed using Anton Paar modular compact rotational rheometer MCR302 using a parallel plate probe (plate diameter 9975 mm). The temperature was ramped from 6 °C to 50 °C.

4.8. Scanning Electron Microscopy (SEM)

The Scanning Electron Microscope used is a JEOL, JSM 6510 LV instrument. The samples were swelling twice and frozen with liquid nitrogen, lyophilized (the materials were lyophilized in a Christ freeze-dryer alpha 2–4 LSC at −85 °C and 0.37 mbar), sectioned to visualize the hydrogel morphology, and were covered by a golden film before examination. Images of the hydrogels were analyzed using ImageJ Software Version 1.54 (National Institutes of Health (NIH) and the Laboratory for Optical and Computational Instrumentation (LOCI), University of Wisconsin, Madison, WI, USA).

4.9. Swelling Ability

The swelling capacity of hydrogels (SA) was measured by the conventional gravimetric method in distilled water (pH 6.5), saline (pH 7.0), and model exudate (pH 7.5). A total of 1 × 0.5 cm hydrogel samples were weighed and incubated at 20 °C and 37 °C in 50 mL of distilled water/saline/exudate/Ringer solution/PBS. After specified incubation times (1, 2, 4, 8, 12, and 24 h), excess liquid was removed, the sample was weighed, and then placed back in 50 mL to continue the incubation at the respective temperatures. The swelling of the scaffolds was calculated as the ratio of the swollen sample's weight ($W_W - W_P$) to the dry polymer weight (W_P). Each measurement was performed in triplicate, and the results were averaged. The swelling ability (S_A) was calculated using the following Equation (2):

$$S_A = (W_W - W_P)/W_P, \quad (2)$$

where W_W and W_P—the weight of the swollen sample and dry polymers, respectively.

4.10. The Enzymatic Degradation of Hydrogel Samples

The enzymatic degradation of hydrogel samples was conducted in a PBS solution (pH = 7.4) with the addition of protease (ORBAproteo P1200, ORBA Biokimya, İstanbul, Turkey). Hydrogels weighing 1–1.2 g were weighed with an accuracy of 0.0001 g and transferred into 10 mL of the protease solution at a substrate concentration of 2.5 mg/mL in PBS. They were then incubated at 20 °C in an air thermostat chamber for predetermined time intervals (3, 6, 16, and 24 h). After incubation, the samples were collected, dried, and weighed to determine the residual polymer content. The percentage of degradation was calculated using the following Equation (3):

$$\text{Degradation (\%)} = (W_d - W_t)/W_d \cdot 100\% \tag{3}$$

where W_d—polymer weight, [g] and W_t—polymer weight that remains after incubation at time t, [g].

4.11. Drug Release Studies

Gelatin–alginate hydrogel samples were initially saturated with medication solutions as follows: lidocaine hydrochloride—55–60 mg/g; novocaine—4–5 mg/g; sodium diclofenac—20–25 mg/g; and chlorhexidine bigluconate—15–20 mg/g. The release of lidocaine, novocaine, diclofenac, and chlorhexidine from hydrogel samples (diameter ~15 mm, height ~4.0 mm, and weight ~1 g) was carried out in model environments: phosphate-buffered saline (PBS) and a 0.001% protease solution in PBS. A vertical Franz diffusion cell with a diffusion area of 3.53 cm^2 and an acceptor compartment volume of 5 mL (PermeGear, Bechenheim, Germany) was used for the study. The hydrogel sample and release medium were separated by a highly porous regenerated cellulose membrane, 200 μm thick, with a pore size of 0.45 μm and a total porosity of 80–85% (Sartorius Stedim Biotech, Göttingen, Germany), providing unobstructed contact with the receptor medium. The experiments were conducted in triplicate at 37 °C with stirring at 480 rpm for 24 h. At specific time intervals (2, 4, 6, and 24 h), 0.2–0.4 mL samples were taken from the acceptor compartment of each sample, and an equivalent volume of aqueous medium was added to maintain absorption conditions in the system by keeping a constant initial volume. The concentration of the drugs in the collected samples was measured by UV spectrophotometry using a UV3100PC instrument; quartz cuvettes with Teflon stoppers type PCS8501, thickness 10 mm (Malvern, Malvern, UK) were used. Measurements were taken at the following wavelengths: λ = 262.8 nm for lidocaine; λ = 291 nm for novocaine; λ = 276.4 nm for diclofenac; and λ = 255.5 nm for chlorhexidine. The concentration (C, %) of the drug in the solution was determined using calibration equations: C = (Abs + 0.0063)/16.0 for lidocaine; C = (Abs + 0.0041)/653.0 for novocaine; C = (Abs + 0.001)/310.0 for diclofenac; and C = (Abs + 0.0304)/297.0 for chlorhexidine. The release degree (Rd) was determined by Equation (4) [25]:

$$R_d = \frac{m_{dss} \times C \times m_{ss}}{m_s \times m_{ch}} \tag{4}$$

where m_{dss}—the mass of the diluted sample solution, [g]; C—the concentration of the diluted sample solution [%]; m_{ss}—the mass of the starting solution, [g]; m_s—the mass of the sample, [g]; and m_{ch}—the mass of the drug in the hydrogel sample, [g].

4.12. Antibacterial Studies

The effectiveness of the bactericidal agent chlorhexidine in the hydrogel samples was studied using the methods specified in [60]. Cultures of Staphylococcus aureus, Pseudomonas aeruginosa, Escherichia coli, and Aspergillus niger were grown on a solid medium (containing 4% glucose, 1% meat-peptone broth (MPB), and 2% agar in 100 mL H_2O). For cultivation, 20 mL of the medium was poured into 90 mm Petri dishes, to which 2 mL of inoculum was added (ensuring that the medium temperature did not exceed 40 °C). To prepare the inoculum, microorganisms were transferred with a loop into test tubes containing 10 mL of the liquid nutrient medium and incubated for 24 h at 32 ± 2 °C. Hydrogel samples saturated with chlorhexidine were prepared with a diameter of 5.0 mm and a thickness of 2 mm, creating a chlorhexidine bigluconate concentration of 1% in the hydrogel. To study the effect of chlorhexidine on microorganism growth in the dishes, hydrogel discs containing the agent were placed on the nutrient medium with microorganisms. The inhibition zone of microorganism growth was recorded every 24 h. The area of microorganism growth inhibition (cm^2) was determined using TotalLab TL120 Software Version 2009.

4.13. Statistical Analysis

All of the experiments were conducted in triplicate, and the results were reported as mean ± SD. Mean values were compared using an independent samples Student's *t*-test, with *p*-values less than 0.05 considered statistically significant. The spectra obtained in the characterization of the hydrogels were processed using Origin 2019 Software Version 9.6 (OriginLab Co., Northampton, MA, USA).

Supplementary Materials: The following supporting information can be downloaded at: https://www.mdpi.com/article/10.3390/gels11030174/s1, Figures S1 and S2: Pore size of gelatin-alginate hydrogels; Figure S3: SEM image PEGDE 500-GelAl blend (1 to 1); Figure S4: SEM image PEGDE 500-GelAl blend (1 to 2); Figure S5: SEM image PEGDE 500-GelAl blend (1 to 3); Figure S6: SEM image PEGDE 500-GelAl blend (1 to 5); Figure S7: SEM image PEGDE 500-GelAl blend (1 to 8); Figure S8: SEM image PEGDE 500-GelAl blend (1 to 15).

Author Contributions: Conceptualization, S.V. and V.S.; investigation, O.M., D.O. and O.K.; data curation, O.M. and N.N.; resources, D.O., P.P. and O.K.; writing—original draft, O.M. and S.V.; administration, S.V. and N.N.; writing—review and editing, O.K., P.P. and S.V. All authors have read and agreed to the published version of the manuscript.

Funding: This publication has emanated from research supported in part by MSCA4Ukraine, grant number 1232572; a grant from Research Ireland under grant number 13/RC/2073_P2; and by the National Research Foundation of Ukraine, state registration number 0124U003832.

Institutional Review Board Statement: Not applicable.

Informed Consent Statement: Not applicable.

Data Availability Statement: All data and materials are available upon request from the corresponding author. The data are not publicly available due to ongoing research using a part of the data.

Acknowledgments: The authors would like to express their gratitude to the Armed Forces of Ukraine for providing security to us and our families and the possibility to perform this work.

Conflicts of Interest: The authors declare no conflicts of interest.

References

1. Sheokand, B.; Vats, M.; Kumar, A.; Srivastava, C.M.; Bahadur, I.; Pathak, S.R. Natural polymers used in the dressing materials for wound healing: Past, present and future. *J. Polym. Sci.* **2023**, *61*, 1389–1414. [CrossRef]
2. Zhang, X.; Qin, M.; Xu, M.; Miao, F.; Merzougui, C.; Zhang, X.; Wei, Y.; Chen, W.; Huang, D. The fabrication of antibacterial hydrogels for wound healing. *Eur. Polym. J.* **2021**, *146*, 110268. [CrossRef]
3. Ndlovu, S.P.; Ngece, K.; Alven, S.; Aderibigbe, B.A. Gelatin-Based Hybrid Scaffolds: Promising Wound Dressings. *Polymers* **2021**, *13*, 2959. [CrossRef]
4. Market Research Report. 2022. Available online: https://www.fortunebusinessinsights.com/industry-reports/chronic-wound-care-market-100222 (accessed on 1 April 2022).
5. Ribeiro, M.; Simões, M.; Vitorino, C.; Mascarenhas-Melo, F. Hydrogels in Cutaneous Wound Healing: Insights into Characterization, Properties, Formulation and Therapeutic Potential. *Gels* **2024**, *10*, 188. [CrossRef] [PubMed]
6. Mogoşanu, G.D.; Grumezescu, A.M. Natural and synthetic polymers for wounds and burns dressing. *Int. J. Pharm.* **2014**, *463*, 127–136. [CrossRef] [PubMed]
7. Dron, I.; Nosova, N.; Fihurka, N.; Bukartyk, N.; Nadashkevych, Z.; Varvarenko, S.; Samaryk, V. Investigation of Hydrogel Sheets Based on Highly Esterified Pectin. *Chem. Chem. Technol.* **2022**, *16*, 220–226. [CrossRef]

8. Nguyen, M.K.; Alsberg, E. Bioactive factor delivery strategies from engineered polymer hydrogels for therapeutic medicine. *Pr Polym. Sci.* **2014**, *39*, 1235–1265. [CrossRef]
9. Li, J.; Wu, C.; Chu, P.K.; Gelinsky, M. 3D printing of hydrogels: Rational design strategies and emerging biomedical applicatio *Mater. Sci. Eng. R Rep.* **2020**, *140*, 100543. [CrossRef]
10. Ebhodaghe, S.O. Hydrogel–based biopolymers for regenerative medicine applications: A critical review. *Int. J. Polym. Mat Polym. Biomater.* **2020**, *71*, 155–172. [CrossRef]
11. Lee, Y.; Mooney, D.J. Hydrogels for Tissue Engineering. *Chem. Rev.* **2001**, *101*, 1869–1877. [CrossRef] [PubMed]
12. Bryan, J. Moist wound healing: A concept that changed our practice. *J. Wound Care* **2004**, *13*, 227–228. [CrossRef] [PubMed]
13. Priya, S.G.; Jungvid, H.; Kumar, A. Skin tissue engineering for tissue repair and regeneration. *Tissue Eng. Part B Rev.* **2008**, *14* 105–118. [CrossRef] [PubMed]
14. Koehler, J.; Brandl, F.P.; Goepferich, A.M. Hydrogel wound dressings for bioactive treatment of acute and chronic wounds. *Eur. Polym. J.* **2018**, *100*, 1–11. [CrossRef]
15. Bahram, M.; Mohseni, N.; Moghtader, M. *An Introduction to Hydrogels and Some Recent Applications, Emerging Concepts in Analysis and Applications of Hydrogels*; InTech: London, UK, 2016; pp. 9–38. [CrossRef]
16. Chen, Z.; Cao, W.; Liu, Y.; Liu, H.; Ru, J.; Yin, M.; Luo, X.; Zhang, Y.; Chen, F. Hydrogel microspheres encapsulating lysozyme/MXene for photothermally enhanced antibacterial activity and infected wound healing. *Int. J. Biol. Macromol.* **2024**, *279*, 135527. [CrossRef]
17. Hou, X.; Huang, B.; Zhou, L.; Liu, S.; Kong, J.; He, C. An Amphiphilic Entangled Network Design Towards Ultratough Hydrogels. *Adv. Mater.* **2023**, *35*, 2301532. [CrossRef] [PubMed]
18. Shin, Y.; Hu, Y.; Park, S.; Jung, S. Novel succinoglycan dialdehyde/aminoethylcarbamoyl-β-cyclodextrin hydrogels for pH-responsive delivery of hydrophobic drugs. *Carbohydr. Polym.* **2023**, *305*, 120568. [CrossRef]
19. Kolour, A.K.; Shahrousvand, M.; Mohammadi-Rovshandeh, J.; Puppi, D.; Farzaneh, D. Absorbable and biodegradable enzyme-crosslinked gelatin/alginate semi-IPN hydrogel wound dressings containing curcumin. *Int. J. Biol. Macromol.* **2024**, *279*, 134938. [CrossRef] [PubMed]
20. Ma, Y.; Wang, Y.; Chen, D.; Su, T.; Chang, Q.; Huang, W.; Lu, F. 3D bioprinting of a gradient stiffness gelatin-alginate hydrogel with adipose-derived stem cells for full-thickness skin regeneration. *J. Mater. Chem. B* **2023**, *11*, 2989. [CrossRef]
21. Dash, R.; Foston, M.; Ragauskas, A.J. Improving the mechanical and thermal properties of gelatin hydrogels cross-linked by cellulose pavowhiskers. *Carbohydr. Polym.* **2013**, *91*, 638–645. [CrossRef] [PubMed]
22. Demirci, G.; Niedźwiedź, M.J.; Kantor-Malujdy, N.; El Fray, M. Elastomer–Hydrogel Systems: From Bio-Inspired Interfaces to Medical Applications. *Polymers* **2022**, *14*, 1822. [CrossRef]
23. Abourehab, M.A.S.; Rajendran, R.R.; Singh, A.; Pramanik, S.; Shrivastav, P.; Ansari, M.J.; Manne, R.; Amaral, L.S.; Deepak, A. Alginate as a Promising Biopolymer in Drug Delivery and Wound Healing: A Review of the State-of-the-Art. *Int. J. Mol. Sci.* **2022**, *23*, 9035. [CrossRef] [PubMed]
24. Razzak, M.A.; Kim, M.; Chung, D. Elucidation of aqueous interactions between fish gelatin and sodium alginate. *Carbohydr. Polym.* **2016**, *148*, 181–188. [CrossRef] [PubMed]
25. Maikovych, O.; Nosova, N.; Bukartyk, N.; Fihurka, N.; Ostapiv, D.; Samaryk, V.; Pasetto, P.; Varvarenko, S. Gelatin-based hydrogel with antiseptic properties: Synthesis and properties. *Appl. Nanosci.* **2023**, *13*, 7611–7623. [CrossRef]
26. Maikovych, O.; Nosova, N.; Yakoviv, M.; Dron, I.; Stasiuk, A.; Samaryk, V.; Varvarenko, S.; Voronov, S. Composite Materials Based on Polyacrylamide and Gelatin Reinforced with Polypropylene Microfiber. *Vopr. Khimii Khimicheskoi Tekhnologii* **2021**, *1*, 45–54. [CrossRef]
27. Branco da Cunha, C.; Klumpers, D.D.; Li, W.A.; Koshy, S.T.; Weaver, J.C.; Chaudhuri, O.; Granja, P.L.; Mooney, D.J. Influence of the stiffness of three-dimensional alginate/collagen-I interpenetrating networks on Fibroblast Biology. *Biomaterials* **2014**, *35*, 8927–8936. [CrossRef] [PubMed]
28. Bukartyk, M.; Nosova, N.; Maikovych, O.; Bukartyk, N.; Stasiuk, A.; Dron, I.; Fihurka, N.; Khomyak, S.; Ostapiv, D.; Vlizlo, V.; et al. Preparation and research of properties of combined alginate/gelatin hydrogels. *J. Chem. Technol.* **2022**, *30*, 11–20. [CrossRef]
29. Catoira, M.C.; Fusaro, L.; Di Francesco, D.; Ramella, M.; Boccafoschi, F. Overview of natural hydrogels for regenerative medicine applications. *J. Mater. Sci. Mater. Med.* **2019**, *30*, 115. [CrossRef] [PubMed]
30. Kolimi, P.; Narala, S.; Nyavanandi, D.; Youssef, A.A.A.; Dudhipala, N. Innovative Treatment Strategies to Accelerate Wound Healing: Trajectory and Recent Advancements. *Cells* **2022**, *11*, 2439. [CrossRef] [PubMed]
31. Díaz-García, D.; Filipová, A.; Garza-Veloz, I.; Martinez-Fierro, M.L. A Beginner's Introduction to Skin Stem Cells and Wound Healing. *Int. J. Mol. Sci.* **2021**, *22*, 11030. [CrossRef]
32. Tottoli, E.M.; Dorati, R.; Genta, I.; Chiesa, E.; Pisani, S.; Conti, B. Skin Wound Healing Process and New Emerging Technologies for Skin Wound Care and Regeneration. *Pharmaceutics* **2020**, *12*, 735. [CrossRef]
33. Suberlyak, O.; Grytsenko, O.; Baran, N.; Yatsulchak, G.; Berezhnyy, B. Formation Features of Tubular Products on the Basis of Composite Hydrogels. *Chem. Chem. Technol.* **2020**, *14*, 312–317. [CrossRef]

34. Mortier, C.; Costa, D.C.S.; Oliveira, M.B.; Haugen, H.J.; Lyngstadaas, S.P.; Blaker, J.J.; Mano, J.F. Advanced hydrogels based on natural macromolecules: Chemical routes to achieve mechanical versatility. *Mater. Today Chem.* **2022**, *26*, 101222. [CrossRef]
35. Panouille, M.; Larreta-Garde, V. Gelation behavior of gelatin and alginate mixtures. *Food Hydrocoll.* **2009**, *23*, 1074–1080. [CrossRef]
36. Goudoulas, T.B.; Germann, N. Phase transition kinetics and rheology of gelatin-alginate mixtures. *Food Hydrocoll.* **2017**, *66*, 49–60. [CrossRef]
37. Varvarenko, S.; Samaryk, V.; Nosova, N.; Puzko, N.; Taras, R.; Tarnavchyk, I.; Voronov, A.; Kohut, A.; Voronov, S. Prediction of interfacial interactions between polymer layers. *Macromol. Symp.* **2010**, *298*, 72–78. [CrossRef]
38. Dou, L.; Li, B.; Zhang, K.; Chu, X.; Hou, H. Physical properties and antioxidant activity of gelatin-sodium alginate edible films with tea polyphenols. *Int. J. Biol. Macromol.* **2018**, *118*, 1377–1383. [CrossRef]
39. Grytsenko, O.; Bratychak, M., Jr.; Dulebova, L.; Ivan, G. Thermophysical Properties of Composite Metal-Filled Copolymers of Polyvinylpyrrolidone. *Chem. Chem. Tech.* **2024**, *18*, 37. [CrossRef]
40. Mulik, J.; Sukumaran, S.; Srinivas, T. Factors structuring spatio-temporal dynamics of macrobenthic communities of three differently modified tropical estuaries. *Mar. Pollut. Bull.* **2020**, *150*, 110767. [CrossRef] [PubMed]
41. Cooke, M.E.; Jones, S.W.; Horst, B.; Moiemen, N.; Snow, M.; Chouhan, G.; Hill, L.J.; Esmaeli, M.; Moakes, R.J.A.; Holton, J.; et al. Structuring of Hydrogels across Multiple Length Scales for Biomedical Applications. *Adv. Mater.* **2018**, *30*, 1705013. [CrossRef] [PubMed]
42. You, G.; Niu, G.; Long, H.; Zhang, C.; Liu, X. Elucidation of interactions between gelatin aggregates and hsian-tsao gum in aqueous solutions. *Food Chem.* **2020**, *319*, 126532. [CrossRef] [PubMed]
43. Lin, L.; Xie, J.; Liu, S.; Shen, M.; Tang, W.; Xie, M. Polysaccharide from Mesona chinensis: Extraction optimization, physicochemical characterizations and antioxidant activities. *Int. J. Biol. Macromol.* **2017**, *99*, 665–673. [CrossRef] [PubMed]
44. Meka, V.S.; Sing, M.K.; Pichika, M.R.; Nali, S.R.; Kolapalli, V.R.; Kesharwani, P.A. Comprehensive review on polyelectrolyte complexes. *Drug Discov.* **2017**, *22*, 1697–1706. [CrossRef] [PubMed]
45. Grover, C.N.; Cameron, R.E.; Best, S.M. Investigating the morphological, mechanical and degradation properties of scaffolds comprising collagen, gelatin and elastin for use in soft tissue engineering. *J. Mech. Behav. Biomed. Mater.* **2012**, *10*, 62–74. [CrossRef] [PubMed]
46. Wang, C.S.; Virgilio, N.; Wood-Adams, P.M.; Heuzey, M.C. A gelation mechanism for gelatin/polysaccharide aqueous mixtures. *Food Hydrocoll.* **2018**, *79*, 462–472. [CrossRef]
47. Lin, D.; Xiao, L.; Wen, Y.; Qin, W.; Wu, D.; Chen, H.; Zhang, Q.; Zhang, Q. Comparison of apple polyphenol-gelatin binary complex and apple polyphenol-gelatin-pectin ternary complex: Antioxidant and structural characterization. *LWT* **2021**, *148*, 111740. [CrossRef]
48. Oyen, M.L. Mechanical characterisation of hydrogel materials. *Int. Mater. Rev.* **2013**, *59*, 44–59. [CrossRef]
49. Ning, X.; Huang, J.; A, Y.; Yuan, N.; Chen, C.; Lin, D. Research Advances in Mechanical Properties and Applications of Dual Network Hydrogels. *Int. J. Mol. Sci.* **2022**, *23*, 15757. [CrossRef] [PubMed]
50. Teixeira, M.O.; Marinho, E.; Silva, C.; Antunes, J.C., Felgueiras, H.P. Pullulan hydrogels as drug release platforms in biomedicine. *J. Drug Deliv. Sci. Technol.* **2023**, *89*, 105066. [CrossRef]
51. Stoica, A.E.; Chircov, C.; Grumezescu, A.M. Hydrogel Dressings for the Treatment of Burn Wounds: An Up-To-Date Overview. *Materials* **2020**, *13*, 2853. [CrossRef]
52. Avila-Rodríguez, M.I.; Meléndez-Martínez, D.; Licona-Cassani, C.; Aguilar-Yañez, J.M.; Benavides, J.; Sánchez, M. L Practical context of enzymatic treatment for wound healing: A secreted protease approach (Review). *Biomed. Rep.* **2020**, *13*, 3–14. [CrossRef] [PubMed]
53. Järbrink, K.; Ni, G.; Sönnergren, H.; Schmidtchen, A.; Pang, C.; Bajpai, R.; Car, J. Prevalence and incidence of chronic wounds and related complications: A protocol for a systematic review. *Syst. Rev.* **2016**, *5*, 152. [CrossRef] [PubMed]
54. Goodarzi, H.; Jadidi, K.; Pourmotabed, S.; Sharifi, E.; Aghamollaei, H. Preparation and in vitro characterization of cross-linked collagen–gelatin hydrogel using EDC/NHS for corneal tissue engineering applications. *Int. J. Biol. Macromol.* **2019**, *126*, 620–632. [CrossRef]
55. Rizwan, M.; Peh, G.S.L.; Ang, H.P.; Lwin, N.C.; Adnan, K.; Mehta, J.S.; Tan, W.S.; Yim, E.K.F. Sequentially-crosslinked bioactive hydrogels as nano-patterned substrates with customizable stiffness and degradation for corneal tissue engineering applications. *Biomaterials* **2017**, *120*, 139–154. [CrossRef] [PubMed]
56. Naomi, R.; Bahari, H.; Ridzuan, P.M.; Othman, F. Natural-Based Biomaterial for Skin Wound Healing (Gelatin vs. Collagen): Expert Review. *Polymers* **2021**, *13*, 2319. [CrossRef] [PubMed]
57. Stetsyshyn, Y.; Raczkowska, J.; Harhay, K.; Awsiuk, K.; Shymborska, Y.; Nastyshyn, S.; Ohar, H.; Vasilyev, V.; Ostapiv, D.; Sharan, M.; et al. Grafted polymer brush coatings for growth of cow granulosa cells and oocyte-cumulus cell complexes. *Biointerphases* **2020**, *15*, 031006. [CrossRef] [PubMed]
58. Rezvaniana, M.; Ahmad, N. Optimization, characterization, and in vitro assessment of alginate-pectin ionic cross-linked hydrogel film for wound dressing applications. *Int. J. Biol. Macromol.* **2017**, *97*, 131–140. [CrossRef]

59. Nosova, N.; Samaryk, V.; Varvarenko, S.; Ferens, M.; Voronovska, A.; Nagornyak, M.; Khomyak, S.; Nadashkevych, Z.; Voronov, S. Porous polyacrylamide hydrogels: Preparation and properties. *Vopr. Khimii Khimicheskoi Tekhnologii* **2016**, *5–6*, 78–86. Available online: http://vhht.dp.ua/wp-content/uploads/pdf/2016/5/Nosova.pdf (accessed on 22 December 2016).
60. Kozak, M.; Stasiuk, A.; Vlizlo, V.; Ostapiv, D.; Bodnar, Y.; Kuz'mina, N.; Figurka, N.; Nosova, N.; Ostapiv, R.; Kotsumbas, I.; et al. Polyphosphate Ester-Type Transporters Improve Antimicrobial Properties of Oxytetracycline. *Antibiotics* **2023**, *12*, 616. [CrossRef] [PubMed]

MDPI AG
Grosspeteranlage 5
4052 Basel
Switzerland
Tel.: +41 61 683 77 34

Gels Editorial Office
E-mail: gels@mdpi.com
www.mdpi.com/journal/gels

www.ingramcontent.com/pod-product-compliance
Lightning Source LLC
Chambersburg PA
CBHW041400210926
51850CB00013B/3139
* 9 7 8 3 7 2 5 8 3 8 2 3 3 *